FORESTRY AND ITS CAREER OPPORTUNITIES

McGraw-Hill Series in Forest Resources

Avery and Burkhart: Forest Measurements
Brockman and Merriam: Recreational Use of Wild Lands
Brown and Davis: Forest Fire: Control and Use
Dana and Fairfax: Forest and Range Policy
Daniel, Helms, and Baker: Principles of Silviculture
Davis: Forest Management
Davis: Land Use
Harlow, Harrar, and White: Textbook of Dendrology
Heady: Rangeland Management
Knight and Heikkenen: Principles of Forest Entomology
Nyland, Larson, and Shirley: Forestry and Its Career Opportunities
Panshin and De Zeeuw: Textbook of Wood Technology
Panshin, Harrar, Bethel, and Baker: Forest Products
Sharpe, Hendee, and Allen: An Introduction to Forestry
Stoddart, Smith, and Box: Range Management

Water Mulford was Consulting Editor of this series from its inception in 1931 until January 1, 1952.

Henry J. Vaux was Consulting Editor of this series from January 1, 1952, until July 1, 1976.

Under proper management forests of the world will continue to provide many resources to help sustain our lives. *(U.S. Forest Service.)*

FORESTRY AND ITS CAREER OPPORTUNITIES

FOURTH EDITION

Ralph D. Nyland

Professor of Silviculture
State University of New York
College of Environmental Science and Forestry
Syracuse

Charles C. Larson

Professor of Forestry and former Dean, School of Forestry
State University of New York
College of Environmental Science and Forestry
Syracuse

Hardy L. Shirley

Dean Emeritus
State University of New York
College of Environmental Science and Forestry
Syracuse

McGRAW-HILL BOOK COMPANY

New York St. Louis San Francisco Auckland Bogotá
Hamburg Johannesburg London Madrid Mexico Montreal New Delhi
Panama Paris São Paulo Singapore Sydney Tokyo Toronto

This book was set in Times Roman by Cobb/Dunlop Publisher Services Incorporated.
The editor was Marian D. Provenzano.
The production supervisor was Diane Renda.
R. R. Donnelley & Sons Company was printer and binder.

FORESTRY AND ITS CAREER OPPORTUNITIES

1 2 3 4 5 6 7 8 9 0 DOCDOC 8 9 8 7 6 5 4 3

ISBN 0-07-056979-7

Library of Congress Cataloging in Publication Data

Nyland, Ralph D.
Forestry and its career opportunities.

(McGraw-Hill series in forest resources)
Rev. ed. of: Forestry and its career opportunities / Hardy L. Shirley. 3rd ed. [1973]
Includes bibliographies and index.
1. Forests and forestry. 2. Forests and forestry—Vocational guidance. 3. Forests and forestry—United States. 4. Forests and forestry—Vocational guidance—United States. I. Larson, Charles C. II. Shirley, Hardy L. (Hardy Lomax), date . III. Shirley, Hardy L. (Hardy Lomax), date . Forestry and its career opportunities. IV. Title. V. Series.
SD373.N95 1983 634.9 82–21652
ISBN 0–07–056979–7

CONTENTS

PREFACE

When Hardy L. Shirley wrote the first edition of *Forestry and Its Career Opportunities* in 1952, he took a view of forestry that was fairly broad for that time. He included traditional forestry activities as well as the allied fields of wood processing and use. In those days some schools offered programs in wood utilization, paper science, and wood products manufacturing to complement the timber production and land management courses. But forestry had a fairly narrow focus. It gave only limited attention to disciplines such as hydrology, watershed management, soil science, recreation management, resource economics, planning, urban and world forestry, biometrics, forest engineering, chemical ecology, and the many other disciplines that now are an integral part of forestry.

As the nature of forestry became more sophisticated, college programs broadened and became more advanced technologically. The schools began to offer more diverse opportunities for study in forest science and practices, as well as among the several fields allied to forestry. As a result we felt a need to revise the text once more, to provide an up-to-date interpretation of the profession. We wanted to share with students the modern challenges of forestry and to review the great growth of technology and science that supports it.

While we recognize the continuance of traditional forestry and land management as an essential and growing profession, we view it from a new perspective. We see forestry as drawing more upon related biologic sciences, engineering disciplines, economic and planning fields, policy and legislation interests, and business management areas. Consequently we have broadened the text to review these fields as well, and to consider the ways they interact with the management and use of forests. We also describe the importance of different resources to human welfare, and outline opportunities for careers. To this end we have included with each chapter a brief discussions of educational preparation focused upon differenct aspects of forestry. These will give students some ideas to discuss with their faculty advisers in selecting elective courses. The final chapter provides additional information about preparing for careers in forestry, and outlines where professionals find employment.

The current revision resulted in considerable change within the book. We merged some chapters, deleted others, formed new ones, and reoriented the focus of our discussion. We have taken a broader approach to interpreting forestry and have strengthened the technical content. Still the discussions give only a broad overview of the different subjects. The chapters represent a starting point, but will not substitute as a detailed text for those who seek a more penetrating study of the sciences and technologies related to forestry. Such readers should refer to the literature cited in the references for each chapter and use those listings to guide further study.

Many members of the faculty at the State University of New York College of Environmental Science and Forestry, and from other academic institutions, provided technical review, contributed ideas, and gave advice about different chapters in this book. Their help proved extremely valuable in our attempts to provide a balanced and accurate coverage of the various topics. We particularly acknowledge the contributions of Dr. Peter E. Black, Professor Harry W. Burry, Dr. Carl H. deZeeuw, Dr. Allan Drew, Dr. Deborah B. Hill, Dr. Miklos A. J. Gratzer, Dr. Bengt Leopold, Dr. Paul Manion, Dr. John B. Simeone, and Dr. Edwin White. In addition, Dr. David W. Robinson, Dr. Craig E. Schuler, and Dr. Raymond A. Young provided a critical review of the manuscript and offered many valuable suggestions that helped to strengthen and improve the book. We also thank representatives of the U.S. Forest Service, the U.S. Bureau of Sport Fisheries and Wildlife, the U.S. Bureau of Land Management, the Weyerhaeuser Company, and several other corporations and individuals who allowed us to use photographs to illustrate parts of the book. Finally, we acknowledge the secretarial help of Shirley Bratt, Tari Pittenger, and Shirley Farricy in preparing the manuscript. For all of these contributions, and for the many ways in which other people aided in our research and deliberations, we express sincere thanks.

We have found much excitement, challenge, and pleasure in the practice of forestry. We believe that as a profession it has important contributions to make to the future. This book, it is hoped, will help others to find in forestry as much promise and excitement as we have enjoyed.

Ralph D. Nyland
Charles C. Larson
Hardy L. Shirley

CHAPTER 1

THE PROFESSION OF FORESTRY

Oceans, lakes, and seas cover 361 million square kilometers (km^2), or 71 percent of the earth's surface. The remaining area serves as a source of land-based resources, including space for croplands, forests, grasslands, home sites, businesses, and industrial developments. According to the general type of land, this continental area of the earth can be divided as follows:[11]

Type of area	Millions of square kilometers
Ice, rock, and sand	24
Desert and semidesert	18
Cultivated	14
Grasslands, savanna, alpine, and tundra	32
Swamp and marsh	2
Lakes and streams	2
Forest, woodland, and shrubland	57
Total continental	149

Source: Based on Whittaker and Likens [13]. Used by permission.

Forests and woodlands of various types cover by far the greatest proportion of the land, and croplands relatively little. Areas generally considered of low productivity for food and fiber crops account for slightly more than 40 percent. The forests by virtue of their extent also outrank other types of land area in the total production of plant organic material, called plant biomass (Table 1–1). They are responsible for 47 percent of the worldwide production. Oceans add only 32 percent of the total net production. Other types of land rank far below these.

DEPENDENCE UPON RESOURCES

People can support themselves in highly organized societies even on artificial islands surrounded by seas, at the polar ice caps, and in wastelands and deserts. But they must draw upon goods from remote croplands, grasslands, and forests to survive. The same holds true for land-based urban centers throughout the world, where inhabitants must look to distant croplands for food and for grains to feed domesticated animals. They also draw upon grasslands for meat products, and look to forests to provide materials for hundreds of goods. However, if people lacked access to one of these types of land or derived only limited amounts of products from them, lifestyles would change importantly. Society as we know it today would disappear.

People throughout the world cherish cropland and often take extraordinary measures to safeguard its productivity. Fortunately, such land need not deteriorate if treated with respect and understanding. The wheat fields of Anatolia that supplied grain to the Roman legions still yield generous crops to modern Turkish farmers. In other regions, however, fields given less careful treatment have eroded. The topsoil today lies under water in the harbors of Ephesus, Nicodemia, and Tarsus. In America as well, careless tilling of steep slopes has led to serious erosion in Appalachia, and along the eastern coastal plain. In addition, vast

TABLE 1–1
PRODUCTION OF ORGANIC MATERIAL BY PLANTS IN DIFFERENT TYPES OF ECOSYSTEMS THROUGHOUT THE WORLD

Surface cover	Production of organic material by plants, billions of metric tons of dry matter per year	Percent of total
Forests and woodlands	79.9	46
Savanna, grasslands, alpine, Tundra	20.0	12
Cultivated lands	9.1	5
Desert and semidesert	1.7	1
Swamps, marshes, lakes, Streams	6.8	4
Oceans	55.0	32
Total	172.5	100

Source: Based on Whittaker and Likens [13]. Used by permission.

quantities of fertile soil also blew off farmlands in the Great Plains during the early 1900s. Yet, through careful methods of cultivation and soil conservation, the wheat fields of the Great Plains will continue to yield grains for future generations. Other croplands of the world also will last, if people respect them and manage them wisely.

Grasslands and forests deserve attention as well. Through careless use, their productivity can decline. The herding of sheep, goats, and cattle, for example, has long provided a livelihood and sources of food and clothing in many parts of the world, and it remains important in the Mediterranean region, Afghanistan, Pakistan, India, and Africa. But much of the land has suffered from overuse, which has led to serious soil erosion, reduced yields of forage, and lower returns to the herders. At the same time, in places where people kept grazing within the capacity of the land to support good growth of vegetation, little soil deterioration has resulted and the lands continue to support good herds of domesticated animals and wildlife.

Forests, too, can yield continuous crops of timber without loss of soil or reduction of productivity, if managed according to good practices. Proper management requires planning and control to keep harvests within the capacity of the land to grow and regenerate new supplies. It also means taking measures to secure new regeneration promptly to replace trees that have been removed, and implementing practices that ensure soil stability. With these, forests will provide renewed resources of timber and support sustained production and harvest without interruption.

THE FOREST AS A PROVIDER OF GOODS AND SERVICES

Within the United States, forests cover about 298,201,781 hectares (ha). Alaska alone has 48,236,802 ha. Rangelands extend over another 331,984,737 ha throughout various parts of the nation. Between 1920 and the early 1960s, forest area in the United States increased by 20 million hectares as a result of reversion of lands once cleared for agriculture. However, substantial land clearing for crops in the south and middle west has again reversed the trend. Recent projections indicate that by the year 2030 forest area may decline by 1 percent and rangeland by 3 percent, compared with 1977.[12]

Of the total forest area, only two thirds has the capacity for commercial-scale production. This means it will produce at least 1.4 cubic meters (m^3) of wood annually per hectare.[10,11] Lands with that minimum potential are called commercial forests. The name implies nothing about who owns the land or the actual use to which it is put. The term simply conveys information about its minimum productive capacity. In fact, private nonindustrial owners control about three fifths of the commercial-quality forest area of the United States, and use it for a variety of purposes. Many have little direct concern about producing wood for industrial use.

Altogether the commercial forests of the United States produce about 619,411,113 m^3 of wood each year. About 66 percent of that volume is harvested for various uses. Another 18 percent dies from natural causes without being recovered.[11] Major commodity uses include wood for lumber, paper, fiberboard, poles, pilings, particle board, veneers, and the many goods made from these. Increasingly wood and wood wastes also serve as fuel to heat homes and produce steam for industrial application and power generation. In addition, forests provide many other products, such as chemicals for pharmaceuticals, resin, maple syrup, and berries. These represent the material goods that we harvest from forests regularly.

Forests also supply many benefits that we do not necessarily classify as forest products, even though they are important to our welfare. Water probably heads the list; forests serve as a major source of water that has purity and clarity unmatched by that from other land areas. They provide a habitat for many species of animals, birds, and fish that some people prize for hunting and fishing and others simply for watching and photographing. They afford recreational facilities as a place to hike, camp, and picnic. In many parts of the world, forests and associated grasslands also provide much forage for domestic cattle, sheep, and goats. Most interesting, no forest provides only one of these benefits. Instead each tract of forested land will offer multiple values as a source of wood, forage, water, wildlife, and recreation potential.

Of the goods and services we derive from forests, wood has the greatest dollar value. It brings considerable income to individuals, corporations, and government. And landowners derive cash from its sale. Others add value by processing the trees into lumber, pulp, and veneers. Additional industries remanufacture the primary products into furniture, building materials, paper, and a myriad of other items derived from wood. With each step, the value multiplies through wages, salaries, and income generated by sales. Landowners benefit directly from the sale of logs, bolts, poles, and other roundwood products. But those sales represent only the first step in a series of commercial activities that progressively compound the economic value of forest products to those who manufacture consumer goods and sell them throughout the world.

Nationally forage and grazing generate the second greatest amount of income, but seldom to individual landowners. Most fee grazing takes place on government-owned land. The income realized goes into the national treasury. Water, too, can generate revenues if a water company owns rights to the water that flows out of the forest. Otherwise water comes freely from our forested lands to serve many domestic and industrial needs. Most wildlife and recreational benefits are available without charge from lands controlled by federal, state, and local governments. But some landowners lease rights to hunt, fish, and recreate on their lands. In the process, they shut their forests to public use by posting them against trespassing. Increasingly this leasing of lands for recreational use provides landowners with revenues to pay the costs of ownership, taxes, and administration.

THE SCOPE OF FORESTRY

At one time, most people lived in the forest, and trees covered much of the landscape. The earth's population remained limited and people settled in relatively small communities surrounded by woodlands. They had easy access to nearby resources in abundant quantities. But over time things changed. The population grew. People clustered into larger and larger communities that became big towns and cities. They cleared more and more land for their agriculture, industry, and housing. Demand upon nearby forests increased, and in some localities the pressures overtaxed the resources available. This happened first in the more crowded portions of Europe and around the Mediterranean, then in the new world, and more recently in the tropics. Under the worst of circumstances, the forests disappeared and the people substituted other resources where they once had used wood.

As the distance between people and the forest increased, communities began to import wood products from other regions of the continent. In some cases, they even had to reach out to other parts of the world where forests still dominated the landscape. In response, some individuals organized themselves into enterprises for harvesting and processing the logs and roundwood pieces. Others formed industries and trades to work with the lumber and boards. Another group began to distribute the finished goods and lumber to willing buyers at other places. And people who had once gone into the forest to procure and process various products of the forest for themselves came to depend upon others to supply these needs. A new era had arrived with new opportunities and challenges.

Eventually people realized that by managing the resources and controlling use within limits of the land to produce the goods over time, they could sustain the benefits of the forest in perpetuity. Gradually they began planning their strategies and developed new technologies to improve management and use. By stages, demand arose for skilled people to implement the programs, develop the plans, and supervise application of the technologies to accomplish the job.

The first people to oversee the forest undoubtedly relied upon their personal observations, experience, and ideas shared with others through rather informal means. With time, these people probably drew upon the biologic sciences for information, and applied their new scientific understanding in devising better and more effective ways to manage the forest and control its growth and development. They melded the practical experience of informed woodsmen with ideas and understanding from the biologic sciences. Eventually a unique discipline of forestry evolved and with it the advent of more scientific management of forest resources to better serve human need.

Forestry Defined

For many decades, even up to fairly recent times, foresters cared for most matters related to the forest. They managed the land, planned for use and harvest of its products, supervised recreational activities unique to forested environments, attended to wild animal populations, oversaw the taking of water and minerals off the land, worked within many operations that processed timber into primary products for further remanufacture and use in other industries, and more. Foresters served as managers, planners, and guardians of the land and trees. In short, all matters related to forests, their use, and care came under the purview of foresters. They functioned in all these various capacities, supported by a few specialists who helped them to address special problems that arose from time to time.

The growth of the natural, physical, and social sciences and the heightened understanding they brought to us led to specialization. Unique skills became available to apply to aspects of forest care and use. At the same time, increased demands upon natural resources brought a new urgency to develop more complex schemes of management. It also forced the profession to seek a better understanding of the nature and growth of forests, to learn how they differ from and complement other ecosystems, and how they influence conditions around the earth. As a result, the scope of forestry broadened. People with rather specialized skills became involved. Eventually forestry took on new significance and scope.

Modern forestry, as it emerged in the process, has come to encompass a multitude of activities and responsibilities. It embraces the science, business, art, and practice of managing forests and the lands that support them. It strives to

Figure 1–1.
Management of forested areas and imaginative use of the associated resources will ensure adequate supplies of essential goods and services for the future. *(U.S. Forest Service.)*

provide for continuing use of forest resources.[2] These resources include trees, animals, various plants and their by-products, and minerals. Foresters also have concern about many in-place values people derive through recreation and other noncommodity uses (Figure 1–1). Overall forestry includes any activity of using and managing the natural resources that occur in association with forests so that these goods, values, and services provide human benefit.

In this sense, forestry involves professional persons educated and engaged in many forest-related activities. It includes those with expertise in allied fields of biology, business administration, economics, ecology, hydrology, meteorology, soil science, political science, wood products engineering, and wood chemistry. By its modern interpretation, forestry has wide bounds that embrace these many fields of science and practice. They are united by a common interest of attending to, managing, and using the forests and their associated resources.

We call anyone involved in the profession of forestry a forester. Such people have responsibility for the practical application of scientific, economic, and social principles to administer forestlands for specific purposes. In this sense, forestland means any place that supports forest growth, plus any other area bearing evidence that it once had forest cover and has not been dedicated to some other use. The practice of forestry, then, involves all the activities for protecting and managing renewable natural resources in accordance with principles that assure their optimum economic and social benefit, and that maintain the productive capacity of the forest over the long run. Such a philosophy carries with it concern for both the present and the future. It puts upon the profession of forestry the responsibility to use resources and forests in such a way as to ensure their renewability and availability for future generations while satisfying the immediate needs of society.

The Concerns of Forestry

Foresters manage both ecosystems and people. The term ecosystem means the community of plants and animals found in a particular place, plus the physical environment that supports their growth, development, and reproduction. Although the earth supports several types of ecosystems, foresters deal primarily with the ones where trees make up the principal plant component. Foresters also become involved with range ecosystems, especially where range and forestlands interface.

Foresters and their colleagues primarily work to influence the development of vegetation. This may include limiting the numbers of species, increasing their diversity, altering the spacing and arrangement between trees, reducing hazards of many natural and people-caused destructive agents, scheduling the times for harvest and use, controlling the intensity of these activities, and undertaking many other related functions. All of these attempt to influence the forest character and development in a way that makes it and the associated resources more useful.

Forestry also manages people. At least it influences the ways in which people utilize or affect the plants, animals, and physical components of forest and range ecosystems. In many respects, this really means facilitating forest use so people can take advantage of the resources according to the principles of conservation. Described either way, forestry gives attention to social values and needs, and to the types of uses and goods people seek from forests. It includes resolving conflicts over potential uses to provide the broadest benefits to a wide segment of our population. It entails the processes of discerning societal needs and interests, and the development of policy and programs to bring those to fruition. It involves interpreting for people the characteristics of forest-use opportunities and describing realistically the consequences of exploiting those values. It means administering the people and programs that bring use to reality. In these and many other ways, forestry encompasses a variety of activities that help people derive value from the forest. Forestry also helps people to use resources in a way that provides for their continuing availability for future generations.

In its broadest sense, forestry also includes the many activities needed to extract, process, manufacture, and market forest products. Thus forestry has concern for both the growth and the harvest of products, and for the ways the wood is handled to turn the trunks and other parts of trees into items used in homes, in manufacturing, and in many service industries. Foresters work most closely with primarily processing plants such as sawmills, pulp mills, and veneer mills. Yet they also have concern for and interest in the many technologies applied to the remanufacture of solid wood pieces and fiber into secondary products that consumers will purchase and use. Such interplay between those who manage the forests and those who convert the resources into consumer goods helps the profession of forestry to perform its functions, and to ensure that the fruits of its efforts will have maximum value to society.

The Ecosystems Approach

Foresters take an ecosystems approach in managing forest resources. That means giving attention to the forest as a complete living community of plants and animals. It also means taking cognizance of the water, soil, atmosphere, and other physical resources that influence and sustain life in the forest. Thus foresters must develop an awareness of ways in which the manipulation of one component of an ecosystem can affect others. They must have an understanding of the interactions that occur between living and abiotic elements of the ecosystem, and how changing one or more of these elements would affect the others. That forces the forester to acquire a broad reserve of scientific information to draw upon in devising realistic ways to manipulate the forest and its resources.

Appreciation of the complexity of forest ecosystems encourages foresters to work with other professions in managing forests for various purposes. They utilize the services of wildlife and fisheries biologists, soil scientists, range managers, hydrologists, watershed managers, recreation specialists, ecologists, botanists, entomologists, pathologists, landscape architects, chemists, physiologists, general biologists, limnologists, biometricians, photogrammetrists, engineers, atmospheric scientists, geneticists, wood technologists, paper scientists, and others schooled in many different scientific disciplines. Foresters also depend upon business managers, administrators, planners, legislators, political scientists, economists, marketing specialists, sociologists, communication specialists, urban specialists, and others who devote themselves to the many social sciences and aspects of business and administration. They become involved with all these people as a rather routine matter in the course of daily activity. Some foresters even pursue advanced studies to prepare themselves to work in one of these specialties, and to merge their basic understanding of forestry with that of a field associated with it.

Education for careers in forestry involves the study of basic sciences as well as the disciplines of management. It also provides opportunity to strengthen understanding in some aspect of forestry science or practice. Consequently foresters receive a broad-based education that interfaces with many related disciplines. This preparation enables them to communicate effectively with other professionals, and to work jointly with them in charting programs for the management and use of forest resources. Forestry education also recognizes that professional skill comes with experience offered through eventual employment. Hence individual development of each forester involves the formal education obtained in a university or college, plus the maturing and experience that comes with actual practice.

THE NATURE OF A PROFESSION

By common definition, the word profession means an occupation requiring extensive education in a branch of science or liberal arts, and usually involving mental rather than manual work. Though this meaning remains somewhat nebulous, certain attributes characterize a profession and distinguish it from other occupations.[5]

1. It supports its practice by a systematic body of theory

2. It has status and prestige among its clientele

3. It receives broad community support and approval as an authority in some field of knowledge and practice

4. It maintains a code of ethics to control relations of its members with clients and colleagues

5. It has an identity or culture sustained by a formal professional association or society

As a discipline satisfies these criteria, it becomes recognized as a profession and maintains that distinction within a society.

Attributes of a Profession

A profession depends upon advanced knowledge of a body of information. Many nonprofessional occupations, however, also require access to many facts and demand a high degree of skill at various tasks. This alone does not make an occupation a profession. Rather the understanding and skills required for a profession come from a fund of information accumulated over time through experience and systematic research. Further the knowledge has been organized into a consistent system called a body of theory. The practitioners must master this underlying theory to use as an intellectual model for decision making, and in explaining and integrating practical experience. In addition, the understanding does not come from on-the-job training or repetition of a fixed set of practices in order to develop certain functional skills, but through formal education whereby individuals study theory and underlying science over an extended period in an academic setting. The practical skills then serve as a means for drawing upon the body of theory and using it as a basis for solving problems and performing specific functions of the occupation.

The depth of knowledge required of a profession serves to highlight the lack of understanding by other persons, generally referred to as lay people. This difference sets the profession apart as an authority on some matter and practice. The clients recognize a need to call upon the profession for advice and counsel. They come to depend upon a professional to determine the needs and identify the best alternatives for meeting them. Clients also lack the capability factually to judge the professional's capabilities, and so turn to the profession to make judgments about its own members.

All professions gain public sanction by persuading society to grant them certain powers and privileges. At least, they strive to have people tacitly recognize their expertise. That gives the profession some degree of authority. The powers assigned to a profession commonly include control over its professional schools by an accreditation process, and the right to determine who enters the ranks of the profession. In some cases, a legislative body may have given the profession legal status. Its practitioners generally administer this law and interpret the qualifications for licensure. Community sanction also comes by recognition of the official professional organization, and in acknowledging its authority to control its members. The professional organization will hold its members responsible for meeting a standard of practice and require self-imposed concern for public welfare that transcends self-interest. Through these powers and privileges granted by society, the profession accepts an obligation to serve public interests.

To help guide the practice of its members, a profession will adopt a specific code of ethics and require allegiance to it. The code will guide the ethical behavior of practitioners and place the profession on public record as upholding a commitment to protect its clients from exploitation. This code will lie at the heart of the professional practice. It serves as a means to instill public trust in the authority of the profession relative to the body of theory that underlies its practice. A failure to endorse the code will threaten the public trust and destroy confidence in the profession.

Above all, a profession distinguishes itself by its culture. This becomes embodied in the social values that guide the behavior of its members in their relationships with clients and colleagues. It instills in the members an attitude of career service and devotion to duty, making one's work and life inseparable. The work of a professional becomes an extension and an integral part of life in general. Another part of the profession's culture includes the symbols by which people recognize it—distinctive dress, insignias, and emblems. This culture develops through a network of formal and informal groups that the profession operates, including the offices and firms of its members, its educational and research centers, and the organized association or society estabiished to promote the profession's vital interests and aims.

The Role of Professional Education

The aforementioned qualities make professionalism as relevant today as in the past, and will sustain it into the future. Other requirements may emerge from time to time and place new demands upon professional persons, and will probably include the ability to respond to the growth of technology and the increasing complexity of problems that confront society. The future seems to call for more multidisciplinary collaboration on a widening scale. It also will require more highly sophisticated and costly methods and procedures. This will take more team effort where professionals and highly skilled technicians of several discrete disciplines work together to develop solutions and apply them. Consequently tomorrow's professionals, and today's as well, will need to develop the capacity to join into and enhance group effort, and to encourage interplay between individuals with different skills and capabilities.

In summary, the key elements of modern professionalism include:

1. A mastery of special skills and methods
2. Knowledge of a body of theory embracing scholarly, historical, and scientific principles
3. Long and intense preparation, and a commitment to continuing study and career service
4. High standards of achievement and conduct
5. A commitment to personal freedom, independence of ideas, and objectivity
6. A strong sense of dedication to public service
7. An ability to blend one's own special skills with those of others, and to find satisfaction and fulfillment in the success of group undertakings

These attributes begin to develop during the college years as the faculty guides a student's preparation for professional service.

Professional education will provide qualified persons needed to help a profession fulfil its obligations to society adequately and efficiently. This must include a mastery of essential technical subject matter, balanced with practical experience in applying the theoretic concepts and principles. It will emphasize those basic concepts and interrelationships to provide a durable frame of reference that a person can later build upon in adding new technologies to replace older ones or obsolete information. In addition, the practitioner must develop a social understanding to use in relating the technical skills to relevant problems and in appreciating potential impacts of different actions upon people. It is hoped that, in the process, a student will develop basic professional qualities, and a dedication to the career that lies ahead. That attitude, in essence, determines the level of professionalism a person will have and exhibit in the daily practice of business.[4]

FORESTRY AS A PROFESSION

Forestry has distinction as a profession of long standing. Its roots run deep, through centuries of time. Over the years, it gained recognition by its dedication to the scientific management of forests, and its concerns for human welfare. It requires skill in using that knowledge when functioning within some specific role in forestry activity. Forestry also has high standards for performance and for fulfilling responsibilities that an individual assumes when engaging in forestry work. The success of practitioners in carrying out these roles and serving the needs of society has earned forestry this status as a profession.

Development in the United States

While forestry traces its history to rather early dates in Europe, American forestry practice began only during the last decades of the 1800s. When Bernard Fernow headed the fledgling Divison of Forestry in the U.S. Department of Agriculture in the late 1880s, he held distinction as the only person in governmental service who had obtained professional education in forestry. Uncertainties of the young profession seemed sufficiently real to Dr. Fernow that he advised Gifford Pinchot (Figure 1–2) to fit himself for employment in landscape gardening, nursery practice, and botany as well. This seemed a wise course of action in case he could not find employment in forestry. Pinchot went to France, and there studied forestry. He later returned to the United States, eventually to become the nation's Chief Forester when Fernow moved to Cornell University in 1898 to establish the first forestry school in the country.[5] By that time, forestry had received recognition as a unique discipline. The profession had taken root.

Early emphasis in forestry was largely on acquisition, caretakership, and management of forestlands for timber values. This work also included scheduling and supervision of logging, road building, tree planting, and other activities involved in growing and harvesting timber. Over time, responsibilities of foresters

Figure 1–2.
Gifford Pinchot, who studied forestry in France, served as the first chief of the U.S. Forest Service from 1905 to 1910. *(U.S. Forest Service photo taken February 7, 1909.)*

expanded. They gave attention to water, animals, grazing, recreation, and protection. Schools of forestry also developed programs in wood products utilization and engineering. They began offering education in entomology and pathology related to trees. Eventually they included other fields of biologic sciences that support forestry practice.

Opportunities for Practice

Professional opportunities for careers related to forestry today include a multitude of functions involving the management of forest plants and animals, influencing the physical environment of forested areas, and guiding the ways in which people use the resources that forests provide. This opens career opportunities that range widely in character and demand the services of people schooled in such divergent disciplines as:

Administrative functions to plan, support, and implement management programs and to facilitate use

In-woods activities directly related to the manipulating of forest ecosystems and harvesting or using the indigenous resources

Business management activities needed to keep enterprises functioning and to bring the products and uses of forests to a marketplace

Development and application of many technologies related to manufacturing and processing of forest products and their use in homes and industrial applications, and for many other purposes

Scientific activities involved in acquiring information and monitoring the condition of forest ecosystems and their component parts

Study and development of policy and legislation to direct and enable forest use for the public welfare

The types of activities a modern forester may have responsibility for vary greatly around the country and with the type of employer. Even people working for the same agency or corporation may do different types of things as part of their daily routine. Approaches to management will differ among forests, depending upon the indigenous species and the quality and sizes of trees present. The primary mission of the employer will also determine what the forester does, as will the nature of the nearby wood-using industry and the ways in which the local population depends upon the forest for various values and commodities. Demands upon forests for outputs of water, wildlife, and forage will also affect priorities given to different duties. And pressures for recreational use in some areas may involve the forester in activities that would have little importance in more remote places to which people do not have easy access. Further, as a forester gains experience, responsibility may shift from directly managing the resources to greater emphasis upon supervising other foresters and work crews who implement the management programs. Consequently a forester, or group of foresters working together, may experience great diversity in their professional responsibilities, and these will change from time to time.

Jobs of the Forester

What you do as a forester depends in large measure upon whom you work for. Foresters for public agencies or private firms may participate in activities of managing land and range, and controlling the use of resources thereon. This involves inventory work to determine the amount, kind, and condition of resources available. Foresters also plan, lay out, and oversee the harvesting of timber and other products. They may deal with recreation programs, work with wildlife managers, address concerns for water quality and yield, help protect the plants and animals against destructive agents, lay out and supervise the construction of roads and trails, and involve themselves in many other types of activities that comprise resource management programs.

Other foresters work for governmental agencies administering or supervising the conduct of various programs mandated by law or public policy. Some help to implement government regulations, such as the many fostered by the environmental movement that mushroomed in the 1970s. Foresters even work for legislative groups as staff persons to give advice on technical matters and aid in the drafting of legislation and policy. Public foresters also have responsibility for many types of government-assistance programs to private landowners, which may include timber stand improvement, tree planting, soil stabilizaton, watershed management, forest resource assessment, and other activities by which governments at all levels interface with private landowners who control so much of the nation's woodland resources.

Industrial management and procurement programs usually place greatest emphasis on timber production and give less attention to noncommodity uses of forests. Foresters employed with private corporations as land managers do most of the things described. Others will work with private landowners, buying the timber the company needs and arranging the harvest of those products. These foresters spend time assessing the amount and value of timber offered for sale. They determine how best to harvest the products and often supervise the logging crews. They will also arrange to transport the logs and bolts to a primary processing plant and may even have responsibility for some phases of mill operations. Foresters working with private industry also participate in many business management activities and contribute to decisions about how the corporation will operate.

Foresters employed both by public agencies and private firms may also work in a consulting capacity to help private landowners arrange and carry out their resource management programs. These foresters develop a working relationship with the individual landowner, determine the interests and objectives for managing the land, ascertain the potentials the land offers, prescribe ways to accomplish the goals, and provide a wide range of technical services needed to implement the plans. Some wood-using industries even maintain consulting programs as a part of their efforts to develop and maintain a lasting supply of materials for their mills. Other private consulting firms work directly for the landowners and serve as their agents in negotiating with perspective buyers. Public agency cooperative management foresters often offer their services as a part of some government-assistance program to help landowners get started in a management program. The actual types of activities a forestry consultant carries out depend upon the employer, the interests of the landowner, and the available resources.

Surprisingly large numbers of foresters work in educational programs of various types, including both traditional classroom instruction and educational outreach efforts. For example, some high schools and technical training schools hire foresters, as do public agencies for various public service programs that involve educational efforts. Corporations employ foresters for public information and public relations activities. Land grant colleges employ foresters as cooperative extension agents to help disseminate forestry information to many different publics. Publishers of books, periodicals, and newspapers employ foresters for writing and technical consultations. Special-interest groups, industrial associations, professional societies, and groups representing lay people hire foresters in many different types of education-related activities. Foresters also work in colleges and universities teaching and doing research. All these have a common interest in helping to share ideas and inform others about various types of forestry potentials and activities.

These few illustrations depict only some of the employment opportunities open to foresters. The different roles foresters play in society, business, industry, and government would number in the thousands. In fact, diversity characterizes the forestry profession. Further, the combinations of activities any one forester may become involved with during a career seem almost endless. They will change as a person matures from a novice forester into a seasoned professional. Most experience a general trend toward greater supervisory responsibility and less hands-on work in actually managing the resources. Hence forestry as a career offers frequent new challenges and opportunities. It provides many choices for shaping a career to one's own interests.

METRIC UNITS IN FORESTRY

For decades, many nations have relied upon the International System of Units (SI) and the metric system for measuring objects, weights, distances, volumes, and land area. Some researchers and laboratory scientists have traditionally used metric units, even in the United States. The United Nations reports all of its statistics in metrics, and during the 1970s Canada converted to the system as official national practice. However, since Colonial times people in the United States have used English units in industry, commerce, business, and government. Currently available information about the forest and range resources of the country do not include metric units. In fact, only in the 1970s did serious discussion begin about changing to SI units and making the system official as a national standard. The Metric Conversion Act of 1975 established a national policy of encouraging increased use of metrics, but in a voluntary program of conversion.[3]

A change to metrics will require a major national effort, and take a period of transition. We experienced some of those changes when the food industry began to label packages in metric weights and volumes during the 1970s. A more widespread use of metric units will require government and industry to convert their old statistics from English units and gradually to begin taking new measurements in metric ones. They also must retool machinery, equipment, and parts. Foresters likewise will need to adjust in a variety of ways. All of this may take at least a decade of transition, and under the all-voluntary system of conversion designated by the Congress will move ahead at varying rates in different states, industries, and businesses. In the interim, foresters can learn to think in both English and metric units. Those newly entering forestry certainly will need to speak the language of both systems and to develop a facility to convert back and forth in anticipation of the changeover. This book thus uses metric units throughout in the hope it will help students make the transition. The conversion factors presented in Table 1–2 should aid in the process.

Forestry depends upon several units of measurement for which most people have little need. These pertain to units of land, and ways to express quantities of different tangible products of the forest and range. Most serve as units of measure for amassing statistics for management planning, in keeping records about effects of management, in describing growth and change of the natural resources, and in determining the quantities and values of products available for sale. Table 1–2 lists several of the terms used throughout the book, and shows the English equivalents for each metric unit. For additional ones, readers can consult any of the metric conversion tables available in the published literature. Some organizations have printed booklets listing a full range of conversion factors to use in changing English to metric units, and vice versa. Table 1–2 utilizes two of these books that include forestry terms.[6,8]

LEARNING ABOUT THE OPPORTUNITIES

Over the years, U.S. forestry has blossomed into a complex profession with much to offer in career opportunities. Generally education in forestry provides for preparation in four broad fields: forest science, forest resources management, forest engineering, and wood products technology. Each has many subdivisions

TABLE 1–2
METRIC UNITS AND ENGLISH EQUIVALENTS FOR MEASUREMENTS COMMONLY USED IN FORESTRY

Metric unit	English equivalent
1 hectare (ha)	2.47 acres (ac)
1 square meter (m^2)	10.76 square feet (ft^2)
1 cubic meter (m^3)	35.31 cubic feet (ft^3)
1 kilometer (km)	0.62 miles (mi)
1 square kilometer (km^2)	0.39 square miles (mi^2)
1 liter (l)	0.26 gallons
1 centimeter (cm)	0.39 inches (in)
1 millimeter (mm)	0.04 inches (in)
1 kilogram (kg)	2.20 pounds (lb)
1 metric ton (mt)	1.10 tons (2000 lb)
Common conversion factors	
1 ha	10,000 m^2, or 107,639 ft^2
1 m^2/ha	4.36 ft^2/ac
1 m^3/ha	14.29 ft^3/ac
1 kg/ha	0.89 lb/ac
1 mt/ha	0.44 tons/ac

Source: Based on Johnson and VanOsdol [6] and Rennie [8]. Used by permission.

with almost endless opportunities for specialization. Among the forest sciences, for example, one may study in such areas as pathology, entomology, zoology, botany, and ecology. Forest resources management involves disciplines such as silviculture, watershed management, wildlife management, soil science, forest administration, forestry economics, mensuration, and biometrics. Study in forest engineering includes aspects of civil engineering, logging, materials handling, forest structures, photogrammetry, and related fields. Wood products technology encompasses primary timber processing, wood products engineering, paper science, wood chemistry, materials science, and associated areas. Choices available to students vary with individual colleges. Most forestry curricula, however, have optional courses that students may elect to focus upon a field of particular interest.

Education at a forestry college will acquaint students with a variety of career choices and alternatives programs of preparation. By 1980, the Society of American Foresters had accredited 44 professional forestry programs at colleges and universities in 35 states.[9] Another seven colleges listed nonaccredited curricula,[9] and 72 institutions offered two-year programs in forest technician training at the nonprofessional level.[1] Many of the schools also provide opportunity for study and research leading to the master of forestry, master of science, and doctor of philosophy degrees in several disciplines of forestry and forest science.

Available Resources

Students will find many resources available to help them explore career potentials and to understand the opportunities available through each. Taking advantage of these depends upon individual initiative. For example, each school has among its faculty people who appreciate what forestry offers and the preparation needed to practice it. Each has developed a specialty and pursues it through teaching and research. They welcome opportunities for discussion and will give advice and provide many types of encouragement. They can suggest courses to take, books to read, and names of others to contact.

Most schools also help students to find summer employment. This gives you another more informal way to find out what forestry practice amounts to and how different specialists operate within it. The summer jobs allow students to explore different parts of the country and to learn how forestry practice differs among different regions. In addition, summer employment provides a chance to learn important skills, to become better acquainted with the forest, and to experience different work environments that you might like to elect for your future career. Then, as you progress in your studies, the college will give you assistance in career placement, and many schools take active roles in helping students to link up with potential employers. These programs should be considered opportunities upon which you can capitalize. You must take the initiative and make the choices so that you end up in a situation to your own liking.

Students will gain increased appreciation for career opportunities by participating in the activities of professional societies. These organizations provide direct contact with working professionals and offer many educational and career guidance services. They hold technical meetings and usually have active programs in the state or region around your college. Many have organized clubs or chapters within the school. Student participants elect officers and plan programs of a professional nature. Often these focus upon career opportunities or look at the different types of activities foresters engage in during their work. Overall they provide excellent ways for students to become directly involved with their chosen profession and to become acquainted with others with whom they will work upon graduation from college.

Many schools, agencies, and corporations offer students opportunities for internships during their college years. These may substitute for formal courses in some cases, and students earn college credit for their participation. Generally the internship provides an experience of learning by working in a functioning forestry or resource-related program. Interns will learn by doing, by their contact with experienced professionals, and by independent study in developing solutions to problems given them to work on. These internships may take place during summer months and substitute for some type of employment. Others may provide for off-campus study during a semester. They may open up to either undergraduate or graduate students. For some they provide an excellent means for learning about a specific career field and for developing an improved capability for functioning within it.

The program of study provided by the college or university serves as the foundation for forestry. The faculty has developed it on the basis of their own professional experience, and through consultations with others. The program will involve a basic core of courses that all students take. These will introduce you to basic biologic and physical science, to mathematics and economics, to communication skills and social sciences, and to other subjects that serve as a base for more specialized forestry courses. Later you will study such subjects as silvics, forest management, mensuration, biometrics, aspects of business management, forest soils, aspects of physical environment, specialized economics, and aspects of forest protection. You will also find time for elective courses and can use them to strengthen your appreciation of some aspect of forestry science or practice, or even to develop more generalized skills with a working appreciation of several disciplines. Here again your faculty adviser and other members of the staff will help you to seek out the different options and to piece together an individual program that best suits your wants and needs.

Using This Book

The chapters that follow provide an overview of how forestry developed in the United States. They describe some important events that shaped its character and explain forest ecosystems and how they function. They identify the different types of forests in North America and outline how foresters manage the timber. Other chapters discuss concerns about soils and their effects on forest development, the nature of water as a resource and its management, wildlife and its management, the protection of forests against destruction and harm, development and management of recreation resources, and the organization of programs to manage forests for multiple benefits. The last section of the book explains the nature of wood products and their manufacture. Each chapter includes a discussion of background and technical information and how foresters and other specialists put that knowledge to use, together with a description of the discipline, and shows how interested individuals might plan their educational program to capitalize upon personal interests in different aspects of forestry practice.

While the many chapters provide considerable information and much detail about forestry and related matters, they present only a brief survey of several related topics. Readers should not look to the discussion for more than an introduction to the subjects covered. However, each chapter will provide sufficient coverage to give an understanding of the types of technical knowledge needed to practice in that field. The discussion should help you to identify an area of forestry you will find intriguing and might want to pursue. With that background, you can select courses and other means of study to develop the more detailed understanding and skills needed to practice in some chosen career. Reference to the listings of literature cited in each chapter will provide a starting point for additional readings about each of the subjects discussed.

REFERENCES

1. J. E. Coufal, "Forest Technician School Enrollment and Employment of Graduates, 1978–79," *Journal of Forestry* **78**(2):89–90 (1980).
2. F. C. Ford-Robertson (ed.), *Terminology of Forest Science, Technology Practice and Products,* Society of American Foresters, Multilingual Forestry Terminology Series no. 1, p. 710
3. P. F. Ffolliott, D. W. Robinson, and J. C. Space, "Proposed Metric Units in Forestry," *Journal of Forestry* (**2**):108–109 (1982).
4. A. Flexner, "Is Social Work a Profession?" *Proceedings of the National Conference of Charities and Correction, Chicago,* Hildmann Printing, 1915, pp. 576–590, as cited by H. S. Becker, "The Nature of a Profession," *Education for the Professions, The Sixty-first Yearbook of the National Society for the Study of Education, Part II,* University of Chicago Press, Chicago, 1962.
5. E. Greenwood, "Attributes of a Profession," *Social Work,* 2 (**3**): 45–55 (1975). Adapted from E. Greenwood and printed with permission.
6. D. M. Johnson and B. M. VanOsdol, *Handbook for Metric Conversion,* Idaho Research Foundation, Moscow, Idaho, 1973.
7. G. A. Pinchot, *Breaking New Ground,* Harcourt, Brace, New York, 1947.
8. P. J. Rennie, "Measure for Measure," Canadian Department of Forestry and Rural Development, Forestry Branch, Department Publication no. 1195, 1967.

9. Society of American Foresters, "Professional Instruction Offered in the U.S.," *Journal of Forestry* **78**(3):175 (1980).

10. Ibid., *Forest Policy Guidebook,* Society of American Foresters, Washington, 1980.

11. U.S. Forest Service, *Forest Statistics for the U.S., 1977,* U.S. Department of Agriculture, Forest Service, Washington, 1978.

12. Ibid., *An Assessment of the Forest and Range Land Situation in the United States,* U.S. Department of Agriculture, Forest Service, Washington, FS–345, 1980.

13. R. H. Whittaker and G. E. Likens, "The Biosphere and Man," in H. Lieth and R. N. Whittaker (eds.), *Primary Productivity of the Biosphere,* Springer-Verlag, New York, 1975.

CHAPTER 2

DEVELOPMENT OF FORESTRY IN THE UNITED STATES

When Columbus arrived in the new world in 1492, native Indian peoples in North America numbered about 900,000. They lived both in open prairie at the center of the continent, and among forests of the east and west. Some followed agrarian lifestyles, cultivating beans, squash, maize, tomatoes, potatoes, and tobacco. Tribes hunted buffalo, deer, and other game animals. But the Indians lacked iron tools and could make little impress on the forests or range, except by fire. Generally their clearings were small and scattered, and their settlements limited. The continent essentially remained a vast wilderness that offered seemingly unlimited resources coveted by the settlers who later arrived from a relatively crowded and urbanized Europe.

White settlers who surged into North America brought two major concepts of land ownership. Those who settled in the north subdivided the land into small owner-operated properties. Along the Hudson, Connecticut, and Delaware rivers the Dutch did establish large estates operated under a feudal system. These soon faltered as a result of Indian raids, of mismanagement, and of revolt by the other settlers who had immigrated to escape such domination. In the south, owners acquired large tracts of land for plantations, which were worked by sharecroppers or slaves. Consequently the landscape became highly fragmented into limited-acreage ownerships along much of the upper Atlantic coast, but partitioned into vaster holdings at many places in the south. The west remained unexplored and unsettled.

Settlers brought with them to the young nation two dominant goals. First, they wanted to occupy the land and use what it offered. Initially they struggled primarily to survive, probably impelled by the potential of the land and its vast resources. Later successful individuals began to exploit these resources for personal gain. Second, they sought to free the creative spirit of individualism, though still guiding it in a common interest. They wanted to control their own destinies and to break free of religious, political, and economic repression.

FORESTRY IN COLONIAL TIMES

American forestry had its foundations in the struggles and indomitable spirit of those early pioneers to make homes in the wilderness of the new world. The early colonists found almost unbroken tree cover that stretched from the eastern seaboard westward to the prairies. Extensive forests also greeted their arrival in the Rocky Mountains, along the west coast, and northward into Alaska. Altogether the eastern hardwood forest occupied about 163.2 million hectares. Coniferous ones spread across another 217.4 million, mostly in the west. In addition, open woodland covered 37.7 million hectares, making a grand total of 418.3 million of forest and woodland. This amounted to 46 percent of the total land area that became the new nation. With such resources, the forest offered great potentials for goods of trade and received great attention in policy, laws, and business, even from the beginning of the settlement era.

Most early colonists really considered the forests an enemy to conquer. The large trees formed dense stands that blocked out sunlight. They felled them to make room for houses, croplands, and pastures. This task proved formidable. And even when felled, the hard and durable trees could be disposed of only by burning. Transportation was limited and few markets or sawmills existed. They had little impetus to convert the fallen trees into lumber. Besides, the seemingly endless expanse of forest lying to the west gave little reason for alarm about future shortages. Fire became a natural ally in the land-clearing process, and as the population gradually increased, the settlers used fire along with their iron tools and work animals to change the face of the landscape to fit their purposes better.

Rough-hewn logs taken from trees felled to clear the land served as materials for the first crude cabins and other buildings (Figure 2–1). Eventually some enterprising people did cut boards by using whipsaws, and in 1625 and 1631 sawmills began to operate in Virginia and Maine. The colonists harnessed waterpower from swiftly flowing streams to drive the machinery, and settlements developed around these mills. Soon others proliferated along the coast and lumber became a major export item to Europe in exchange for weapons and other manufactured goods.

Lack of facilities to transport heavy materials limited lumbering to the lands along the coast and the rivers that ran to the sea. There the colonists loaded the lumber on ships for transportation to the essential European markets. The ocean provided settlers with an important link to their homelands, and shipping thrived. Coastal trade also developed among the colonies and with the West Indies to the

Figure 2–1.
The early Colonists depended upon the forest as a source of building material, such as the rough-hewn logs used to recreate the old French missionary settlement near Syracuse, N.Y.

south. Again the sea provided a major transportation link for this trade and as a means for communication. This soon gave rise to a fledgling shipbuilding industry that rapidly grew into prominence in New England. With it came increased demands upon the forests. The shipbuilders needed oak and pine for timbers, planking, and masts. By the time of the Revolution, Colonial shipyards were producing 100 ships per year,[4] supported by a growing lumber industry to supply the materials.

Settlers along the frontier continued to dispose of timber by burning it. But rather early along the settled seacoast concern began to develop about the oak and pine needed in the shipbuilding industry and for other construction. As the population increased, more and more land was cleared for farming. Readily accessible supplies of lumber began to dwindle. In fact, just six years after the Pilgrims landed, the Plymouth Colony enacted a law forbidding the sale or transport of timber out of the colony without the approval of the governor and the council. By 1668, the Massachusetts Bay Colony had adopted other restrictions on cutting ship timbers. In 1681, William Penn incorporated a provision in his land purchase contracts "that in the clearing of ground, care be taken to leave one acre of trees for each five acres cleared, and especially to preserve mulberry and oak for silk and shipping."[3] A later Massachusetts Bay Colony charter, in 1691, strengthened the early restrictions, and in it the king reserved the best white pine trees for ship masts for the Royal Navy. He also ordered his surveyors to scout the forests of New England for suitable white pine and to carve into their bark a broad arrow as a sign of the king's property. This practice later was extended into New York and New Jersey.

The king's broad-arrow policy proved difficult to enforce. Trespass commonly occurred despite severe punishment to anyone found guilty of cutting a broad-arrow pine, and such punishment actually seemed only to encourage defiance of the king's authority by the independent-minded colonists.[9] Later, during the Revolution, patriots in Maine, Massachusetts and New Hampshire did their utmost to prevent the Royal Navy from getting the mast and spar timbers so critical to the British shipbuilding program. Free use of these trees became a major grievance against the king in the colonists' fight for independence.

Early needs to transport timber and other goods by water also led governments to designate many streams large enough to float logs as navigable waters. These became free of restriction and control by riparian owners. Colonists also passed laws to protect forests from fire and timber trespass on both private and communal lands. Other laws set standards for wood products exported from the colonies in the hope of protecting the reputation of its forests as a source of high-quality lumber for European markets.[5] Thus, from rather early times, public action began to shape a forest policy through legislation that would eventually guide the development of U.S. forestry activity.

Such measures had purpose. Colonists faced shortages of fuel for iron furnaces, glass factories, home heating and cooking, building, and many industrial uses. Quite early in their history, cities such as Providence and Philadelphia faced problems with easy access to fuel to satisfy a growing demand by their expanding populations. These shortages resulted from problems in transportation rather than from scarcity. Nearby resources dwindled and suppliers had to reach out farther and farther to get their materials. Benjamin Franklin commented in 1774 that: "Wood, our common fuel, which within these three hundred years might be had at any man's door, must now be fetched near (160 kilometers) to some towns, and makes a considerable article in the expense of families."[4] Three hundred years had carved up the seemingly endless eastern wilderness. The gradual urbanization of the new world created problems in supplying the wood needs of the fledgling nation.

THE PUBLIC DOMAIN AND ITS DISPOSAL

Conflict over control of land arose among the colonies even before they won independence. In large measure, these problems resulted from the vagueness of the charters. For example, Virginia's charter covered land "throughout from Sea to Sea, West and Northwest." Massachusetts claimed territory even beyond the Delaware River.[10] Also, Connecticut and Pennsylvania both had rights to rich farming lands in the Wyoming Valley of Pennsylvania. As a result of these disputes armed conflict occurred at times. Tension developed, and the colonies seemed unable to resolve their claims.

To reconcile the disunity, the Continental Congress in 1780 passed a resolution urging the new states to cede their western lands to the Union. By this means, from 1781 through 1802 the federal government acquired jurisdiction over about 94.3 million hectares.[5] The public domain began to accumulate, supplemented by vast areas acquired in the west and south through the Louisiana Purchase, various acquisitions from Mexico, the purchase of Alaska from Russia, and in several other ways.[5] By 1867, the public lands covered just over three fourths of the contiguous United States and all of Alaska, a total of 731.6 million hectares of land and 12.2 million of water.[5]

These federal lands became a source of wealth to the developing nation, but irregularities in their management embarrassed its elected officials. The federal government sold land to obtain public revenues, granted it to soldiers in lieu of cash payments, gave it to companies to promote the building of railroads, and granted some to the states to further public education. An act of 1807 forbade squatting on the public domain in an effort to preserve it for public revenues and national development. But embarrassment arose from the lack of trained land administrators to oversee its use and control. The government had no police agency to enforce the laws and prevent trespass.

Attitudes of the Congress and many federal officials ran contrary to the desires and actions of settlers along the frontiers. Those stalwart people occupied any land not already taken by others, and felt free to use it for their own purposes. The concept of a public domain closed such land to their use and conflicted with their interests and needs. Strong demand arose that the public lands be opened for private settlement, and these pressures eventually prompted the Preemption Act of 1841. This act granted homesteaders the right to squat on public land, and eventually to purchase it. In 1862, during the administration of Abraham Lincoln, the Congress passed the Homestead Act, which authorized the granting of about 65 ha free to settlers who lived on the land and developed it. In 1873, the Timber Culture Act empowered the government to grant 65 ha to anyone who planted trees on 16 of them and kept the new forest healthy for 10 years. Even later, the Timber and Stone Act (1878), Desert Land Act, and Carey Act (1894) all liberalized conditions and made it easier for settlers to acquire title to public land through homesteading.[17] In addition, the Swampland Act of 1849 and 1850 resulted in the transfer of millions of hectares of swamplands from the public domain to private individuals. Through these decisions by the Congress, portions of the vast public ownership became available for private occupancy and helped to foster the move to westward settlement.

To encourage railroad construction and expand that critical transportation network, the Congress also gave land to the railroad companies. These included alternate sections of land (1.61 km^2) as far back as 32–64 km from either side of the right-of-way. This created a checkerboard pattern of ownership along the route of the rail line. Later the Forest Reserve Act of 1897 included a provision for any legal claimant to swap previously acquired lands for others of equal size anywhere in the open public lands. Many companies and individuals promptly used this authority to improve the quality of their holdings, and to consolidate them. In the process, they acquired some of the best timber and rangelands in the country. Altogether, by the various means of homesteading, railroad grants, and other authorizations, the government had transferred about 325 million hectares of public domain to the private sector as of 1953. Another 91.7 million hectares went to states for various purposes.

The spread of settlement across the continent consumed great amounts of lumber and timber, but the government still controlled most of the forest resources and had no legal means for selling them to lumber companies. Further, administration of the public domain and the numerous laws related to it remained weak. These factors contributed to misuse, plus extensive trespass onto the government lands to steal the timber.[5] As a means of acquiring the land and timber, some companies even hired bogus entrymen to file for title under the Timber and Stone Act, and then transfer their claims to the company.[16] Later the company could exploit the timber at a great profit. The blame fell upon the Congress for not taking appropriate action to make the timber legally available through sales and other means, and for failing to provide a mechanism to control the disposal of the public lands in a better fashion.

BEGINNINGS OF FEDERAL FORESTRY

During the Colonial period and the first 100 years of the Republic, many prominent citizens and groups urged public action to reserve the nation's forests for sustained crops of timber and to protect watersheds. None of these pressures evoked congressional measures, but their arguments helped to convince many other citizens of the importance for action. The spark that eventually started federal efforts in forestry as a governmental program grew out of the American Association for the Advancement of Science (AAAS) in 1874. In a message to the Congress, AAAS included excerpts from a talk by Franklin B. Hough of Lowville, New York. He wrote:

> The preservation and growth of timber is a subject of great practical importance to the people of the United States, and is becoming every year, of more and more consequence, from the increasing demands for its use; and while this rapid exhaustion is taking place, there is no effectual provision against waste or for renewal of supply . . . besides, the economical value of timber for construction, fuel, and the arts . . . question of climate . . . the drying up of riverlets . . . and the growing tendency to floods and drought . . . since the cutting off of our forests are subjects of common observation . . .[15]

At the urging of AAAS, in 1876 the Congress authorized a survey of existing forest resources and of demands placed upon them by consuming industries. It also included a study of possible influences of forests on climate, and a review of forest practice in Europe. The Commissioner of Agriculture employed Dr. Hough to carry out the task. He prepared three voluminous reports. His successor, N. H. Eggleston, issued a fourth. These contained the best information then available about America's forest resources and industries. The study constituted the first national forest assessment.[15]

In 1881, Congress acted again and established the Division of Forestry under the Commissioner of Agriculture. That agency later became the U.S. Forest Service. Numerous other bills brought before the Congress sought to reserve the nation's timberlands under public control. These suffered defeat because of pressures from special-interest groups that preferred to have the lands under private ownership. In 1891, however, a rider attached to an act amending the land laws authorized the president to set apart from the public domain reserves of land bearing forests, whether commercial or not.[10,11] This act proved critical to the development of federal forestry. With its authority, Presidents William Henry Harrison and Grover Cleveland began to reserve large blocks of public land for federal ownership. These became the National Forests and remain today under federal management by the U.S. Forest Service.

At the request of President Cleveland, the National Academy of Sciences set up a special forestry commission in 1896 to report on the forest reserves and to recommend measures for their proper use. He appointed Charles Sprague Sargent of Harvard University as chairman. Under Sargent's leadership, the commission recommended increasing the federal reserves by an additional 8.6 million hectares. It offered no advice for managing them. A storm of protests quickly arose throughout the western states over what local people called the locking up of public lands.[10,15] This protest led Congress to suspend all but two of the reserves created by Cleveland. In addition, it adopted the Sundry Civil Appropriation Act of 1897, which provided for managing existing and future reservations. The law stated that: "No public forest reservation shall be established except to improve and protect the forest, secure favorable conditions, and furnish a continuous supply of timber for use and necessities of citizens of the United States."[15,16] The decision received the support of the western residents, and laid the groundwork for a rational policy of managing and protecting the forest reserves. It served as the basis for federal land management until 1976, when the Congress passed the National Forest Management Act as a new authority for managing those public forest holdings.[16]

The Roosevelt-Pinchot Era

By the turn of the century, the nation seemed ready for a major conservation effort. It just seemed to await effective leadership from the White House. This came with dramatic suddenness on September 14, 1901, when Theodore Roosevelt succeeded to the presidency following the assassination of William McKinley. Roosevelt had become imbued with principles of conservation and an appreciation of the importance of a national forestry program. These grew out of his interests in wildlife and his long friendship with Gifford Pinchot, the nation's Chief Forester. Knowing of this interest, Pinchot and F. H. Newell, who later became the first director of the Reclamation Service, urged the president to include a strong conservation statement in his first message to Congress. The president agreed, and added these paragraphs[2,11]:

> The fundamental idea of forestry is the perpetuation of forests by use. Forest protection is not an end in itself. It is a means to increase and sustain the resources of our country and the industries which depend upon them. The preservation of our forest is an imperative business necessity. We have come to see clearly that whatever destroys the forest except to make way for agriculture threatens our well-being.
>
> The practical usefulness of the National Forest Reserves to the mining, grazing, irrigation, and other interests of the regions in which the reserves lie has led to a widespread demand by the people of the West for their protection and extension.

He went on to advocate uniting the administration of the nation's forest reserves into the Bureau of Forestry.

Under Pinchot's leadership, the Bureau of Forestry had gradually grown in activities, size, and influence. Through its service to owners of large private forest properties, many prominent citizens had come to learn and appreciate the meaning of forestry. At the same time, the forest reserves under the jurisdiction of the Department of the Interior had received little protection and no management. Hence the president's proposal to transfer jurisdictional control of the forest reserves to Pinchot's bureau received widespread public support.

Throughout the same period, the American Forestry Association (AFA) had gained an important leadership role in national conservation programs. Organized in 1875 as a lay forestry group, the AFA worked to complement the efforts of scientists and scholars in advocating development of a federal forestry program. It had helped to organize the first American Forestry Congress in 1882. That gathering brought national attention to forestry.[2] The second Forestry Congress convened in Washington, D.C., in January 1905. Participants included the Secretaries of Interior and Agriculture, other government officials, and conservation-minded citizens. Association leaders from the lumber, railway, and cattle industries also participated. It resolved to urge formation of a dynamic federal forestry program, and called for immediate transfer of forest reserves to the Bureau of Forestry. The Congress responded, and in 1905 passed the Transfer Act and President Roosevelt signed it into law. A month later, the Department of Agriculture assumed full legal authority for regulating the forest reserves.

Altogether the Roosevelt administration arranged the transfer of nearly 61 million hectares from public domain to what would become the National Forests.

Opposition developed in the west and eventually Congress withdrew from the president the right to set aside new National Forests from the public lands. But before signing this bill, President Roosevelt first designated even more land as National Forest. This included much of the area he considered the more important public domain.[5]

Up to this point, questions of public domain and National Forests referred to the west. The Weeks Act of 1911 changed the situation. It authorized acquisition of land for National Forests wherever necessary to protect navigable streams, thus providing the basis for expanding the National Forest system to the eastern states. The authorization later included lands for the growing of timber. As a result, the Forest Service broadened its holdings to include 43 National Forests and to purchase other units, with lands in 23 states east of the Rocky Mountains.

Regulating Use of the National Forests

Following passage of the Transfer Act, Secretary of Agriculture James Wilson promptly sent a letter to the chief, proposing administrative policies to guide management and use of the new National Forests. This letter, from which the following is excerpted,[11] became a classic in federal forest policy.

> In the administration of the forest reserves it must be clearly borne in mind that all land is to be devoted to its most productive use for the permanent good of the whole people, and not for the temporary benefit of individuals or companies. All the resources of forest reserves are for use, and this use must be brought about in a thoroughly prompt and business-like manner, under such restrictions only as will insure the permanence of these resources. The vital importance of forest reserves to the great industries of the western states will be largely increased in the near future by the continued steady advance in settlement and development. The permanence of the resources of the reserves is therefore indispensable to continued prosperity, and the policy of this department for their protection and use will invariably be guided by this fact, always bearing in mind that the conservative use of these resources in no way conflicts with their permanent value.
>
> You will see to it that the water, wood, and forage of the reserves are conserved and wisely used for the benefit of the home builder first of all, upon whom depends the best permanent use of land and resources alike. The continued prosperity of the agricultural, lumbering, mining, and livestock interests is directly dependent upon a permanent and accessible supply of water, wood, and forage, as well as upon the present and future use of their resources under business-like regulations, enforced with promptness, effectiveness, and common sense. In the management of each reserve local questions will be decided upon local grounds; the dominant industry will be considered first, but with as little restriction to minor industries as may be possible; sudden changes in industrial conditions will be avoided by gradual adjustment after due notice; and where conflicting interests must be reconciled, the question will always be decided from the standpoint of the greatest good to the greatest number in the long run.

Soon after this jurisdictional change, the Bureau of Forestry became known as the U.S. Forest Service, a subdivision of the Department of Agriculture. Pinchot threw his dynamic energy into building it into a strong agency and strengthening its program for National Forest administration. The task was not easy. Cattle and sheep ranchers had grazed their animals on the public domain without permits or control and resented federal intervention. To win their cooperation, the forestry officers had to exercise a great deal of diplomacy and to rely on sheer courage in many cases (Figure 2–2). By operating with fairness, honesty, and devotion to the ideals set forth in Secretary Wilson's letter, the Forest Service established itself as a responsible public agency. Through its research it brought new philosophies and methods to the attention of administrator and user alike. Gradually ranchers, lumber companies, water users, and a large segment of the general public began to see the Forest Service as a partner in resource management and use.

Federal-State Cooperation

Gifford Pinchot recognized that practices on farms, rangelands, and forests have an impact upon soil erosion, flooding, inland navigation, water power generation, fish and game supplies, and other long-term societal needs.[11,16] He sought to bring

Figure 2–2.
Early forest officers such as these shown breaking camp in Colorado's White River National Forest in 1916 worked hard to win the support of local people and to establish the credibility of the U.S. Forest Service in resource management and use. *(U.S. Forest Service.)*

this idea to national and international attention, and arranged a White House Conference of Governors in 1908 to create state and national commissions of conservation. In response to the conference, the president created a National Conservation Commission to give attention to the nation's basic resources of water, forest, land, and minerals. When William Howard Taft succeeded Roosevelt as president of the United States in 1909, he appointed Richard Ballinger Secretary of the Interior. The strong-willed Ballinger soon clashed with the equally stubborn Pinchot. It became apparent that Taft could not keep both men in the government if he wanted harmony to prevail, and so he dismissed the lower ranking Pinchot. The president wisely selected as his new Chief Forester Henry Solon Graves, dean of the Yale School of Forestry. The choice pleased both Pinchot and the staff of the Forest Service. Graves proved a good chief, and helped the advance of forestry by giving it the backing of the federal Forest Service. Federal forestry programs soon began to broaden.

From this early foundation, the Forest Service expanded its programs to build cooperation with the states, forest owners, and forest industries. It broadened its research to cover both forest management and forest products. Results influenced new state approaches to taxation that have provided relief to many owners. In addition, the Forest Service initiated a nationwide forest inventory to provide a better appreciation of the nature and extent of the nation's forest resources and their capacity to supply present and future needs. This inventory continues with periodic updating to the present.

The Morrill Act of 1862 and subsequent similar acts had granted land and money to the states for establishing colleges of agriculture and mechanic arts. Education, research, and extension of information to users in agriculture thus received national support. Initially this did not include work in forestry. However, the Forest Service had enlisted the aid of the states to protect nonfederal lands from fire and pests, and to encourage other forestry practices on private lands. That effort gained momentum as a result of the Weeks Law of 1911. It authorized federal matching funds for fire protection. Later the Clark-McNary Act of 1924 extended cooperation to include forest extension, tree planting, assistance to forest owners, control of forest pests, help to landowners and primary wood-using industries, research, public education in fire prevention, acquisition of land for recreation, and the development of forests for paid recreational use. By 1966, all 50 states had established cooperative programs with the federal government in these areas, facilitated by the infusion of federal funds to help defray costs.

Public Works Programs

The great economic depression that began in 1929 caused failure of corporations and banks, foreclosures on farms and homes, wide-scale loss of jobs, and general paralysis of industry and trade. Businesspeople, politicians, government officials, and unemployed veterans thronged to Washington to lobby for decisive steps to reverse the downward drift of the national economy. As a result, President Herbert Hoover in 1932 requested, and the Congress approved, a public works program to create employment. Under the later leadership of President Franklin D. Roosevelt, this program took effect in 1933. It included river valley development, forestry practices on federal and state lands, park expansion, soil conservation and flood control works, and construction of much needed public facilities. These developments proved important to the nation's forestry activity.

As one program through the public works bill, the Congress created the Civilian Conservation Corps (CCC) in 1933. It established work camps across the nation and employed people for tree planting, care of plantations, timber stand improvement, construction of forest roads and trails, building of lookout towers, and various other forestry measures on public lands. At its peak, the program included 2600 camps, each organized to handle 200 persons. Until it was closed in 1941 at the onset of World War II, the CCC program acquainted many different people with forestry work through their employment in it. The CCC also initiated deliberate management of forest stands on national and state lands. In many ways, it marked the beginning of active public forest management.

Through 1939, cooperative federal activity with states grew as a result of these emergency programs, plus a general expansion of regular activities. Land purchases for National Forests continued, especially adding cutover lands that private individuals and firms could not afford to hold. This same era saw the initiation of a prairie–plains shelterbelt program. Through this cooperative endeavor, federal and state officials hoped to solve problems that had plagued farmers during the dust bowl days of the early 1930s. It emphasized tree and shrub planting, and resulted in establishment of 29,800 km of windbreaks in the prairie and plains states by 1942. Also, the Taylor Grazing Act of 1934 and subsequent measures placed about 59.5 million hectares of the remaining public domain into federal grazing lands. This also resulted in the formation of the Grazing Service within the Department of the Interior to manage this program. Later, in 1939, the Congress appropriated funds to initiate fire control in Alaska, in an effort to protect the public domain in that territory.

Developments Since 1945

The national economy had reached record levels by 1945. Demand for consumer goods and homes caused a boom following the war. Forest products were in great demand, and the improved market activity encouraged forest industries to look ahead to long-term growth and development. In turn, this stimulated the industries to take more deliberate actions to protect their raw material supplies.

Through this expansion era, timber supplies of the nation remained strong. Even in 1963, annual timber growth exceeded the cut by a substantial margin. Yet growth of the nation's population and a projected increase in demand for timber products to supply the needs of these people seemed to presage a shortage of timber by 1980.[12]

Improved prosperity resulted from the postwar expansion. With it came improved mechanisms for economic development, bringing new wealth, more leisure time, and increased potential for recreational pursuit. Transportation facilities had also improved vastly. And, along with the national prosperity, personal income increased to give more people the opportunity to own automobiles. These factors acted as a catalyst upon each other. By the mid-1950s, people began looking to the outdoors for recreation. They had the capacity to travel to areas that had been inaccessible to a majority of them in earlier days. With this development came increased pressures upon forests to serve nontraditional needs. Demand for outdoor recreation rose sharply. Attention shifted to the managing of lands for multiple benefits, and conflicts over priorities arose.

The Forest Service had recognized its task of providing adequate recreational facilities and needed enabling legislation to authorize the several uses public forests served. In response, Congress adopted the Multiple-Use Sustained Yield Act of 1962. That law declared that National Forests should provide for outdoor recreation, range, timber, watershed, wildlife, and fish purposes.[16] The act set no priorities, but left this to local determination. By determining public interests and balancing the needs in relation to available resources, each forest supervisor could develop a program of management and use best suited to satisfy local demands.

The Outdoor Recreation Resources Review Commission appointed by President Dwight D. Eisenhower in 1958 established guidelines for outdoor recreation by both public and private agencies. In response, the Department of the Interior formed a Bureau of Outdoor Recreation to initiate cooperative federal, state, and private planning and development of outdoor recreational facilities. The work of the bureau gained impetus through the establishment of the Land and Water Conservation Fund in 1964. With these monies the bureau could finance acquisition and development of lands for recreational use in accordance with statewide outdoor recreational plans. In response, states undertook the requisite planning. For the first time, this gave comprehensive attention to the nation's growing recreational demands.

Interest in wilderness also grew rapidly after World War II, and so executive orders established specific wilderness areas within the National Forests. These represented administrative areas set aside in roadless portions of the National Forests, mostly in the west. The idea for these came from the forestry profession, which institutionalized it by administrative orders. Foresters watched after the wilderness as an ongoing part of the routine forest management for government-controlled lands. But such groups as the Wilderness Society and the Sierra Club pressured the government to protect these set-aside areas more permanently. In the end, the political pressures motivated the Congress to enact specific legislation to replace the administrative orders. The Wilderness Preservation Act of 1964 did this by establishing additional areas in the National Forests, National Parks, National Wildlife Ranges, and National Monuments. By 1976 the coverage totaled about 5.1 million hectares, with another 10.6 million earmarked for further review by the Congress. However, to create these wilderness areas, Congress overrode provisions of the Multiple-Use Sustained Yield Act and ordered exclusive use as nonmanaged wilderness.

One other movement had provoked major changes in federal natural resource programs. Like the wilderness policy, this one also resulted from an upsurging of public pressure upon the Congress. It began in 1962 when Rachel Carson's book *Silent Spring* appeared in bookstores across the nation. In simple, convincing language she argued that in an effort to curb insects and other pests, people had also poisoned the environment. Her arguments stirred public awareness and concern about the way in which careless acts of many kinds had resulted in a degrading of natural ecosystems. As public pressure mounted, Congress acted, and in 1969 passed the National Environmental Policy Act. It set forth a national environmental policy and established the Council on Environmental Quality, which subsequently led to the creation of the Environmental Protection Agency.[15] The agency's tasks include research, enforcement of certain environmental standards set by the Congress, monitoring of environmental conditions across the nation, and coordination of environmental protection activities. Although directed at all types of potential threats to air, water, and land resources, the agency's programs also affect the ways in which we use and manage these natural resources. The act has had profound effect upon the programs and activities of private individuals, forest industries, and municipalities and other divisions of government whose actions might adversely affect the quality of our natural environment.

Recent attitudes by some of the public toward management and use of forest resources and attitudes toward wilderness and environmental quality have affected many forestry operations, largely in relation to timber harvesting. In particular, the pressure placed upon such federal agencies as the U.S. Forest Service by various special-interest groups warrants consideration. One example shows the conflicts that can arise in trying to satisfy several public interests.

In 1968, Congress adopted a national goal of stimulating the building of new housing, and acted to provide aid in financing home ownership. Builders responded to the program, but soon ran short of lumber. At the same time, Japan launched its own housing program and sent buyers to the United States to purchase logs and lumber for export. As a result of these two demands, shortages became exaggerated and prices rose. By the early 1970s, both the lumber industry and the president were urging the U.S. Forest Service to increase timber sales from the National Forests. At the same time, presidential restrictions on federal hiring prevented employment of additional foresters to lay out and supervise the increased harvests.

Just prior to this time, the Forest Service had adopted clearcutting as a method for regenerating mature stands of timber in the eastern hardwood forests. It had used the technique successfully among western conifer forests since the 1930s. Clearcutting fitted the biologic characteristics of many tree species. Further, it permitted harvest of designated timber with a minimum of supervision once the forestry staff had determined that clearcutting appeared appropriate. Hence the U.S. Forest Service moved ahead with its program for even-aged management on a national scale. It began more widespread use of clearcutting to remove mature stands and create new ones for future crops. The apparent simplicity of setting up timber sales by clearcutting fitted well with the shortage of staff and the pressure to increase harvesting operations. However, the program began to create conflicts with the growing number of people who looked to the forests for recreational uses.

Though it based the decision to use clearcutting upon results from extensive research, the method met growing public criticism. Citizens, awakened to a new awareness of natural resources and concerns for protecting them for the future, complained loudly. People in places as widespread as Montana and West Virginia voiced their opposition. Many argued that clearcutting limited the multiple-use options of the areas and some wondered if it might prove harmful to the productivity of forestlands. Special-interest groups such as the Isaac Walton League brought suit against the Secretary of Agriculture for violating provisions of the Sundry Civil Appropriation Act of 1897, which had given the U.S. Forest Service authority to sell timber. That law specifically authorized the sale of large, mature, or dead tress, if individually marked or designated. In selling timber by clearcutting, the agency had not individually marked the trees within its clearcuts.

Both the trial and appeal judges concurred that clearcutting did not strictly follow the letter of the law. They stopped timber sales that violated exact provisions of the law. In response, the U.S. Forest Service asked for broader authority to sell timber from the National Forests. As a result, Congress passed the National Forest Management Act of 1976, permitting more flexibility in managing timber resources of the National Forests. This act provides the current authority for the management of our federal forest resources.[18]

Developments in Review

Certain milestone legislation seems to characterize the administration of each past chief of the Forest Service. These men helped channel the nation's forest policy development to respond to the needs of their day. To illustrate, Gifford Pinchot (1898–1910) brought about the transfer of timber reserves from the Department of the Interior to that of Agriculture, and their organization for effective administration. Henry Graves (1910–1920) saw the Weeks Law passed, and implementation of cooperation with the states in protecting and managing forests not under federal jurisdiction. William Greely's administration (1920–1928) encouraged the Clark-

McNary Act, which greatly broadened cooperation with states and forest industries. Robert Y. Steward (1929–1933) worked for passage of the Knutson–Vandenburg Act, which provided funds for restocking lands after timber harvesting. Ferdinand Silcox's (1933–1939) administration initiated the shelterbelt project and the Cooperative Farm Forestry Act. Earle H. Clapp (acting, 1939–1943) helped in developing the first national plan that served as a basis for later public acquisitions. Lyle Watts (1943–1952) brought provisions of the Forest Pest Control Act to bear upon all forests. Richard McArdle (1952–1962) worked successfully for passage of the Multiple-Use Sustained Yield Act. Edward P. Cliff (1962–1972) was serving as chief when Congress passed the Wilderness Act. John McGuire (1972–1980) helped to guide passage of the Forest and Rangeland Renewable Resources Planning Act, and its amendment to embrace the National Forest Management Act.[18] These achievements testify to the cooperation between various chiefs of the Forest Service and the Congress, and to the many people who helped convince a growing nation of the importance of forest care and use in meeting its increasing demands.

Study reports presented in 1973 by the President's Advisory Panel on Timber and the Environment, in the *Outlook for Timber in the United States* prepared by the U.S. Forest Service, and in the *Report of the Nation's Material Resources* have highlighted the current status of our national wood and energy situation.[12,19,20,22] They served as one impetus for Senator Hubert Humphrey to begin work on drafting the Forest and Rangeland Renewable Resources Planning Act that passed the Congress in 1974. This act requires that the Forest Service more broadly assess the nation's forest resources and keep the inventory current by regular updating. It also calls for more intensive effort to regenerate National Forest lands and for high-level timber production. Further, it gives the U.S. Forest Service responsibility for keeping the nation informed about the status of forest resources as these change as a result of use and regrowth. In 1976, the Congress also incorporated these provisions within the new National Forest Management Act. That action supplanted the authority originally given to the government by the old Sundry Civil Appropriation Act of 1897.

These and many other developments in U.S. forest policy awakened public awareness to the value of the nation's forest resources. At the same time, they encouraged restraint in dedicating National Forest lands exclusively to timber uses. But just as the nation seemed well on its way to solidifying its attitudes toward forest resources, an embargo by petroleum-producing countries in the late 1970s also awakened public appreciation concerning our dependence upon overseas energy sources. This in turn has focused some attention on the potentials for using wood as one substitute for many oil-based products, and as a source of energy. Almost overnight, the forests of the nation seemed to gain new importance as a supplier of materials that we could regularly renew through good management. How much the country will turn to its forests for these purposes remains unclear. The situation warrants watching as a future issue.

DEVELOPMENT OF STATE FORESTRY

State forestry in the United States began in Colonial days with early ordinances in Delaware, Pennsylvania, and Massachusetts. All recognized the importance of forests and regulated their use. Agricultural societies formed in Philadelphia as early as 1791, in New York in 1795, and in Massachusetts in 1804 soon recognized the value of tree planting and urged steps to encourage it. These represented some of the earliest organized drives to promote good forestry practices in the nation.

In 1819, the Massachusetts Legislature asked the state's Department of Agriculture to promote the growth of oak suitable for ship timbers, and in 1837 authorized a survey to determine measures that might encourage private forestry practice. Earlier, in 1788, New York had passed laws to regulate lumber grading for national and international trade.[23] In 1867, Michigan created a forestry commission to investigate timber operations, land clearing, and proper care of timberlands within the state. Wisconsin appointed a similar group shortly thereafter. Activity included shelterbelts and proper management of forested areas. Later New York, Connecticut, New Hampshire, Vermont, Ohio, Pennsylvania, and North Carolina also formed forestry commissions.[9] These early developments set the stage for programs that spread across the nation and persist today.

In 1872, New York set up a commission to investigate the preservation of forests in its Adirondack region. This mountainous area had served both as a source of timber and as the playground of many affluent people who resided in larger eastern cities (Figure 2–3). The study led to the establishment of the Adirondack Forest Preserve in 1885. Some nine years later, in 1894, citizens dissatisfied with the management of these lands pushed for and obtained a state constitutional amendment prohibiting the cutting of trees within the preserve. That control still remains in effect. But New York's example of acquiring land for permanent forest use triggered acquisition activities in Pennsylvania in 1895, in Minnesota in 1899, and in many other states thereafter.

In many cases, state ownership came as a result of the default of owners on taxes and of land abandonment, rather than because of deliberate state policies.[6] The attraction of more fertile lands in the middle west, and the increased capacity to ship goods back east to the population centers following the building of the transcontinental railroad, lured farmers from the hill farms of the eastern states. Marginal farming operations in the east and north collapsed, and once active farms became idle. The depression that followed World War I intensified the situation. By the early 1900s, agricultural use of land in the east had declined sharply, leaving many hill farms unused and unwanted.

Farmers and rural bankers recognized the economic and human waste in farming poor land. States began to take action, and in 1929 the New York legislature authorized the purchase of marginal farmlands for conversion to forests. By 1978, that state held about 288,090 ha and had planted the old farm fields as tree plantations. The Federal Resettlement Administration that became

Figure 2–3.
Creation of the Adirondack Forest Preserve by the New York State Legislature in 1885 represented one of the earliest programs of state ownership to promote forest conservation, with the states acquiring about 7.53 million hectares of forestland by 1975. *(State University of New York, College of Environmental Science and Forestry.)*

active in 1934 began purchasing similar types of land, and turning them over to states to administer. Through these measures and various types of acquisitions, as of 1977 the states owned about 9.3 million hectares of commercial timberland distributed as follows:[21]

Region	Commercial timberland owned by the states, millions of hectares (1977)
New England and mid-Atlantic states	1.95
Lake States and Central states	3.36
South Atlantic and Gulf states	0.73
Rocky Mountain states	0.89
Pacific Coast states	2.35

In most instances, the authority to administer these lands calls for multiple-use management to include timber production, recreation, watershed protection, public hunting, and other purposes.

The Weeks and Clarke-McNary acts proved a boon to state forestry departments. Fire control had become important and represented the largest budget item in many state forestry programs. Federal aid under these laws provided essential financial support to ensure continuation of other types of forestry activity as well. Additionally the Smith-Lever Act of 1914 allotted funds through the state agriculture colleges for extension work in agriculture, including forestry. These initial forestry extension programs emphasized tree planting and field demonstrations. But by 1924 the need for additional effort became evident. Passage of the Clark-McNary Act enabled states to distribute tree seedlings to farmers and to strengthen other farm forestry extension efforts.

The experience gained through these programs encouraged passage of the Norris-Doxey Farm Forestry Act in 1937. It provided technical advice to farm owners who wanted to grow and market timber. The Cooperative Forest Management Act of 1950 extended these benefits to all woodland owners. Other federal and state programs to stimulate good forest practice on limited-acreage nonindustrial private holdings included those of the Soil Conservation Service, the Prairie-Plains Shelterbelt Program, the Agricultural Stabilization and Conservation Programs, the Food and Agriculture Acts, and the Pitman-Robertson Game Management Program. Most of these acts gained strength through parallel programs enacted by the states. As of 1970, all states were participating in one or more of these or related programs.

PRIVATE AND INDUSTRIAL FORESTRY

Early records show that in 1820 Zachariah Allen planted some abandoned farmland in Rhode Island to oaks, chestnut, and locust.[7] During the following 57 years, he sold fuel and timber for $4948. After allowing for interest on the land, taxes, and planting costs, the venture netted $2543, or 6.93 percent per year for the 57-year investment period. Certainly many other private landowners took an interest in their forest holdings, harvested the products, and attempted to protect the young growth. Some farmers, plantation owners, loggers, and lumber companies even set aside specific forest areas where they kept logging to a minimum or excluded it entirely. By using common sense and judgment, many practiced good forest management over a long period of time with little or no aid from anyone technically trained in forestry.

Deliberate Management Started

Accumulation of forest lands in large family holdings had begun in Maine as early as 1850,[13,14] and included very large properties such as the Coe and Pingree estate. The founder, David Pingree, cut only spruce trees at least 36 cm in diameter. Other early large-scale forestry ventures included those on large estates owned by the Vanderbilt, Webb, and Whitney families. However, George Vanderbilt had employed Gifford Pinchot as a consulting forester for his estate of just over 3000 ha in North Carolina, later named the Biltmore Forest. Pinchot brought in a sawmill and started timbering operations. The venture proved financially successful, at least initially, and attracted the interest of other forest owners to do the same.

Ne-Ha-Sa-Ne Park in New York, owned by Stewart Webb, adjoined the Whitney estate. Gifford Pinchot and Henry Graves developed plans for managing these two properties, and included the first silvicultural rules ever written into a logging contract in the United States.[5] About 1912, Finch Pruyn and Company, which manufactures lumber and paper products, also initiated a forestry program on its nearby Adirondack holdings. Foresters began marking the trees for harvest and the company developed a cutting budget for sustained yield. Modern versions of those early programs remain in continued use by the company today.

Mobilization of the nation's resources during World Wars I and II made heavy demands upon forest resources and industries. Lumber, fiberboard, and plywood for construction of military quarters, overseas bases, and packaging of war materials consumed one half of the national output. Expansion of war industries and housing used most of the remainder. Efforts to expand the output of wood products helped stimulate a recovery of economic health in the forest industry. Additionally research efforts to find ways to substitute wood for scarcer materials provided new product ideas upon which the forest industry could capitalize. These included laminated beams and arches, new types of chip and particleboard, and more efficient ways of converting logs and bolts into usable secondary products. Many of these still offer promise as substitutes for nonrenewable materials in current and future markets.

Gradually many corporations began to acquire and manage forestlands for sustained production to keep their mills supplied with wood. The rapid expansion of industrial activity and the need for new housing following World War II encouraged both paper and solid wood products companies to purchase forestlands on a large scale, especially in southern and western states. Also, development of methods for successful pulping of southern pines gave major impetus to paper manufacture in that region and triggered intense industrial expansion within the great pinery of the south.

Although industrial management takes place across the country, the heaviest concentrations of corporate lands are in the southeast and northwest, and to a lesser extent in the upper Lake States and northern Atlantic Coast states. These include industrial lands managed both for pulpwood and sawlogs. Most corporations employ extensive forestry staffs to manage their lands and several major corporations maintain active forestry research programs. In addition, many companies also hire foresters to work with private landowners to promote good forestry practices and to assist in developing and implementing management plans for these noncorporate properties.

Impetus for Industrial Management

Three major factors encouraged forest products companies to acquire forestlands and manage them for sustained yield production. These also stimulated a fairly intensive management to capitalize fully upon the potentials of resources. First, their large mills require great amounts of wood year-round, as much as 1500 m^3 or more per day. Hence the firms must ensure a dependable supply within easy reach of the mill. Also, the cost of mill construction and machinery acquisition usually runs high, reaches into the hundreds of millions of dollars for a large plant, and requires a long period for amortization to pay off the investment. This, too, necessitates a steady long-term supply of raw materials to sustain production at a full level. Further, companies have demonstrated that under good management they can both reduce losses to mortality and shorten the time it takes to grow trees to usable sizes. For example, in the southeast, proper management may yield a crop of pulpwood in as few as 20 years. Consequently, through good management, the companies could realize greater yields from each property and reduce the amount of land they must depend upon to supply their mills adequately. Also, by acquiring and managing their own timber resources, the corporations could assure themselves of regular and continued supplies of wood of the kinds they need to serve the demands of the markets they serve.

Once top management across the forest products industry became committed to good forestry, they entered upon it with imagination and enterprise. Logging, tree planting, nursery operations, management of immature stands, and other forestry operations became mechanized to improve operating efficiency and control the results better. Industrial owners began to inventory their forest resources to determine the amounts and kinds of products available, the volumes added annually by growth, and those lost to mortality by the many destructive agents that affect forestlands. With this information, foresters could better plan their silvicultural operations and keep harvests in balance with production from the land. In all of this, companies quickly recognized the benefits of technological innovation resulting from research in tree growth and physiology, entomology, pathology, tree genetics and improvement, and many other forest sciences and disciplines. And they applied these through their silviculture and management programs to improve the quantities, quality, and kinds of yield from their lands. As a result, forest industry management has come to rank among the best in the world, and its promptness to adopt research findings has spurred forest research to new levels of achievement.

INFLUENCE UPON PROFESSIONAL FORESTRY AND EDUCATION

Early forestry education trained people in field practices needed to organize and manage forest properties dedicated primarily to timber production. Course work included heavy emphasis on timber cruising, surveying, tree planting, stand treatments for harvest cuttings, forest inventory, planning of allowable cuts, forest installations, logging engineering, and similar matters that field foresters would deal with daily. Colleges supplemented these forestry technology courses with ones in literature, mathematics, economics, chemistry, physics, writing, and other basic college subjects common to science-oriented curricula. However, most students concentrated the bulk of their time on working in forest technology and the basic biologic sciences that support forestry practice. These included botany, entomology, zoology, pathology, dendrology, and physiology. Social and managerial sciences stood at the periphery of forestry education.

Gradually the nation developed greater awareness and concern for its natural environments and began looking to forests and rangelands to serve a broader spectrum of needs, including outdoor recreation. The profession realized the necessity for greater stress upon social and managerial science to complement forestry science and practices courses. Abilities to communicate effectively also received greater attention, along with business management skills. And with these developments, forestry curricula began to involve greater stress upon interdisciplinary studies that addressed a wider complement of forest values and professional functions. These included increased recognition of outdoor recreation, water quality and yields, social values, environmental quality, and other nontraditional concepts for forestlands. Attention also turned to preparing people who could better manage both public and private organizations to fulfill their increasingly complex missions.

Forestry programs available today reflect the changes that have taken place in forestry as it has emerged from a fairly narrow practice devoted to timber production to a profession that considers the full range of values offered by the forest and its associated resources, and emphasizes modern techniques for managing them. In this transition, forestry education changed from offering a relatively standardized curriculum to all students, to providing a more diverse educational opportunity for individuals to develop personal interests within the many associated disciplines.

The development of forestry education has somewhat paralleled changes in the profession and in the national interest in forestlands and their resources. On the one hand, colleges and universities have tried to keep abreast with technologic innovation and to introduce students to modern ideas and approaches to forestry science and practice. And, at the same time, they have broadened the concept of forestry. It has become an interdisciplinary field offering opportunities within a host of related subject areas. This in turn places new demands upon the students themselves. They must readily accept the challenge to master new ideas and practices. In addition, they must learn to appreciate the many opportunities available and carefully select combinations of courses that will provide the preparation needed to enter some particular discipline.

Undoubtedly the nature of forestry will continue to change as human perceptions of forest values and use mature. As demands upon commercial forestlands increase with the swelling of human populations and the shrinking of the base of the world's forested lands, forestry professionals will face challenges that require new skills and technologies. So increasingly, it appears that the future of forestry education will involve both continued change within the resident curricula and development of opportunities for experienced professionals to renew their education by postgraduate pursuit. This should involve both degree and nondegree programs, but the form and methods remain unclear at present. Still the notion that foresters must seek new opportunities for continued professional growth and development seems clear, if they expect to keep pace with future developments in their chosen field. Forestry and forest use will certainly change in the years ahead. Through education and continued learning, forestry professionals will find the means to fulfill their role as leaders to help guide the wise development and use of the nation's forest resources.

REFERENCES

1. American Forest Products Industries, *Progress in Private Forestry in the United States,* American Forest Products Industries, Washington, 1961.
2. O. Butler, "70 Years of Campaigning for American Forestry," *American Forests* **52**(10):465–469, 512 (1946).
3. J. Cameron, *The Development of Governmental Forest Control in the United States,* Johns Hopkins Press, Baltimore, 1928.
4. G. Chinard, "The American Philosophical Society and the Early History of Forestry in America," *Proceedings of the American Philosophical Society* **89**(2):444–488 (1945).
5. S. T. Dana, *Forest and Range Policy, Its Development in the United States,* McGraw-Hill, New York, 1956.
6. S. G. Fontanna, "State Forests," in *Trees, The Yearbook of Agriculture,* U.S. Department of Agriculture, Washington, 1949, pp. 390–394.
7. M. L. Holst, "Zachariah Allen, Pioneer in Applied Silviculture," *Journal of Forestry* **44**(7):507–508 (1948).
8. J. S. Illick, *An Outline of General Forestry,* 3d ed., Barnes & Noble, New York, 1939.
9. R. G. Lillard, *The Great Forest,* Knopf, New York, 1947.
10. F. L. Paxson, *History of the American Frontier 1863–1893,* Houghton-Mifflin, Boston, 1924.
11. G. Pinchot, *Breaking New Ground,* Harcourt, Brace, New York, 1947.
12. President's Advisory Panel on Timber and the Environment, *Report of the President's Advisory Panel on Timber and the Environment,* U.S. Government Printing Office, Washington, 1973.
13. W. B. Sayers, "To Tell the Truth. 25 Years of American Forest Products Industries, Inc.," *Journal of Forestry* **64**(10):657–663 (1966).
14. H. L. Shirley, "Large Private Holdings in the North," in *Trees, The Yearbook of Agriculture,* U.S. Department of Agriculture, Washington, 1949, pp. 255–274.
15. W. N. Sparhawk, "The History of Forestry in America," in *Trees, The Yearbook of Agriculture,* U.S. Department of Agriculture, Washington, 1949, pp. 702–714.

16. H. K. Steen, *History of the Forest Service,* University of Washington Press, Seattle, 1976.

17. U.S. Bureau of Land Management, *Brief Notes on the Public Domain,* U.S. Department of the Interior, Bureau of Land Management, Washington, 1950.

18. U.S. Congress. *The National Forest Management Act of 1976,* 94th U.S. Congress, 2d Session, Washington, 1976.

19. U.S. Forest Service, *Timber Trends in the United States,* U.S. Forest Service, Forest Resources Report no. 17, 1965.

20. Ibid., *The Outlook for Timber in the United States,* U.S. Department of Agriculture, Forest Service, Washington, 1973.

21. Ibid., *Forest Statistics of the U.S., 1977,* U.S. Department of Agriculture, Forest Service, Washington, 1978.

22. U.S. National Committee on Materials Policy, *Materials Needs and the Environment Today and Tomorrow,* Final Report, U.S. National Committee on Materials Policy, Washington, 1973.

23. R. E. Wolf, *Partial Survey of Forest Legislation in New York State,* M.S. Thesis, New York State College of Forestry, Syracuse, 1948.

CHAPTER 3

FOREST POLICY

The preceding review of history and the development of forestry helps to depict how people have used forests for personal benefit and to support the growth of nations and civilizations. But focusing upon forest policy as a part of history involves more than just a recall of past events. Instead it helps to show how use and abuse of forests affected them as viable ecosystems and had an impact upon societies, human welfare and aspirations, and the future of life. Forest policy involves more than just laws, regulations, and processes of government. It goes beyond administrative measures to protect the physical environment, trees, and rare or endangered species of plants and animals. A review of forest policy reflects the attitudes of people toward the care and husbandry of all life forms that directly and indirectly contribute toward making the earth a wholesome place to live for people, and for the many species of plants and animals that inhabit it. Study of forest policy, then, focuses upon the ways people have banded together in common action to help maintain viability of the forests in serving the needs of future generations.

Even a brief review of history shows that people derive many benefits from forests. These give the plants, land, water, and wildlife a distinctive status. Individual landowners may own the timber and can manage and sell it for personal profit, but many of the associated resources belong to society at large. No one owns the wildlife, for example, that move about independent of property boundaries, and often migrate over vast distances at different times of the year. No one owns the esthetic appeal of the forest landscape. In some regions, individuals can own rights to water, but for the most part it freely serves the needs of many. As a result, over the centuries society has taken corporate action to enhance and safeguard its general interests by exercising a variety of controls, incentives, and influences. These affect the ways individual landowners may use many of the associated resources of the nation's forestlands.

The organized actions of society first began centuries ago. Many of the oldest persist in concept today. Others developed in more recent times, in response to changing perceptions and needs. For example, the depletion of timber supplies led Chinese priests, Roman emperors, and Egyptian pharoahs to regulate tree cutting even before the time of Christ. They sought to protect resources for shipbuilding and domestic uses, and for the construction of temples and palaces. Centuries later, the need for a navy to match the growing sea power of Britain during 1669 led Colbert in France to develop his celebrated policies. These later became a model for protecting and using the forests of many other nations.[18]

During its early history, Britain likewise established special forests of oak to supply timbers for ship hull construction. Later it acted to reserve American white pine trees for spars and masts of its ships. Domestic needs, as well as those for national defense and commerce, also moved nations into action, such as the development of community forests around the city-states of Germany, Switzerland, and France. More selfishly, love of sport led feudal landlords to protect forests in Britain, France, Germany, Russia, and Sweden as hunting grounds for their personal pleasure.[18]

When European settlers colonized North America, their fear of shortages of fuel wood and timber soon prompted the formulation of Colonial forest policies. Even as early as 1626, the Plymouth Colony forbade the selling and transporting of timber. Many other laws followed as the governments acted to protect their forests from exploitation. This concern persisted, and in 1799 Congress appropriated $200,000 to purchase federal forest reserves to ensure a supply of oak timbers for shipbuilding.[12] Ironically later realization that oaks occurred in abundance throughout the eastern United States allayed concerns of the new nation to the degree that Congress failed to protect the timber it had earlier purchased for such a high price for the day.[5]

Awakening awareness of the richness of the nation's natural resources throughout the 1800s evoked complacency among the people and leaders alike. With an abundance of forests and a scarcity of federal money, the government began to give away forestland in lieu of payment for its debts. Interest in promoting settlement across the continent compounded the situation and led to neglect by the Congress to reserve timberland for public use until 1891. Even then, it failed to provide for the protection and management of those reserves until 1905.[1] In fact, after the national frontier pushed west, the forests of the continent attracted little attention, except from people and firms who could exploit the wealth of those natural resources. Persons in a position to formulate and implement policy to protect and enhance the national good saw no reason to act. Consequently, throughout the period of expansion during the 1800s, the nation developed no coherent forest policy. It felt no need.

Therein lies a basic premise. Policy develops in response to need. People must first perceive a reason to act, then they will begin to pressure government to develop a policy. In part, it involves understanding what demands protection or control, and that individuals must sacrifice some personal freedom to ensure necessary benefits for society as a whole. Persons of authority within the legislatures and structures of government must feel the pressure from public demand. In response, they will make a commitment to take action on behalf of others they serve. However, critical issues must first gain widespread endorsements. Then government and legislative leaders will act to formalize that public commitment into new policies and organize efforts to implement them.

THE NATURE OF FOREST POLICY

Forest policy as a discipline involves the laws, regulations, and other settled courses of action pertaining to management and use of forested lands and their associated resources.[5] These arise from policy issues provoked by various special-interest groups. Eventually they may become enacted into law or issued as regulations by formal governmental action. Forest policy, then, sets forth objectives and directions for forest management and use within the purview of the agency having jurisdictional responsibility for their implementation. These policies can take many forms. In effect, they stimulate or regulate actions of landowners and resource users. Or they delineate ways that government spending and programs will channel actions of public and private agencies and firms to satisfy the interests and welfare of society.

Forest owners, professionals who work for them, and people who directly use our forests have special concern about emerging policy issues. These matters often portend directions forest management and use may take if the issues capture sufficient attention and public acclaim to move government and legislative bodies into action. Landowners, industries, and other users whose livelihood depends upon the forest and its resources are especially concerned about the implications of new forest policy. They must take cognizance of how it might affect their capacity to manage the forest and utilize its goods and benefits. They must know who will reap the benefits, what that will cost, and who will foot the bill for any added expenses imposed through the new policy. In fact, emergence of a new forest policy often raises serious concern about how the costs and benefits of owning, managing, and using forests will be distributed among the citizens.

Policy acts like a two-edged sword. In effect, it cuts at matters of critical interest to society and helps ensure the public welfare. It provides citizens with the mechanism, through government, to influence matters of critical importance to their future. On the other hand, policy often limits the freedom of individuals to exercise absolute rights over the property and resources they own. When policy requires promulgation of rules and regulations to control use or require action of some type, it ultimately means that an individual landowner or user must bear a cost associated with doing something extra or exercising restraint in the manner of use or development. Fair policies provide essential assurance that the needs of society at large are met without undue burden on those who own the land and pay the costs of maintaining and developing its resources.

BUILDING OF A NATIONAL FOREST POLICY

Forest policy does not develop in quick, clear steps. Rather it forms gradually from its roots in our perceptions of needs. It grows through interaction and compromise between interest groups and policy makers. Then the issue stimulates debate and action of legislators and elected government officials. Through this evolutionary process, ideas take shape and ramifications become recognized. Finally the issue becomes transformed into a policy through the laws, rules, and regulations of government. It remains for the responsible agencies to design and implement appropriate programs that translate the new policy into activities to address concerns that provoked action in the first place. A few examples from American forest history illustrate the process.

Establishing the National Forests

Throughout its first hundred years after independence, the nation took no organized action to provide a comprehensive plan for continued supply of forest goods and services. When Franklin B. Hough addressed the AAAS on this subject in 1873, his statement convinced the members to approach the Congress and the state legislatures about the importance of forests to future timber and water supplies. The AAAS urged Congress to establish an office of Commissioner of Forestry with responsibilities to ascertain the extent and use of the nation's forest resources, the influence of forests on climate and agriculture, and methods to perpetuate forest values and resources. But before the act passed in 1876 and Hough became Commissioner of Forestry, prominent members of the association had to spend much time with their senators and representatives lobbying for votes to pass the legislation. Even in developing these early forest policies in the face of great national need, the Congress moved slowly and took steps only after citizen concern pressured it to act.

The actions of special-interest groups working to influence Congress and state legislatures historically has proved a major factor in the development of U.S. forest policy. Creation of the National Forest System by the Congress shows the importance of that public pressure. To illustrate, the Department of the Interior assumed jurisdiction over the federal forest reserves established under the Forest Reserve Act of 1891. Yet it lacked trained personnel to develop and implement forestry programs as could the nucleus of trained foresters in the Department of Agriculture. When Theodore Roosevelt sent his first message to the Congress in 1901, he urged transfer of those reserves to Agriculture with its technical capability to put the lands under effective management.[5,13] Congress failed to act, even after three successive annual messages.

In 1905, in response to this situation, the American Forestry Association convened the Second American Forestry Congress, expressly to pressure Congress into transferring the federal forest reserves. They had support from the Secretaries of the Interior and of Agriculture, and the heads of the Reclamation Service, Geological Survey, and General Land Office. From outside government, the AFA received support from the presidents of the National Livestock Association, National Wool Growers Association, National Lumber Manufacturers Association, the Union Pacific and Great Northern Railroads, and from the head of the Weyerhaeuser Lumber Company.[10,12] Resolutions of the American Forestry Congress, fortified by the many supporters of the AFA, carried sufficient weight that Congress passed the Transfer Act on February 1, 1906. Once Congress saw the upswelling of support for President Roosevelt's proposal, it moved effectively to implement the policy into law. Without that showing of support, the president and his staff officials seemed powerless to bring the transfer to fruition in response to an acknowledged national need.

The Wilderness Act

A second series of events in Forest Service history illustrate another aspect of forest policy development and deserves attention. Two things are involved. First, they show the response by persons outside government, who can pressure for transformation of administrative action into law to gain more lasting assurance that a policy will continue in effect even if government programs change with time. Second, they illustrate the interaction between government officers and the pressure groups to mold an issue into a workable policy that serves the broad needs of society as a whole. In this case, the issue involved wilderness and actions by the Forest Service beginning in 1919 to set aside areas for wilderness within boundaries of the National Forests. Through administrative orders, these covered 40,486 ha by 1950. The Secretary of Agriculture also established special rules and regulations to govern their quality, protection, and use.

The Forest Service wilderness areas attracted little attention at first. In 1935, however, a small nucleus of Forest Service employees and former officers led by Robert Marshall organized the Wilderness Society and sought to make people aware of the status of the U.S. wilderness and the need to protect it (Figure 3–1). As time passed, the society grew in membership and influence, perhaps reflecting an increased interest in wilderness across the country. In 1950, its executive secretary, Howard Zahniser, approached the Forest Service to propose transformation of the wilderness rules and regulations into legal status. The Service tried to dissuade him, but Zahniser persisted.

With the support of the Sierra Club, he persuaded Senator Hubert H. Humphrey of Minnesota and Congressman John Saylor of Pennsylvania to introduce bills to

Figure 3–1.
As chief of the Division of Recreation and Lands from 1937 to 1939, Robert Marshall developed Forest Service policies that led to preservation of natural, roadless, and wilderness areas within the National Forest System. *(U.S. Forest Service.)*

solidify a national wilderness policy. Congress took no action at that time, or in response to 60 other wilderness bills introduced over the next eight years. In fact, as pressure for a wilderness policy increased, the U.S. Forest Service recognized that none of the emerging bills seemed workable. It decided to join with proponents to draft more realistic legislation. Simply defining wilderness proved a stumbling block. Also, the pressure groups wanted a Wilderness Preservation Council with no administrative powers or responsibility to any governmental authority. This seemed undesirable, since it could accomplish little and might stand between the Secretaries of the Interior and of Agriculture in all subsequent dealings with the Congress over wilderness.

Formulation of a wilderness bill drew strong opposition from forest industries, mining interests, cattle owners, and others who depend upon the use and development of natural resources around potential wilderness areas. Forest Service staff, the two societies, and congressional supporters had to seek a fair compromise among the factions in order to proceed in developing appropriate legislation. At the same time, wilderness support grew nationwide, and through lobbying the proponents gained more backers in Congress. As debate revealed needed points of compromise, these became incorporated into the bill, eventually enabling its passage in 1964.[8]

Effectiveness of standards written into the Wilderness Act still awaits the proof of time. For one thing, the Act limits flexibility of the National Park Service in managing park lands under its jurisdiction. This may cause the Service problems in addressing public demands for increased use of the National Park System. Withdrawals of lands from the National Forests into wilderness status also reduces considerably the supplies of available timber the Forest Service can draw upon in supplying increasing demands for wood products. Nevertheless the Wilderness Act gives evidence of concern among citizens for protection of natural ecosystems. Through the process of debate and compromise by pressure groups, the Wilderness Act developed as a compromise tempered by conflicting interests of divergent special-interest groups that claim to represent the complexity of U.S. society. It is hoped that the process has helped to temper congressional action in formalizing a forest policy reflective of the emerging needs of the entire nation.

Renewable Resources Planning Act

Previous examples depict how emerging issues became formulated into policy that changed government programs through legal mandate of a legislative body. At times, however, solutions to complex problems have not developed so clearly, requiring a different type of policy decision and action. One such case arose following a 1968 decision by Congress to set a national goal of erecting 26 million new housing units by 1978. This policy attempted to anticipate housing requirements of a growing population. To stimulate construction, Congress also provided for liberal financing of new homes.

Nationwide the building industry responded to an extent that demand rapidly exhausted available lumber supplies, especially in eastern yards. Suddenly prices doubled and attempts to control them had little effect. The many sizes and shapes of wood products and the dispersed nature of the industry made price regulation impractical. Heavy buying by Japan further compounded the problem. The hoped-for housing program began to falter, and the nation's capacity to satisfy its domestic need for wood products seemed uncertain.

As shortages and prices of lumber increased, President Richard Nixon, with backing of the forest industries, urged immediate increased cutting on the National Forests. These lands held 51 percent of the nation's softwood supplies. He also assigned a special task force to deal with immediate problems, and appointed an advisory panel to assess alternative long-range strategies for meeting the United States' lumber needs. The panel's report of April 1973 recommended several measures applicable to both federal and private forestlands. It urged government to develop comprehensive plans for managing and using the nation's forests to realize their many benefits, to set standards for efficient and safe harvesting, to apply the best practices available and accelerate management efforts, to encourage greater management efforts of private lands, to strengthen research and development relative to timber production and use, and to pursue several other measures aimed at increasing the growth and yield of timber and optimizing forest multiple-use potentials.[16]

As an immediate response, the Forest Service reappraised the national timber resources and reevaluated its own harvesting program and ways to make available more timber to satisfy immediate shortages. The Council for Environmental Quality, Office of Management and Budget, Housing Administration, and other federal and private agencies also began seeking additional ways to provide both immediate and lasting solutions. In the midst of this disarray, Senator Humphrey, with cosponsorship of Congressman John Rarick of Louisiana, conceived of the Forest and Rangeland Renewable Resources Planning Act. It took two years to perfect as the legislators and their staffs organized public hearings at which forest industries, the Forest Service, and others could offer advice and debate ideas contained in the legislation.

The resulting bill called upon the Forest Service to prepare inventories of all National Forest resources, and to develop plans for their use and management. It also had to coordinate with private ownership plans to ensure a comprehensive approach for meeting the expanding wood products and other resource use needs of the country. Thus a policy evolved whereby government, through an appropriate agency, would take deliberate efforts to maintain awareness of the nature and supply of natural resources. In addition, the U.S. Forest Service must recommend to the Congress programs needed to satisfy demand as it emerges.

In many ways, the Forest and Rangeland Renewable Resources Planning Act brought no immediate change in timber supply or harvest. Nor did it set forth specific programs to provide for them in the future. Rather it established a deliberate planning effort to help keep government well informed about the country's resources and the changing demands for them. It established a policy that the nation should remain prepared and sufficiently knowledgeable to anticipate likely problems and take early action to circumvent potential difficulties. It exemplifies the emergence of forest policy through a process of deliberate, continual study and review. It provides for concerned groups, agencies, industries, and individuals to pool their capabilities in seeking effective ways to attack important problems that confront the nation. Unfortunately Congress failed to appropriate funds to pay for the extra activity called for by the law. The money to support the planning came instead out of timber sales revenues that the U.S. Forest Service had traditionally used for other purposes.

In response to the Forest and Rangeland Renewable Resources Planning Act of 1974, the U.S. Forest Service has identified both high- and low-level use potential for public and private forest and range.[17] These cover the period through the year 2030 and suggest changes that might occur under varying assumptions of growth. Table 3–1 includes a listing of several for the National Forests. It contrasts annual

TABLE 3–1
PROJECTED NATIONAL FOREST PROGRAM DIRECTIVES FOR 1978–2020 TO SATISFY HIGH-LEVEL DEMAND*

Program element	Units†	1978 level	2011–2020 levels
Developed recreation	Million RVD	79.6	141
Dispersed recreation	Million RVD	130.2	225
Wilderness	Million hectares	6.2	17
Livestock grazing	Million AUM	9.9	10.3
Timber sales	Thousand cubic meters	28.3	38.7
Roads	Kilometers	1104	1460
Total cost	Million dollars	1666	2404

**Source:* Based on U.S. Forest Service [17]. Used by permission.
†Abbreviations: RVD - recreation visitor days; AUM - animal unit months.

outputs in 1978 with those forecast for high-level demands in 2011–2020. These reflect a likely upper limit for programs on the basis of the best information available. If demand does materialize as projected, use will increase substantially and necessitate more intensified management. The need for funds will also mushroom.

Clearly Congress and the federal agencies involved with managing and administering natural resources face many challenges. So do their counterparts in state governments. Both must find ways to satisfy probable increases in demand and make wise choices about how to apportion available funds to pay the costs. However, as a result of this early planning under the act, government officials now recognize the magnitude of the work that lies ahead. They can identify matters that seem to demand the greatest attention in shaping the development of forestry for the future.

National Forest Management Act

A final example of policy development illustrates the action of citizens through the courts to force abrupt changes in policy and bring the Congress to resolve an important national issue. Again the matter involves the U.S. Forest Service, but this time focuses upon the authority it had to cut and manage timber under the Sundry Civil Appropriation Act of 1897. That law permitted the sale of dead, physiologically mature, or large trees, if individually marked. The law did not recognize needs for thinning of young forests as a method for culturing their development. Nor did it provide a means to manage forest vegetation to improve wildlife habitat or water yields, or for many other desirable purposes. In 1897, forestry had just begun to gain recognition and need for concern about growing trees to cut in future years seemed remote. After all, the nation had abundant old-growth timber, and terms of the law facilitated its removal.

By the mid-1900s, the Forest Service had begun efforts to improve the growth of young stands to eventually replace the dwindling old-growth forests. In addition, as a result of extensive research and trials, the Forest Service began clearcutting western forests to regenerate new crops of timber to replace the old ones (Figure 3–2). The method was efficient for logging. It resulted in favorable species composition among the new trees, and it created stands that could be readily managed. Once a stand proved suitable for harvest, a sale could proceed. Foresters did not need to take the time to mark all the individual trees within boundaries of the sale area. Yet many special-interest groups felt that neither the thinning of young stands nor this clearcutting of mature ones without marking of individual trees seemed to fit the provisions of the old enabling legislation of 1897.

During the early 1970s, public awareness of natual resources, and use of forests and rangelands for recreation, reached a heightened level. Some who used the National Forests did not like the clearcutting, which suddenly changed the character of an area. Many found this unsightly and a detriment to recreational use. Others worried about possible environmental harm. These concerns became amplified through the protests of a few major preservation groups. A national controversy developed. It started with the Bitterroot and Medicine Bow National Forests of the West, and spread across the country to the Monongahela in West Virginia. A major policy issue had emerged and some resolution seemed necessary.

The Isaac Walton League decided to take action in the courts. It filed suit against the Secretary of Agriculture, claiming that cuttings on the Monongahela National Forest violated the words and purposes of the 1897 law. The federal judge in Virginia agreed, saying, "The law may well be an anachronism, but it is up to Congress to change the law." Later a federal judge in Alaska applied the same reasoning in a suit against the Forest Service in that state. Timber harvesting in many National Forests virtually stopped. Public action through the courts had forced the Congress into prompt action to resolve a major policy issue. Congress had to act to maintain usefulness of the National Forests as a major source of timber supply.

Congress set to work promptly. Senator Humphrey and Senator Jennings Randolph (West Virginia) introduced separate bills. Though similar in some respects, the two differed greatly in the degree of discretion foresters would have in treating a forest stand. Under the Humphrey bill, guidelines for cutting were left up to the Secretary of Agriculture after consultation with an advisory board independent of the Forest Service. Randolph's bill repeated the language of the 1897 act on restricting harvests to individually marked dead, mature, and large trees. It prescribed in detail requirements for every timber sale in the National Forests. The Sierra Club, Audubon Society, and like-minded groups backed the

Figure 3–2.
The old Sundry Civil Appropriations Act of 1897 authorized the cutting of only individually marked mature trees, but clearcutting such as shown in Douglas-fir in Oregon had only a weak basis in law until the National Forest Mangement Act of 1976.

Randolph proposal. The Forest Service, forest industry, and their supporters favored the one by Humphrey. The Senate seemed on the verge of an impasse.

During hearings by the Senate Committee on Agriculture and Forestry, a group of nine professional and technical organizations concerned with forests, wildlife, and fisheries argued for flexibility in allowing forest managers to prescribe practices fitted to local needs. They objected to the strictness of the Randolph bill. These recommendations proved quite influential in the committee's effort to protect the Secretary of Agriculture against the possibility of frequent lawsuits for nonconformity to the law. The Senate wanted to allow flexibility in meeting differing conditions from place to place.[7,14] The resulting version of the Humphrey bill passed the Senate unanimously on August 24, 1976.

Time had run short. With a hotly contested presidential election coming up and final adjournment of the 94th Congress set for October, the House had to decide among seven separate companion bills. Based upon comments obtained during hearings, Representatives Jerry Litton (Missouri) and Stephen Symms (Idaho) designed a compromise bill that cleared the House Committee on Agriculture with some amendments. The entire House acted favorably on September 17. Immediately a House-Senate Conference Committee set to work to resolve differences. Both houses agreed to the conferees' report just one day before Congress adjourned. President Gerald Ford signed it into law on October 22, 1976; thus ending a year of crisis.[7,15]

Events leading up to passage of the National Forest Management Act of 1976 have importance to forestry beyond provisions of the law. A major issue involved the extent to which a law should dictate details of on-the-ground practice in managing forests. It supported the idea of depending upon professional judgment to prescribe appropriate treatments, as long as that work conformed to broad directives established by the Secretary of Agriculture. The issue mobilized foresters, forest industries, representatives of conservation organizations, and professional societies to appear before Congress to state their views. Prominent foresters also came forth to advise the chief of the Forest Service about ramifications of various bills under consideration. They helped to draft language appropriate to conditions at hand.[9]

The resulting law does require public notice before all sales and provides for public input about the planned method for cutting.[7,15] But Congress put responsibility upon local forest officers to take cognizance of these recommendations. They must justify the actions decided upon, and provide convincing reason to undertake the management proposed in light of local needs. In addition, the cuttings must prove consistent with a philosophy of multiple use for the National Forests. The decisions must reflect local advice in identifying needs and opportunities. These provisions will increase the costs of administering the timber sales program, but they do ensure serious attention in planning the management of each area. Clearly, through them, Congress placed its confidence in professional foresters to perform responsibly.[9] Congress also demonstrated its own capacity to address a complex issue, to give divergent views fair hearing and consideration, to develop a compromise that serves the welfare of the whole nation, and to resolve a critical matter without undue delay.

The National Forest Management Act did not provide funds to finance the new program. Initial budgeting came from timber sales receipts. Long-term solutions to funding remain unclear, though revenues for most uses do not cover the costs. For example, of the several resources on the 80 million hectares of National Forests, sale of timber produces most revenue for the national treasury. These exceed expenses for timber sales by a considerable margin. Fees for livestock grazing cover about one half of the costs of that program. Leases for summer homes also cover a substantial portion of administrative costs. Expenditures to promote fish, game, and recreation, however, return negligible income. These programs operate at a large deficit and revenues from other sources must pay their cost. Consequently new provisions of the National Forest Management Act and the broadened emphasis on concerns that generate no revenues demand new approaches to paying costs for administration.[3,4]

That issue remains unresolved. Eventually Congress may face the difficult choice of providing additional revenues to more fully subsidize recreational programs in the National Forests. They may need to demand user fees to help defray costs, or find some reasonable alternative to finance activities mandated under the National Forest Management Act. Even questions about whether the program might reduce timber sales activities may require serious attention. How and when these issues will reach resolution remain unclear.

Major Policy Issues of the 1980s

By the end of the 1970s, the major U.S. forestry-related policy issues centered on the future of use and administration of public lands. Passage of the Alaskan Native Claims Settlement Act of 1971 had sparked national debate about setting aside large areas of federal land in Alaska for national parks, wildlife refuges, and wilderness. As an alternative, the government could encourage development of the mineral, timber, and petroleum resources. Congress failed to settle the issue, so to ensure protection of these lands pending future legislative action, the Carter administration declared them National Monuments for safe keeping. Nevertheless their ultimate use for development, preservation, or a combination of these, remains highly controversial. Congress must eventually decide the question.

Future use of the so-called roadless areas on the National Forests also came into question after 1971 as a result of comprehensive review by the U.S. Forest Service. Called the "Roadless Area Review and Evaluation" (RARE I), the evaluation sought to identify and recommend which lands Congress should designate as wilderness, and those it should thereafter devote to multiple use. Environmentalist groups sought to have most of the roadless areas of the National Forests declared as wilderness. Forest industry and others opposed such an action. The ensuing controversy prompted legal actions against the U.S. Forest Service under provisions of the Environmental Policy Act. It moved the Carter administration to attempt a resolution by implementing a second review called RARE II. When the review was completed in 1979, the U.S. Forest Service sent its new recommendations to Congress, but that body has acted slowly. The wilderness advocates continue to pressure for greater dedication of the roadless areas to wilderness uses. What Congress and the administration will do about the issue remains uncertain. Inaction has left continuing questions about the management of these National Forest lands.

These unresolved issues about public lands in Alaska and the roadless areas of the National Forests have raised considerable concern among Americans faced with increasingly critical demands upon the nation's mineral, timber, and energy resources. Yet their importance may seem small compared with the mounting debate over ownership of the western public lands. This issue, called the Sagebrush Rebellion, has gained momentum in the 11 western states. There state governments seek transfer of ownership of just over 126 million hectares from federal to state jurisdiction. The states also seek authority to manage or even dispose of those lands and the resources thereon. Obviously such a move would have major repercussions for the management and use of natural resources throughout the West. Any action undoubtedly will precipitate a major political struggle with the potential of becoming one of the most controversial and heated natural resource policy issues of all time.

The Lesson of History

The collection of cases cited exemplify the United States' democratic and legislative process in action. They emphasize the important role that concerned citizens play in raising issues of national importance and making divergent opinions known. They also show how the administrative branch of government can act to initiate and guide legislation and work with Congress in forging policy. They illustrate that Congress can move effectively when it hears an articulate message and sees the strength of contending parties in pressing for action. Collectively these selected cases show the dynamic nature of policy and its formation. They highlight the critical importance of interplay between government and the people in identifying and charting a course of action for the use and development of our natural resources.

Study of U.S. forest policy tells the story of the nation's forestry history. History and policy go hand in hand, with policy setting the course that forestry activity will ultimately follow. The two reflect a long-standing national concern for renewable natural resources, for multiple use of the associated products and services, and for attention to the ways that components of natural environments interrelate. History also shows that the evolution of national policies improved the welfare of the American people. It indicates the need for continued planning to ensure the best possible use of the country's natural resources in the years ahead.

The brief review also shows that Americans lack a unified, coherent national policy with regard to forests and other natural resources. Instead the country operates under a diverse collection of separate rules and regulations that Congress has never pulled together into a clear and concise plan for the future. As a result, U.S. forest policy remains fragmented and characterized by an odd collection of governmental reactions to issues raised by many special-interest groups. In some cases, the policies reflect emergency actions to solve an immediate crisis. Perhaps the Forest and Rangeland Renewable Resources Planning Act represents a new direction. It should identify emerging national needs and provide Congress with a way to anticipate future situations that require national attention. It is hoped that, in the process, Congress may take a more unified approach. However, for the present the country will continue to depend upon promulgation of rules and regulations by a diverse collection of governmental agencies to shape forest policy.

STATE, LOCAL, AND PRIVATE FOREST POLICY

Federal laws historically have provided an overall direction for the use and development of forest resources and have shaped the principal form of U.S. forestry. Yet the fabric of that policy has threads of state, local, and private rules, regulations, and procedures intertwined with federal law. Some represent extensions of federal legislation. Others deal with entirely local responsibilities. Each state, for example, has developed an array of laws enabling cooperation with federal programs established by the Weeks, Clark-McNary, and similar acts. These provide uniform approaches to fire control, insect and disease programs, importation of tree seedlings, and much forest research activity. In addition, each state has its own set of laws governing matters of state responsibility, such as the management of state forests, timber harvesting, land use control, and cooperative forestry programs. Further, federal law has no bearing on nonmigratory wildlife, so states have responsibility related to that resource as well. State forest policy, then, usually encompasses cooperative endeavors with federal government, plus matters deemed solely within the jurisdiction of the state's control.

Individual states have complete responsibility for timber harvesting on private properties, even though federal laws such as the Water Pollution Control Act require the states to develop and implement standards relevant to activities that might impair water quality. The degree of control exercised varies. Some states, including Nevada, New Hampshire, Louisiana, Idaho, and New Mexico, adopted laws prior to 1940 to control destructive cutting of timber. Later several other states took action as well.[5] The approaches differ. For example, California requires submission of an operating plan prepared by a state-licensed forester, and posting of performance bonds to ensure compliance. The state determines whether landowners have followed the plan and may charge for reforestation if the owner's regeneration plan fails. Washington and Oregon also have strict cutting laws, and

Maryland restricts timber cutting to state-licensed operators. Several states have no restrictions, but many offer incentives to private owners who follow recommended forestry practices. States commonly offer timber-marking services and help landowners in working out sales agreements with the harvesting contractor. Through these devices, the states encourage use of practices deemed desirable for the types of forests and land conditions common to the region.

Beginning in the early 1970s, when citizen environmental action groups proliferated and pressured for protection of natural resources, several state legislatures responded by passing land-use and development plans for nonurban areas and forestlands. These extended to rural communities the types of restrictions that have governed land use in urban and suburban areas for decades. Principally they limit some landowner alternatives as a means to enhance presumed needs and interests of society at large. Maine, for instance, adopted regulations requiring landowners to leave buffer strips along highways and streams during logging operations. In Vermont, the legislature decided that landowners must obtain permits for timber harvesting and related forestry activities on lands at elevations of more than 763 m. And New York passed a broad land-use and development plan for both private and state-owned lands within its Adirondack Mountain region. This law does not control most forestry operations, but does require a permit for clearcutting on more than 10 ha in a single operational unit.

In addition to these, provisions of Section 208 of the federal Water Pollution Control Act require that each state develop a plan for controlling nonpoint sources of water pollution, including those that might develop from forestry activities. In effect, this program mandates that individual states must take direct action based upon a plan approved by the federal government. Like the separate laws enacted by states over wildland use and development, the federal water quality program has wide-sweeping ramifications for future forestry activity. All represent important policy decisions that directly regulate forestry activities on private lands and mandate landowner compliance. Each adds a cost the landowner must bear or pass on to consumers through prices for the products and services derived from forestlands. They all show the determination of government to control private activity to protect perceived interests of society as a whole.

In recent years, many towns and counties in several eastern states have also enacted local ordinances to control forest use, principally related to timber harvesting. Mostly these involve localities with heavily settled countrysides near urban centers, or where urban-oriented people have built homes in rural environs to escape city living. The ordinances often represent a governmental response to conflicts between people who own land and hope to market forest products from it, and others who feel that timber harvesting destroys esthetic characteristics of their rural neighborhoods. As these local ordinances proliferate, forest industries must adjust to the multiplicity of attitudes toward forest production. They may find supplies of timber from noncorporate lands more difficult to obtain, more costly to acquire, and less dependable as a steady source of supply to their mills.

Issues such as these arise in a host of situations where government may move to restrict the latitude of individual landowners to use and develop private forest resources. Most strive to provide scenic, environmental, or recreational benefits for society. Many owners consider these regulations a form of governmental appropriation without just compensation, since few of the laws provide owners any payment to make up for values lost through the limitations imposed by governmental control. The U.S. Constitution does forbid such taking by government. However, proponents argue that needs to protect the long-term interests of society exceed a landowner's loss. Consequently, through the implementation of one kind of policy that regulates private activity, laws have raised another issue about prerogatives of government to impose the interests of society without compensating the owner. Landowners may need help through tax relief, payments in lieu of values denied, or alternate means to relieve the financial burdens that the legislation has placed upon individual landowners. Such issues may require both court action and legislative decisions to resolve. They seem certain to take prominence as an important forest policy activity for years to come.

Unlike many private individuals, most corporate forest owners hold land so they can grow timber to harvest. Their investments serve principally to ensure a steady supply of raw materials for mills that manufacture the many wood-based products upon which we depend. Historically attitudes and policies of these industries have changed to reflect the national situation. For example, during early times the nation had seemingly endless supplies of large timber. Most corporations entered a region, cut out the accessible merchantable timber, and moved on to a new place. Once their land was cut over, they sold it, abandoned it, or accepted whatever natural regeneration appeared. Investments to maintain a continuing and steady flow of timber from the land seemed relatively unimportant.

Changes in demand for wood products and a decline in the availability of land beginning around World War II triggered a change in these corporate policies. Companies began holding their forested lands, controlling more carefully the removal of timber from them, balancing their harvests with growth of new supplies, and investing in deliberate programs to ensure adequate regeneration for a second and succeeding crops. They employ foresters to oversee woodland operations. And they assign to these professionals the responsibility for controlling regeneration, planning proper cutting schedules, protecting the timber against destruction by natural forces, and otherwise developing sustained yield forestry programs. Economic necessity, amplified by the unavailability of endless resources in new regions, forced companies to alter their attitudes. They had to adopt management policies geared toward ensuring stable supplies of needed raw materials far into the future.

Corporations do not control most private forestland, nor do federal, state, or local governments. Rather nonindustrial owners hold 59 percent of the nation's commercial forest area.[11] These owners have many different policies for management. Each espouses some unique personal reason for owning the land. Some seek revenue through sale of products and services, but many look mostly for recreational benefits, or have a variety of reasons that do not include timber production as a goal. Nationally the private policies of landowners limit the proportion of timber resources available for commodity uses. One study in New York, for example, showed that people who have no interest in products control about one quarter of the land, and others who might sell timber if it did not interfere with other primary interests hold about one half.[2] Such studies clearly indicate that attitudes by private landowners will have a major impact on the availability of resources to satisfy society's needs, especially in the eastern United States.

Various landowner assistance programs offered by the federal government through the state forestry agencies are designed to influence private management activity and increase the production of timber on private lands. These programs offer technical assistance, financial subsidies, landowner education, market development, and other professional services as a means to promote wise management and use of private resources. Through these efforts, governments attempt to influence policies of owners who control about three fifths of the total forestland in the United States. Whether these individuals withhold or make available their personally owned timber resources can influence to an important extent the nation's capabilities for wood product self-sufficiency.

POLICY WATCHDOGS

Even the best public agencies and corporations need supporters and critics, both from within and outside the organization. They must have objectivity, a universal point of view, a deep interest in public affairs, and the courage to speak out about matters needing attention. But this alone may not suffice. Intimidation, threat of dismissal, and fear of transfer often dampen the ardor for critique that develops within an agency or corporation. Outsiders, too, must act as critics by giving attention to activities of government and business to ensure that actions remain consistent with the peoples' interests. In forestry, such organizations as the Society of American Foresters, National Forest Products Association and its affiliates, National Livestock Association, National Woolgrowers, American Forest Institute, and others with economic interests in the policies and work of forestry act as advocates for professional, technical, and business concerns. The American Forestry Association, the Wilderness Society, the Sierra Club, the Isaac Walton League, and thousands of sport and outdoor recreation clubs all have deep concern for use of our forest resources as well. They look after public interests concerning policies that influence use and development relative to preservation and environmental protection. This wide range of advocacy groups provides a system of checks and balances to make sure that no one special interest overshadows the needs of all. They scrutinize government programs to assure compliance with policy. They also bring to public attention new issues of importance, and in the process provide an assessment of the importance of an issue before it reaches crisis stage. In their different ways, the many groups serve as the voice of society in clarifying issues, finding ways to discharge minor tasks without legislation, and airing critical matters of national concern.[3,4]

The management of U.S. forest resources remains with the private landowners, governmental agencies, and industrial companies. Yet many citizens' organizations and forestry associations have worked actively to promote and shape forestry practice. These groups influence the formation of forest policy and champion awareness of attitudes about and concerns for the wise management and use of our forestlands and associated resources. Today such organizations number in the thousands and represent many diverse interests. They have attitudes that range from concern for preservation of open space and natural resources to promotion of active use and development. The organizations serve local, national, and even international purposes. They vary from lay and youth organizations to prestigious scientific societies. In their diversity, they keep before us concern for wise and balanced use of resources to serve human need both now and in future generations.

Since its action to stimulate Congress to adopt a sustained yield federal forestry program in 1876, the American Association for the Advancement of Science has continued its interest in forestry. It publishes feature articles about forestry research in its periodical *Science,* and has included explanations of forest policy issues for its readers, who include the nation's scientific leaders. Its annual meetings have featured broad issues of environment and open-land preservation and use. Many foresters have become members of the AAAS, and some have served on its governing council. In these ways the association continues to influence attitudes, policy, and scientific inquiry important to forestry.

The National Academy of Sciences (NAS) represents another national-level group that affects forestry policy and practice. In particular, it has sponsored research and technologic innovations that have helped to shape U.S. forestry. Support came from the National Research Council during World War I, the Office of Scientific Research and Development during World War II, and more recently through the National Science Foundation. This support to both basic and applied research directed to matters of demonstrated national need has been substantial during most of this century. Through its future funding of research, the NAS should continue to influence the direction of technologic development in forestry-related fields and to help shape the nature of our future forestry policy and practice.

The Society of American Foresters (SAF) and similar organizations in Canada and Mexico represent the profession of forestry in North America. Through these societies, professional foresters work to advance the science and practice of forestry and to provide a technically sound influence on forestry policy formation from local to national levels. As part of its work, SAF testifies before the Congress and state and local legislatures to explain and interpret technical information on forestry-related matters. Through its journals and meetings, it also disseminates information and ideas among foresters, publishes results of research, and promotes excellence in forestry education and practice. The SAF encourages employment of foresters for tasks requiring their professional competence. And through its association in the International Union of Societies of Foresters, it engages in a worldwide exchange of information and participates in joint forestry ventures.

Other scientific and professional groups also influence U.S. forestry in significant ways. As a multidisciplinary field, forestry must draw upon many biologic, physical, and social sciences, and so professionals in these allied disciplines and their organizations have become involved in forestry matters. The Ecological Society, for example, almost from its inception welcomed foresters to membership and service on its council and editorial board. The Wildlife Society and the Society for Range Management came into being to enable researchers and managers to cope with problems concerning National Forests and other public lands. These two groups, along with the American Fisheries Society, have collaborated with the SAF to advance closely related interest and activities. Also, the Forest Products Research Society has influenced forestry activity through its programs to further scientific knowledge of wood and its application in technology. Other societies that contribute to forestry include those for botanists, entomologists, pathologists, soil scientists, recreational managers, photogrammetrists, engineers, geographers, landscape architects, statisticians, hydrologists, chemists, economists, sociologists, and anthropologists. In fact, through this interplay among professionals in many disciplines, we see an example of the nature of forestry itself. It has taken its shape by drawing upon the understanding of many scientific fields to manage the forest and its many interrelated associated resources.

Many different forest industry organizations have also developed throughout the forested regions of the United States. While organized principally to promote industrial concerns and activities, many have contributed importantly to technologic development in forestry practice and resource use. The American Pulp and Paper Association and the National Forest Products Association stand out as most influential. Each has a large membership of affiliate organizations to deal with particular technical and operational areas, and numerous state and regional groups to serve local needs. They collaborate in supporting the American Forest Institute to represent the industrial viewpoint before government agencies and consumers. Through their skilled professional staffs, these organizations have effectively worked with the Congress on such issues as water and air pollution, fire and pest control, and incentives to nonindustrial landowners. At the same time, they promote good practice and improvement among member industries, and have organized programs to stimulate forestry practice on private lands. These and similar organizations have gained a high degree of public esteem for their efforts on behalf of U.S. forestry.

Citizen organizations far outnumber all the professional, scientific, and industrial groups combined. These include such diverse organizations as sports clubs, labor unions, youth organizations, wildflower clubs, and hikers' associations. Most of the citizen groups concerned with nature, outdoor recreation, parks, soil and water conservation, wildlife, natural scenery, and environmental protection have helped to influence forestry or forest resource use in some way, in part by supporting measures related to protecting or conserving forest productivity. The American Forestry Association probably ranks as the most prominent of these nationally. Since 1875, it has mobilized citizen support for broad forest conservation legislation and programs at the national level. Further it has influenced forestry activities in several states. Its American Forestry Congress has highlighted major forestry issues through the years, and developed resolutions that have led to significant legislation by the Congress. Other prominent national groups include the National Audubon Society, Friends of the Earth, Sierra Club, Wilderness Society, Isaac Walton League, National Rifle Association, National Parks and Conservation Association, National Grange, and the Soil Conservation Society of America. All take a strong interest in forest conservation in general, as well as in promoting their own special interests. All are prominent on the forestry scene.

The awakening of environmental concern among citizens during the early 1970s triggered a new impetus and excitement for forestry and the wise use and management of forestland and its associated resources. Many organizations, such as those cited, helped to alert forest managers, agencies, companies, and legislative bodies to the interests and needs of the nation. Above all, this new era has helped us to realize the way in which these many groups represent interests of different people. They help to identify alternative courses of action to deliberate in shaping future forest policy and in translating those attitudes into action. As long as such broad professional and citizen concerns become actively manifest through interplay of the thousands of forestry-related organizations, the best means for wise conservation and use of our natural resources should emerge for our consideration and use.

CAREER OPPORTUNITIES RELATED TO FOREST POLICY

This overview has not fully covered the broad field of forest policy. It serves primarily to review several selected issues typical of the many that have strongly influenced the development of forestry in the United States. It shows the critical role of skilled professionals, political leaders, and concerned citizens in helping to shape and implement policy matters. However, the discussion gives little attention to the future. It ignores any anticipated growth of the world's population, the resultant demand by people for more forest goods and services, and the added pressures these will place upon all land resources.

Such worldwide developments, including shortages of low-cost oil products, will force industrialized nations to turn toward alternative sources for fiber, fuel, and structural materials. The United States and Canada are likely to find their great national endowments of forest resources sought after as a source of wood-based materials to serve increasing world needs. With that demand will come important questions about the best ways to manage and use resources, what amounts to offer in trade, and what priorities the nation should give to the many opportunities for trade and domestic uses that present themselves.

As populations increase around the world, demands for food production will also rise. This may force many nations to convert portions of their existing forests and grasslands to crop production. That in turn will increase even more the worldwide demand for wood products from the forest-rich countries. It also will raise new questions about maintaining around-the-world critical balances of forest for the benefits they offer in water yield, effects of local climate, habitat for wildlife, and the many other values we often consider of secondary importance. Such matters deal with environmental quality and the preservation of major forested ecosystems. To resolve these interrelated issues of forest production, preservation of forest cover, and all that goes with them will take increasingly complex solutions—solutions that will force some important changes in attitudes about the ways in which we use our natural resources.

Future forest policy must reflect great sensitivity to the complexity of natural ecosystems that foresters deal with daily, and the needs for modifying them to meet each changing situation. People all along the chain from forest owner, to manager, to logging contractor, to timber processor, to user must work together. We need solutions to safeguard forest productivity while we more actively manage them to realize greater benefits. We must improve utilization to take fuller advantage of the forest output for construction material, fuel, pulp, and paperboard products. We must ensure continual availability of the water that flows from our forests to preclude limitations on the nation's development and crop production. And we must guarantee perpetual use of forests as wildlife habitats, places for recreation, ameliorators of local climate, and for the many ancillary benefits we derive from them. All contribute importantly to modern life.

When foresters and their colleagues in related professions practice their specialties, they unavoidably become involved with forest policy and the issues that represent public attitudes about using and managing forested lands. Resource managers must respond to the directives, rules, and regulations promulgated by government to implement decisions of the legislative bodies. And within corporations, agencies, and firms of all types, internal policies will provide to the managers guidance for controlling day-to-day operations, as well as the course of management over the longer run. Through these, professional resource managers deal with and help to shape policy and to implement it.

Schools of forestry introduce their students to concepts of forest policy and the social and political forces that influence these national and regional actions. These take the form of courses in political science, sociology, economics, and policy. Individuals interested in pursuing policy-related careers can take elective courses to develop their understanding more fully in these subject areas. Often additional studies in environmental law, public administration, history and government, macroeconomics, policy analysis, planning, and regional development provide the means to enhance capabilities in policy and public affairs. An interest in business or corporate policy can be strengthened by work in corporate finance, accounting, business law, corporate management, and other courses in general business management.

Many schools of forestry and natural resources offer graduate-level studies leading to the master of science or doctor of philosophy degree with specialization in policy and related fields. In some cases, these may involve internship opportunities in government agencies, legislative bodies, or other appropriate organizations that permit candidates to gain first-hand experience by working with lawmakers and seasoned professionals. These opportunities may suffice as a substitute for thesis research. For such advanced studies, individuals may actually do little course work in the traditional forestry subjects, but instead may specialize in areas of law, economics, public administration, and the like.

Persons interested in forest policy will find career opportunities working directly in the development and implementation of policy decisions. Some find places on the staffs of legislative committees, or within executive branches of government. Others may serve on the policy staffs of governmental agencies, or graduates may work with organizations that take active roles in lobbying for the interests of their members. Many with advanced degrees pursue careers in research or teaching in universities or colleges, or in research branches of governmental agencies. In all of these ways, persons schooled in resources management may make valuable contributions to the future of forestry within the United States, or to worldwide efforts to shape policy to ensure the continued benefits of forests in satisfying the needs of people.

REFERENCES

1. J. Cameron, *The Development of Governmental Forest Control in the United States,* Johns Hopkins Press, Baltimore, 1928.
2. H. O. Canham, *Forest Ownership and Timber Supply,* School of Environmental and Resources Management, SUNY College of Environmental Science and Forestry, Syracuse, N.Y., 1973.
3. M. Clawson, *Forests for Whom and for What,* Johns Hopkins Press, Baltimore, 1975.
4. M. Clawson, "The National Forests: A Great National Asset Is Poorly Managed and Unproductive," *Science,* 191:762–767 (1976).
5. S. T. Dana, *Forest and Range Policy: Its Development in the United States*. McGraw-Hill, New York, 1956.
6. J. Giltmier, G. Bergoffen, R. Manthy, W. Hyde, and H. Vaux (eds.), "The Resources Planning Act," *Journal of Forestry* **74**(5):274–387 (1976).
7. D. C. LeMaster and L. Popovich, "Development of the National Forest Management Act," *Journal of Forestry* **74**(12):806–808 (1976).
8. R. F. McArdle and E. B. Maunder, "Wilderness Policies: Legislation and Wilderness Policy," *Journal of Forest History* **19**(4):166–179, 1975.

9. J. R. McGuire, "National Forest Policy and the 94th Congress," *Journal of Forestry* **74**(12):800–805 (1976).
10. G. Pinchot, *Breaking New Ground,* Harcourt, Brace, New York, 1947.
11. R. A. Sedjo and D. M. Ostermein, *Policy Alternatives for Non-industrial Private Forests,* Society of American Foresters, Washington, 1972.
12. W. N. Sparhawk, "The History of Forestry in America," in *Trees, The Yearbook of Agriculture,* U.S. Department of Agriculture, U.S. Government Printing Office, Washington, 1949.
13. H. K. Steen, *The U.S. Forest Service: A History,* University of Washington Press, Seattle, 1976.
14. H. E. Talmage, "In Defense of S.3091," *Journal of Forestry* **74**(7):432–434 (1976).
15. U.S. Congress, *National Forest Management Act of 1976,* Report of the Committee on Agriculture and Forestry, U.S. Senate to Accompany S.3091, Calendar No. 849, 94th Congress 2nd Session, Report no. 94–893, 1976.
16. President's Advisory Panel on Timber and the Environment, *Report of the President's Advisory Panel on Timber and the Environment,* U.S. Government Printing Office, Washington, 1973.
17. U.S. Forest Service, *The 1980 Report to Congress on the Nation's Renewable Resources,* U.S. Department of Agriculture, Forest Service, FS-347, 1980.
18. R. K. Winters, *The Forest and Man,* Vantage Press, New York, 1974.
19. A. C. Worrell, *Principles of Forest Policy,* McGraw-Hill, New York, 1970.

CHAPTER **4**

THE FOREST ECOSYSTEM

The science of ecology helps us to recognize the consequences of our actions. It gives us an appreciation for the interdependence between resource use and environmental protection. It helps us to find ways to keep our forests and grasslands productive and healthy. Overall ecology provides the understanding we need to use our resources effectively while ensuring their renewability for future generations.

Ecology deals with mutual relationships among living organisms, and between them and the environment. It involves study of how living and nonliving elements interact, and how changes in one affect the condition of others. Ecology embraces the soil, water, sunlight, and other life-supporting elements of the physical environment. It also involves the ways in which these affect the myriad life forms that occur on earth. It studies the earth's life systems, how they develop, and what factors influence the types of plants and animals that occur together at various places. In short, ecology deals with ecosystem composition, structure, and function.[11] Forest ecology focuses that study upon assemblages of trees, related plants and animals, and the environments that support them.[26]

ECOSYSTEMS

Life on earth plus those environmental elements that support it constitutes an earthwide system called the biosphere. But the earth as a whole represents something too vast and complex to comprehend or study easily. Instead ecologists usually work with parts of the broad earth system by subdividing it into units that have relatively uniform composition and character. They call these simpler parts ecosystems. The principal ones include the seas, reefs, estuaries and shores, streams and rivers, lakes and ponds, fresh water marshes, tundras, grasslands, savannas, chaparral, and forests.[18] Each of these has features that readily distinguish it from the others. Yet all are similar in the basic ways in which their component parts function and interact.

The living green plants in all ecosystems manufacture food through a chemical process called photosynthesis. Solar radiation provides the energy for this reaction. The sun's energy also drives other chemical processes that result in the production of starch, cellulose, and fats that make up the tissues of plants. In addition, ecosystems have herbivores that feed upon the green plants and derive their energy from the plant tissues. Carnivores eat the herbivores and other carnivorous animals, getting their energy from feeding upon these other life forms. Ecosystems also have fungi, bacteria, and other microscopic organisms that decompose plant and animal remains and reduce them to simple chemical substances like those originally used by green plants in converting incoming solar energy into plant tissues.

Through these and other processes, energy flows and chemical elements cycle through living organisms of the ecosystem.[11] The incoming solar radiation becomes fixed in plants as chemical energy through photosynthesis. As animals eat the plants and in turn are eaten, the energy is transferred from one life form to another. But at the different steps along the way, all release heat. Thus much of the stored energy is dissipated and eventually leaves the ecosystem by radiation to outer space. Through a variety of processes, carbon dioxide, oxygen, and water also are released to the atmosphere. In addition, as decomposers break down plant and animal remains, the mineral elements go back into the soil and become available again for recycling through the system. However, small amounts of chemicals are flushed out of the site in the runoff of water from precipitation. Other elements may be transported by animals in their migrations, or in products harvested by people.[17]

The amount of organic material that accumulates in living plants will vary greatly between different ecosystems. To a large measure, the volume of plant material present depends upon the types of species common to the place. In forests, for example, woody portions of the tree contain vast amounts of stored energy and considerable amounts of chemical materials. These remain in solid form as wood and bark for many decades. By contrast, grasslands have plants that die back to the groundline annually, recycling nutrients in the above-ground plant parts in a rather short period of time. In both cases, however, the general sequence of recycling occurs, but with different patterns over time.

Although forests also function in these ways, they differ from other types of ecosystems by virtue of the trees. Trees make up the primary vegetation and form the bulk of plant substance. The trees also create environments suitable for other characteristic plants, plus animals well adapted to living in a forested environment. In some regions, tree cover forms a dense community; in others considerable space remains between the trees, shrubs, grasses, and other herbaceous vegetation. Depending upon the characteristics of the soil, climate, and other features of the physical environment, the species of trees will also differ. Still, despite the variability in species composition and the arrangement of living organisms present, these forests have common ties with other types of ecologic systems.

FLOW OF ENERGY THROUGH AN ECOSYSTEM

The sun supplies energy to the earth in vast quantities. In fact, every minute our planet receives about as much new energy from the sun as humankind uses in a year.[22] But not all of what reaches the earth's outer atmosphere penetrates to the ground. Approximately one third is reflected back into space by clouds, and another 9 percent by atmospheric dust. Various gases absorb another 10 percent. The remaining 48 percent of the incoming solar energy that arrives at the outer atmosphere ultimately reaches the earth's surface.[11,26] Atmospheric scientists call this the insolation, meaning incoming solar energy. Yet not all insolation will trigger photosynthesis in living plants. About one half of the available energy occurs in the proper light spectrum, and much of that reflects off the leaf surfaces. Other portions of the available energy also become dissipated in the evaporation of water out of the leaves through a process called transpiration. It serves to cool the tissues and regulate heat within the plant. Only about 14 percent of the solar energy that reaches a plant remains available for photosynthesis. And, at most, 2–3 percent will activate the chemical processes that result in the synthesis of sugars and the building of plant tissues.[2,22]

Although living green plants fix incoming solar energy into their tissues, they also lose some in various physiologic functions. In general, the amount lost through growth, maintenance, and respiration differs among ecosystems. For example, in a field ecosystem these functions may take 15–24 percent of the energy initially incorporated into the plants through photosynthesis. For temperate forests, the amounts may run as high as 50–60 percent. Respiration alone may account for 30–40 percent of the energy loss. Thus only up to 60–70 percent of the energy used by plants ends up fixed chemically in the plant tissues.[11] Ecologists call this the plant biomass, meaning the organic material accumulated in all the plant tissues present in the ecosystem.

Herbivores consume some of the plant biomass and thereby derive energy. The parts of plants not eaten by animals accumulate chemical energy in the stems, leaves, and roots until they die. Then decomposing organisms reduce the plant materials to basic chemical components and release heat in the process. During the intervening time, the decomposers work on cast-off parts of plants, such as fallen leaves and branches. Bacteria do the same thing with animal remains. Depending upon the predominant type of vegetation in the ecosystem and its amount, the flow of energy and cycling of chemicals can occur over either a short or long period of time. Nevertheless some changeover takes place annually.

Once fixed in green plants by photosynthesis, energy transfers through the ecosystem by animals as they eat and are eaten. Ecologists call this the food chain. It means that a herbivore initially eats a green plant, is eaten by a carnivore, which in turn may become the prey of some other carnivorous animal. Eventually the last carnivore dies and the decomposers reduce it through decay. In this process, energy flows through the system from one organism to another in the food they eat. Along the way, however, the amount passed on diminishes with each step in the food chain, because each organism uses some of the energy to maintain its own physiologic processes. For example, herbivores expend large amounts of energy in their daily activities. Consequently, of the total energy they take in by eating plants, only a small proportion will remain to be passed on to the carnivore that may eat the herbivore. In turn, the carnivore expends energy, leaving only a small portion of the energy derived from its prey. In fact, the amount may be only 10 percent or so at each step in the food chain.[11]

Actually an ecosystem neither creates nor destroys energy. Rather the amount temporarily fixed in the biomass eventually will dissipate as heat that cannot reenter the system. Only continuous inputs from the sun keep the earth system functioning. In general, solar energy converted or fixed by an average forest ecosystem will amount to only about 0.3 percent of the total reaching the ground level in a sunny climate, and up to 1 percent in a cloudy one. This appears inefficient. But trees have relatively large leaf surface areas in their crowns, and the leaf area may be 3–13 times larger than the horizontal ground space under the tree. Because of this, forests surpass other forms of vegetation in the proportion of solar energy fixed by photosynthesis.[14,20] They also store the energy for long periods in the form of wood, bark, and other plant parts that people later can harvest and utilize in a variety of ways. A redwood stand such as that in Figure 4–1 on page 74 may contain as much as 3200 metric tons of biomass per hectare.[7]

BIOMASS PRODUCTION

Data from monitored watersheds indicate that forests of the northeastern United States convert about 0.7 percent of the incident solar energy into woody tissues by photosynthesis. Several ecologists have also quantified total annual production for various western conifer forests in order to determine the amount of energy fixed into tree tissues annually. The following set of data provide a comparison of several different types of forest.[7]

Type of forest	Forest age, years	Production of biomass, metric tons per hectare per year
Coast redwood	Old	14.3
Western hemlock	26	36.2
Western hemlock/Sitka spruce	110	10.3
Sitka spruce/western hemlock	130	14.7
Western hemlock/Sitka spruce	130	12.3
Douglas-fir	150	10.5
Douglas-fir	125	6.6
Douglas-fir/western hemlock	100	12.7
Douglas-fir/western hemlock	150	9.3
Nobel fir/Douglas-fir	115	10.3
Douglas-fir/western hemlock	450	10.8

Source: Adapted from data assembled from several publications and included in Franklin and Waring [7]. Used by permission.

Figure 4–1.
Forests fix and store energy for long periods of time in the wood, bark, and other plant parts, with the greatest biomass produced by the redwood forests of California. *(U.S. Forest Service.)*

These figures indicate that the type of forest and its age may affect the amounts of material produced annually.

Ultimately the amount of solar energy reaching the earth at a given place determines the maximum amount of plant material the forest can produce annually. But solar energy does not occur in equal amounts at all places, so forests do not have a uniformly high production of biomass everywhere. This difference in energy input depends upon latitude and other factors. Generally tropical regions receive more energy per year with a more uniform distribution of it from month to month. Thus, while tropical forests occupy only 43 percent of the world's forest area, they yield 62 percent of the primary net production of all forests. Overall, forests and woodlands occupy about 42 percent of the earth, and account for almost 46 percent of the total world net primary production of dry matter annually.[13,27]

Tropical rain forests comprise a complex of tree and nontree species, with a dense undergrowth of vines and other vegetation not common to most temperate forests. These plants use space and resources and produce considerable amounts of dry matter that add to total plant biomass. But plantations kept free of the undergrowth develop most of the annual production as growth to the main stems and branches of trees. In such forests, the trees produce more usable volume than do natural areas. They give an indication of the productive potential of the tropical systems. For example, *Pinus caribaea* produced about 20 m^3/ha/yr over a 15-year period in one plantation. *Gmelina arborea* grew about 30 m^3/ha/yr over a 12-year period, and *Albizia falcateria* could yield 40–50 m^3/ha/yr in eight years. In the favorable climate, these fast-growing tropical species develop in height and diameter much more rapidly than do trees in temperate forests.[10] By contrast, Douglas-fir in Oregon has produced up to about 13 m^3/ha/yr over a 90-year period,[23] and loblolly pine in a 16-year-old plantation produced about 19 m^3/ha/yr along the southeastern coastal plain.[27] Even fast-growing cottonwood in plantings along the lower Mississippi Valley produced only 17–18 m^3/ha/yr over a 15-year period.[13] While these figures appear impressive by U.S. standards, the potential of the tropical ecosystems appears considerably greater.

CYCLING OF WATER

Water plays an indispensable role in any ecosystem. Higher plants and animals use water directly in liquid form. This water becomes a constituent of cellulose, starches, sugars, lignin, plant and animal protein, and even of such key compounds as ribonucleic acid (RNA). In fact, water accounts for 70 percent of the weight of living organisms. Essential chemicals and food materials dissolve in it, and water serves as one raw material for photosynthesis. Its high heat capacity and the large amounts of energy required to melt ice and vaporize water make it an important regulator of temperature, both in living organisms and within the physical elements of an ecosystem.

Water enters an ecosystem from the atmosphere mainly in the form of rain, snow, and hail. Some may condense directly onto plants and objects as dew or frost. Also, the leafy canopy in a forest ecosystem may intercept up to 11 percent of the precipitation entering it during a heavy storm, and as much as 80 percent deposited by light showers. The leaf litter and other organic debris on top of the ground will trap additional moisture, preventing it from reaching the soil. Hence, in a forest, only a small proportion of moisture entering the system as precipitation may filter through into the soil to become available for plant use.

Much of the intercepted water evaporates off the leaf and bark surfaces to the atmosphere. Some drips from the plants to the ground. Of the quantity reaching the ground, some flows across the surface or through the soil directly into streams and ponded waters. Then it eventually drains out of the ecosystem. The remaining water enters the soil and remains there between soil particles and among the decomposed organic matter. Here basic chemical elements dissolve in it and may become transported into the plants as water moves from the soil, into the roots, and through the plant tissues. Some of this water leaves the plant through transpiration. The remainder, or about 1.6 percent of that falling to the earth and into the ecosystem, may enter into such chemical reactions as photosynthesis. This may include as much as 11.2 metric tons per hectare during a single year.

Forests affect the hydrologic cycle. First, they intercept important quantities of precipitation, and much of it evaporates directly to the atmosphere off of the foliage. More becomes incorporated into the plant tissues, and trees transpire large amounts of water. In fact, on a warm sunny day, as much as 22–44 metric tons of water per hectare may transpire out of their crowns. Hydrologists usually do not separate the amount lost by evaporation from various surfaces and that transpiring out of foliage. Instead they consider the two a combined loss and call it evapotranspiration. Together these losses greatly reduce the amount of water that remains in an ecosystem, and the quantity that runs out of it as streamflow.

The effect of transpirational loss can be considerable. One elaborate study of a watershed in the northeastern United States indicated that by removing all vegetation, the foresters increased streamflow by 100 to 300 mm during the first year, primarily by reducing losses to transpiration.[14] Within five years, regrowth of vegetation again reduced yields by two thirds, and by 10 years plant use cut yields back to the preclearing levels. On a worldwide scale, this transpiration consumes about 10 percent of the total solar energy reaching the land. Thus vegetation serves as an air conditioner of continental proportions by dissipating heat back to the atmosphere.[24]

Availability of water in large measure determines the type of terrestrial ecosystem supported at different places. The actual amount of precipitation reaching the earth is not the critical factor. Rather the type of plants growing in a particular place will reflect the amounts of water remaining after losses to evaporation from intercepting surfaces, soil, and other sources of storage. In general, if the input by precipitation amounts to less than 16 percent of the water that will evaporate and transpire due to the heat energy present, deserts form. If precipitation supplies

20–60 percent of the water potentially needed for these purposes, grasslands dominate. Forests occur where inputs of water by precipitation accommodate 80 percent or more of the amount potentially needed for evaporation and transpiration.

When precipitation deposits more water than evaporation, transpiration, and plant functions use, the excess leaves the ecosystem as runoff. It forms streams, rivers, ponds, lakes, and the seas, and it serves as the vehicle for leaching basic chemical substances out of the ecosystem. These become available as a result of release by weathering of geologic materials, of decomposition of dead organisms, and as by-products of plants and animals. Then the elements dissolve in water and become transported by it. When the runoff occurs on the surface in large quantities, it may also dislodge and transport organic and soil particles by a process called erosion. Erosion occurs continuously in terrestrial ecosystems, even in the absence of human disturbance. The rate and amount depend upon the amount of excess water, the severity of the topography, and the stability of the soil.

Losses of water from plants, soil, and surface storage actually prove temporary in an ecologic sense. As water vapors rise into the atmosphere, they form into clouds. Eventually droplets develop that again fall to the earth, bringing along nutrients and solid particles dissolved from the atmosphere. Then the cycle begins again with a regular series of input and outflow controlled in broad global pattern by solar energy and influenced by latitude, topography, and other physical features of the earth.

NUTRIENT CYCLING

The movement of water in the system from the soil through plants is a primary mechanism for nutrient transfer. Sulfur, potassium, phosphorus, calcium, iron, and other elements essential to plant growth dissolve in water and thereby become transported into the plant tissues. A review of cycles for carbon, nitrogen, and the mineral nutrients illustrates the general nutrient flow, and exemplifies the complexity of processes involved.

Carbon is an essential element of all plant and animal tissues. It occurs in the atmosphere as gaseous carbon dioxide at a concentration of about 0.03 percent. Through the open stomata, or closable pores in the leaf surfaces, carbon dioxide comes into contact with moist leaf cells and passes into them. Here, by action of chloroplasts energized by light, it combines with water to form carbohydrate substances. Oxygen released in the process passes through the cell walls and out of the intercellular spaces through the stomata into the atmosphere. To a limited extent, by these processes vegetation serves as a regulator of atmospheric carbon dioxide. The rate of photosynthesis increases during periods of high atmospheric concentrations, and decreases at other times.[9,12,17]

Oceans play a role in global carbon cycling. They act as great reservoirs of carbon dioxide.[11] Also, certain water plants such as *Elodeas* precipitate calcium carbonate from seawater. This compound sinks to the bottom eventually to become limestone. If later uplifted by geologic action, the limestone may react with organic and other acids to release carbon. Thus present-day chemical reactions in an upland ecosystem may release carbon previously bound up in limestone under the ocean many geologic ages ago.

Carbon enters into many chemical compounds as it cycles through the system. Eventually it is released to the atmosphere through oxidation. This may occur rapidly as when a wildfire burns over a forest, releasing great amounts of carbon as well as other nutrients. Carbon in wood used by people, and supplies in dead plant parts that fall from the vegetation, also eventually return to the atmosphere as carbon dioxide. This occurs slowly through weathering and decay, or rapidly through combustion.

Nitrogen cycles in a somewhat different way. Paradoxically it makes up about 79 percent of the gaseous content of the atmosphere, but often remains in limited supply for plants. Atmospheric nitrogen occurs in molecular form not directly usable by higher plants. Bacteria must first convert it into nitrate or ammonium. Some of these organisms grow in association with leguminous vegetation and other plants, and others as free-living bacteria. Both kinds fix atmospheric nitrogen into a useful form. Still other bacteria reduce nitrates and nitrites to molecular nitrogen, thereby controlling their levels in the soil. Annually biologic processes fix about 54 million metric tons; industrial ones, 30 million; and photochemical ones, 7.6 million. Denitrification by bacteria recycles only 83 million metric tons.[11] The excess tends to accumulate in such areas as feed lots, sewage effluents, soils receiving runoff from fertilized lands, and undecomposed organic matter.

Plants derive mineral nutrients such as potassium, phosphorus, calcium, sulfur, magnesium, iron, boron, manganese, zinc, copper, and molybdenum directly from the soil.[1,30] These nutrients result from weathering of geologic materials, and to some extent from precipitation and dust. From the stockpile of elements stored in soil, nutrients move into plants in solution through the intake of water. Then the elements are incorporated into tissues through physiologic processes, accumulating mostly in leaves and fruits. Generally concentrations of nutrients remain at low levels in the main woody stem of trees.

As parts of plants die and fall to the ground, flora and fauna in the litter act upon the fallen organic materials and decompose it to basic carbon, nitrogen, and mineral elements. Once released, the gaseous nutrients dissipate into the atmosphere. Or, like the mineral elements, some are held within the soil until leached out in runoff or absorbed by living plants. Some nutrients cycle directly from living plant tissues to the litter and soil after being washed out of leaves during periods of precipitation. These become deposited directly into the litter and soil in soluble form.[1,4]

Collectively these processes result in a cycling of nutrients from soil into plant tissues, and back to the soil. The nitrogen and carbon may move directly to the atmosphere from the plants or organic materials, and later become replenished as noted earlier. Small amounts of nutrients are continually added to the pool of elements in an ecosystem by inputs from dust, precipitation, the weathering of rock and soil, and by further fixation or direct incorporation from the atmosphere by plants through physiologic processes. Small amounts leave the site by leaching through runoff or as wind carries away dust and small bits of organic matter. Also, some carbon and nitrogen become volatilized by microbial activity and transported away by wind. In most ecosystems, however, there remains a balance between inputs and losses, with recycling of the overall pool of mineral nutrients repeatedly from vegetation to soil and back to vegetation. Water acts as the vehicle for transporting nutrients into the vegetation, with carbon and nitrogen continuously added from the atmosphere.

This balance may be disturbed. In fact, many natural phenomena temporarily upset the equilibrium. Wildfire, for example, will consume vegetation and release nutrients through rapid oxidation. Winds may blow down vegetation, suddenly covering the ground with large amounts of dead plants, interrupting the uptake of nutrients, and adding large quantities of dead organic matter to decay and release nutrients. Insects and diseases also kill vegetation and have the same net effect as windthrow relative to nutrient cycling. In addition to these natural events, human intervention may temporarily upset the nutrient cycling. This occurs following logging and land clearing. However, unless maintained by unusual means, the interruption persists for relatively short time periods. The equilibrium of nutrient cycling becomes reestablished as new vegetation replaces the old to form a new forest community.

SITE FEATURES AND THE PHYSICAL ENVIRONMENT

Atmospheric and soil conditions profoundly influence the composition, growth, and development of trees and associated plants at a place. These compose elements of what foresters call the site. Site means the collection of environmental conditions at a place on the earth's surface. It includes both the living and nonliving features that affect the plants and animals that inhabit the locality. Site conditions also influence the way in which living organisms grow and develop.[6,26] Overall many different site factors interact to create a characteristic environment at a place. In forests, these conditions enable the establishment and development of tree communities, plus other plants and the animals that live in forested areas.

Climate

Climate represents one collection of site features that has a strong effect upon an ecosystem. It includes incoming solar energy, called insolation. Both the length of day and the angle of incidence of the sun's rays affect the amount of insolation. For example, the sun appears most directly overhead at the equator, and so that part of the earth receives the greatest intensity and longest daily duration of sunlight. These decline with distance north and south. Atmospheric filtering by water vapor and impurities also influences the quantity and quality of insolation. These losses differ with latitude and with daily changes of atmospheric conditions at any one place. On cloudy days, less total solar energy reaches the earth's surface. Effects of low sun angle and atmospheric attenuation become most pronounced toward polar regions during winter seasons. Along with other characteristics of climate, those reflecting the amount and duration of solar energy are important influences in determining where different plants survive, and how well they grow.

Temperature and Light. Both the quantity of sunlight and its quality influence plant growth and survival. Quantity means intensity and duration. Scientists use temperature to measure intensity at any time, and accumulation of heat over the day length to quantify the amount of energy received. Quality involves the spectrum or wavelength of solar energy, including both the visible and nonvisible portions. Such facets of insolation affect plants in several ways. For example, temperature controls the rate of most chemical reactions, including photosynthesis. It also influences the breakdown of carbohydrates and release of energy through a process called respiration.[4] Few reactions take place to augment stored chemical energy at temperatures below 0 °C, and photosynthesis declines importantly above 25–30 °C. In fact, losses of energy through respiration exceed additions by photosynthesis at these higher temperatures. Consequently the net productivity of a plant community depends upon the number of days when light of suitable quality occurs in a temperature range that sustains net growth of plants.

Both high and low temperatures and their duration may adversely affect plants. While photosynthesis may continue with temperatures up to 50 °C, few plants survive these high levels for very long. Cell material coagulates and death occurs.[4] As temperature increases, more and more water also transpires from the plant. Otherwise the plant dies. The same thing can occur if the soil contains insufficient moisture to supply plant needs. Hence availability of abundant water may compensate somewhat for high temperatures. In these ways, the temperature regime and moisture available affect the kind of plants that survive at a place. Such relationships help to determine the shift of plant species that occurs from the equator to the poles, and from low to high elevations at any given latitude.

Tropical ecosystems usually have the greatest biomass; net productivity will average as high as 22.0 metric tons per hectare per year in a tropical rain forest. But net production may not exceed 1.4 metric tons per hectare per year in arctic, alpine, and tundra communities.[14] Low-incident radiation, low temperatures, variability of climate, and short growing seasons account for this slow growth and the small sizes of trees present in the more northern communities. In arctic and subarctic regions, trees will usually not survive and tundra vegetation dominates the landscape. Even at lesser latitudes, these same vegetation patterns will occur with increased elevation on the highest mountains. There the greater distance from sea level results in lower overall temperatures, which has the same effect as would be true many miles closer to the poles. Distinct timberlines may form on mountains at least as far south as New York and Colorado.

Length of Day. Length of day and regimes of temperature also regulate the seasonal activity among plants and animals. Some physiologic functions depend upon a specific combination of these two climatologic factors. At northern latitudes, for example, the time of bud breaking in the spring and subsequent flowering depend upon the increased temperatures as the days lengthen toward April and May. Hardening of plant tissues and the onset of dormancy in autumn result from the shortening of the days as the season progresses, coupled with lower temperatures common to that season.[26] Animals' migration, breeding, nesting, and hibernation may also reflect changes in day length. Such responses in both plants and animals reduce injury and mortality. Also some species, such as those common to equatorial zones, require minimum periods of light to induce flowering and maturing of seed. When moved to latitudes with shorter days, these plants may fail to reproduce and consequently will not survive beyond a single generation. Such mechanisms profoundly influence species distributions.

Precipitation. Precipitation, too, constitutes a major climatologic determinant of the lushness of an ecosystem, and of the species that survive and grow on a site. The precipitation must supply transpirational and other requirements of plants, and occur at times to support physiologic activity. However, the amount needed to sustain forest growth varies with the temperature regimes of the region. When temperatures remain relatively cool, plants transpire less water, and therefore use less precipitation for various functions. To illustrate, a closed forest in the tropics would use 770–1000 mm/yr, while 500–760 mm would suffice in temperate regions, and 250–500 mm would sustain forest growth toward the subarctic regions. Considerable differences also occur at any single latitude due to important changes in elevation. Higher altitude above sea level results in lower temperature regimes, and less transpirational use. Thus, like other climatic factors, precipitation effects also depend upon temperature.

Wind. Wind also influences forest vegetation, especially along seacoasts, on mountain ridges, and on the plains of the western United States. Wind has varied effects on vegetation, including an influence on the character of individual trees. For example, trees at timberline, ridge tops, and other exposed windy places frequently have distorted shapes (Figure 4–2). The live crown may extend at a sharp angle away from the prevailing wind. Under extreme conditions, some trees even become prostrate from wind-driven snow, ice, and soil particles. In these ways, wind may profoundly alter the character of a forest community.

Winds of hurricane force that occur periodically may blow trees down over wide areas. Though infrequent in most regions, such strong winds may destroy the

Figure 4–2.
Persistent winds from one direction such as those along the shores of Upper St. Regis Lake in New York may distort the shapes of trees, and under extreme conditions wind-driven particles may keep the trees prostrate and bent.

forest cover and add to the ground surface great quantities of dead vegetation that will eventually decompose and settle on top of the soil. The uprooting also exposes mineral soil, mixes the different soil layers, and breaks up the regularity of the ground surface. These may result in a change in the composition of species that germinate on the newly exposed seedbeds, and ultimately over the type of forest that develops there. However, the effects seldom prove permanent, and a new growth of forest trees eventually develops to reestablish the forest ecosystem.[26]

Forests in themselves greatly affect wind. Trees serve as surface obstructions that break up the flow of air across the earth's surface, dissipate the energy of the wind, or divert it above the ground surface to tree-top levels. Within temperate forests, the wind velocity near the ground may measure only 5–30 percent of that over an open field. Tropical rain forests, with their dense growth of vines and other vegetation in the understory and the thick leafy canopy, may slow winds by as much as 99.6 percent of that flowing across the top of the crowns. Foresters put such understanding to use in designing the planting of windbreaks on the Great Plains as a means of reducing soil erosion by wind. The shelterbelts also reduce surface evaporation of precious soil moisture and protect homesteads during winter months.

Not all wind harms the forests. In many ways, wind benefits forest growth. For example, air masses moving inland from the ocean bring fog to the Pacific Northwest coastline and supplement the limited rainfall of the summer months. Winds also move moisture-laden rain clouds across the country, carry fresh supplies of carbon dioxide, and transport mineral particles that augment nutrient supplies of the soil. In addition, they serve to disseminate many tree seeds, and help spread pollen needed to fertilize tree flowers. On the other hand, even gentle winds may move some important tree insects and disease spores over long distances. Thus, even at moderate force, winds may profoundly affect the forest in both beneficial and undesirable ways.

Many other examples would serve to illustrate the ways aspects of climate influence the kinds of trees and associated vegetation that grows and the way the forest develops. The potential list seems almost endless, but these few illustrations help emphasize the importance of climatic effects upon forest ecosystems. Some also indicate that the forest in turn alters the microclimate near the earth's surface, and thereby influences the physical environment. Trees will ameliorate extremes of temperature, light, and wind, and often make conditions conducive to species that will not survive and develop without such protection.

Soils

Soils are another major component of the physical environment. They are characterized by a variety of depths, textures, and nutrient contents. Most soils come from the bedrocks that underlie the surface mantle, and the rocks form an impervious foundation beneath them. Consequently the geology of an area and the geologic forces that tempered the surface conditions also profoundly influence the character of the forest ecosystem that develops at a site. Soil scientists call the features of soil that influence plant growth the edaphic factors. These include its character as related to the soil minerology and physical condition, the biotic organisms and organic materials that occur within the upper soil layers, plus the water and air found between the soil and organic particles. Conditions of these factors within the proportion of the soil penetrated by tree and other plant roots exert an important influence over the type of vegetation that occurs on a site, and how well it grows.

Many factors influence the nature and condition of a soil, its productivity, and the type of plant and animal community that develops or survives in it. As with so many other aspects of natural ecosystems, the interrelationships between soil properties and the biotic community have a complexity that defies adequate description. Rarely does any single edaphic condition determine what will grow in a soil, or how well. Rather different soil conditions interact, often tempered by the temperature and rainfall patterns of a site. These all blend together to create a particular physical environment. Plants and animals that thrive under such conditions survive and grow well. Species not suited to the environment remain absent. In turn, as a biotic community develops, it affects the soil. It adds cast-off plant parts and by-products and acts upon the mineral soil in various other ways. By these means, the living organisms temper the condition of the inert materials, thereby modifying the physical condition of the soil.

In many ways, the soil constitutes a living system. At least it involves an inseparable combination of inorganic materials, and living plants and animals and their dead remains. As such the soil is a dynamic system under continual change. Forest soils differ from others in that they contain large volumes of roots from a variety of tree and nontree species. They also have the bacteria, fungi, actinomycetes, algae, protozoa, worms, arthropods, and mammals that inhabit the forest ecosystem. These form a definite biotic community that influences the development and character of the soil, and tempers the nature of plant and animal communities supported by it.

BIOTIC COMMUNITY DEVELOPMENT AND CHANGE

Given a particular mix of environmental conditions such as the ones illustrated above, and a readily available source of viable seed, a reasonably characteristic type of plant community will develop on a site. Animals that find that community a suitable habitat will gradually inhabit the area. While not totally predictable as to content and organization, the vegetation that develops will generally reflect characteristics of the soil and climate. People, of course, may modify the land, and the types of management practices they use can affect the edaphic factors and make them different from those in other soil nearby. Still, conditions of the physical environment at a site ultimately will limit the array of species that become established and eventually develop to form a characteristic ecosystem that remains somewhat stable over time.

According to human perceptions of time, most natural ecosystems appear highly stable, but few remain that way for long periods. Instead, as a community reaches physiologic maturity, some plants die. That creates space where new ones regenerate. But as one community matures, it modifies the physical environment. It shades the ground and adds organic litter to the soil. It also tempers the amounts of moisture available and influences the temperature near the ground. These changes, in turn, affect the species that becomes established on the site. And they may set the stage for the replacement of the existing community by an entirely different collection of species. The ones best adapted to survive and thrive under the types of environmental conditions present at the time regeneration takes place will replace the old species. As a result, one type of plant community may gradually supplant another. Ecologists call this process succession.[26]

The natural revegetation of an old farm field illustrates this process. If we could revisit the place repeatedly over a long period of time, we would see an amazing transition in the character of the ecosystem. Eventually the kind of community that originally dominated the place before the land was cleared and cropped for food would return. That gradual change over time illustrates succession.

During the first few years, a series of annual plants and grasses would colonize on the exposed soil. Gradually perennials would dominate the ground. In prairie areas, the succession would stop with occupancy by grasses that characterize the region. In forested places, trees would eventually take root among the herbaceous plants. Slowly the number of trees would increase, and they would grow in size. Their crowns would gradually shade out the lesser plants until only trees remained. In time, small plants and shrubs that survive in the shaded environs of the forest would appear. Organic matter from the trees would accumulate on the surface, converting the old field soil to one more characteristic of forested areas.

Among species of the early succcessional stages might be such trees as birches and aspen in eastern regions, or lodgepole pine in the west. These species would dominate the site for a long period of time, but as the trees matured, their crowns would thin out and some would die and fall over (Figure 4–3). In the space opened up, a group of new trees would develop, but because the environment near the ground remained somewhat shaded by the taller, older trees, the new ones would include species that can grow under partial shading. Hence the stand of birch and aspen would gradually disappear and in its place sugar maple, beech, and hemlock, or similar species, would emerge. In the western regions, the lodgepole pine would give way to new stands of spruce and fir. Eventually, through this gradual process, the species composition becomes restricted to ones that can regenerate under shading, leaving a community of the kind that farmers had cleared from the lands decades earlier. Further, the kinds of animals present at the different successional stages will depend upon the food, shelter, and other habitat features prevalent as the vegetation changes.

Figure 4–3.
As trees in a forest mature they eventually die and fall, making way for new regeneration that fills the space and develops into a new forest. *(U.S. Forest Service.)*

In time, the species composition of a community will stabilize, and the same species will occur repeatedly in successive generations. Ecologists call this condition the climax stage, or a climax forest. But once a climax develops, it does not necessarily endure, even if left undisturbed. Rather many natural forces can bring about dramatic changes in the physical environment and create conditions suitable for a subclimax stage of vegetation. This might follow a major infestation by a harmful insect, blowdown by strong winds, wildfires, and other natural catastrophes. All can abruptly open up a forest and may also expose mineral soil and otherwise alter the environment.

Given such a change, those species that grow best under the new more exposed conditions may become established, creating a new forest representative of earlier stages of succession development. Thus, when wildfire or an insect attack destroys a mature stand of lodgepole pine, that same subclimax species may regenerate again. Species composition may not shift to the spruce and fir that might otherwise follow after it. Among the deciduous forests of the East, stands of sugar maple, beech, and hemlock may give way to new forests with black cherry and white ash. These would not normally occur in abundance in climax communities.

Overall forest ecosystems show great resilience in response to disturbance. Otherwise they would have disappeared from this earth long ago. But loss of some critical environmental factor due to repeated catastrophic events, or some major change in the balance of the physical environment, can change the ecosystem. Severe disturbance on steep slopes, for instance, may lead to important erosion and a loss of valuable soil and organic matter. Then the forest might disappear and another type of vegetation community form in its place. In tropical regions, the clearing of a forest to make space for agriculture results in a rapid leaching of nutrients. After only two or three years, the soils become so impoverished that farmers abandon the fields. If so little of the mineral elements and organic matter remains that forest trees may not regenerate the area, however, grasslands supporting only scattered trees take over the site. This creates a new ecosystem called savanna. Fortunately most disturbances do not lead to such severe depletion of site resources and in most places new trees readily regenerate to replace the old forest.

CHEMICAL ASPECTS OF ECOLOGY

Many functions of plants and animals involve chemical reactions or controls. Most commonly, we think about food synthesis, digestion, tissue and organ building, growth, metamorphosis, and reproduction. All involve chemical reactions that have been known about for decades. More recently, ecologists have come to appreciate that chemical mechanisms also influence animal communication, behavior, defense, mating, and rearing of young. Odors attract animals to some foods, and repel them from others.[31] Also, chemicals exuded from or leached out of some plants may determine whether two species coexist at the same place.

All these things have become clearer in recent years as chemical ecology has developed. This discipline embraces the chemistry of forest ecosystems and involves the ways in which chemicals impinge upon living organisms through health and disease, predator–prey relationships, growth, senescence, death, and decay. Key chemical activities also affect air, water, and soil pollution and purification.[25] In fact, the chemistry of ecosystems in many ways may have as profound an effect on their composition and development as do conditions of the physical environment.

The pheromones represent one key group of chemicals that animals and insects emit to communicate with others of the same species.[31] Some pheromones act as danger signals, such as when ants mobilize the colony to fight off invaders. Others may attract bark beetles to concentrate their feeding on a single tree, or entice bees into a swarm around a new queen. In many animals, pheromones attract males to receptive females, and enable young and parents to identify each other (Figure 4–4). Among mammals, the pheromones appear to provide a rather complex system of communication that may influence the social behavior among communities of a species, where they go, and what they do. Similar communications systems may even serve to attract salmon from the ocean to a particular stream to

Figure 4–4.
Pheromones provide a complex system of communication between animals of a species, and may also serve to influence their behavior, defense, mating, and many other social functions. *(U.S. Bureau of Sport Fisheries and Wildlife.)*

spawn in the place of their birth. Such mechanisms have important implications concerning the way in which a wide array of animals function within a particular ecosystem.

Allelochemicals also have important ecologic implications. These are natural substances produced by one plant species that may affect the health, growth, and behavior of another.[28] For example, black walnut trees produce the chemical juglone in their leaves. It leaches out of the foliage during times of precipitation and can harm plants that grow under the walnut tree. Although some species may not suffer ill effects, the chemical impedes the growth of many. As a result, the juglone may limit profoundly the species composition of a plant community that develops around a walnut tree. Several other chemical substances found in many natural ecosystems will also induce this and other forms of abnormal growth, such as the galls that form on some plants. Collectively, therefore, the various chemical mechanisms may have a profound effect upon the survival or adaptability of a species. They may have as much influence over the success, failure, or abundance of some plants in an ecosystem as has traditionally been attributed to the physical environment.

Mycorrhizal fungi serve as a particularly fascinating example of ways in which chemical interaction benefits two species. The exact mechanism still remains unclear, though mycologists and physiologists have recognized for about a century that trees infected with mycorrhizal organisms among the roots grow better than do ones that lack the fungus. Apparently the fungus acts as an extension of the tree roots, aiding absorption of water and nutrients. In turn, the fungus may derive sugars and other food materials from the tree. Recently scientists have discovered that the fungi also produces antibiotics, which may prevent infection by pathogenic fungi, thereby protecting the tree roots against diseases.[16] Hence, in another subtle way, chemical substances produced by one living organism profoundly affect the capacity of another to thrive and develop.

Other examples could serve to illustrate further the increasingly important role of chemistry in ecology. These include many defense mechanisms such as the irritation caused by oils from leaves of poison ivy, the swelling of your skin after a bee sting, and that pungent odor given off by skunks when they sense danger. Chemists also have found several less obvious substances produced by plants and animals that act as defense mechanisms.[5] Increasingly such mysteries capture the interests of the scientists who study the natural regulation of ecosystems or their implication for the evolution of an organism. This study of chemistry in ecology has provided new understanding of the ways in which plants and animals survive and thrive. It has made an important contribution to our appreciation of the complexity of natural ecosystems across the earth.

CAREER OPPORTUNITIES AND THE ECOLOGIC SYSTEM

People make up an integral part of the ecosystem in which they live. They either adapt to their environment, or take measures to alter conditions to suit their capabilities and interests. In this way, human action inevitably alters natural ecosystems. We have an impact in the way we grow crops, build and heat our homes, transport ourselves, manage our water and mineral resources, dispose of our wastes, and much more. Almost everything we do affects the environment in some way—sometimes for the good and sometimes not.

Resource managers and professionals who draw upon the resources of our forests, grasslands, and farms play critical roles in maintaining the long-term productivity of these ecosystems. Through their understanding of ecology, they have learned to temper their activities to prevent destructive impacts. Thereby they can help ensure that society will continually benefit from the many uses of and products derived from the land. But before management can begin, these professionals must consider carefully how the ecosystem functions, how features of the physical environment affect the indigenous plants and animals, how biotic components of the ecosystem interact, and how different treatments might change the environment or the plant community for good or bad. In this process, botonists, wildlife biologists, soil scientists, ecologists, chemists, entomologists, pathologists, planners, administrators, and many other professionals may contribute their skills and advice.

There are few full-time ecologists, though such positions do exist. Many foresters who do research in forest sciences actually function as forest ecologists. They focus attention upon tree biology, reproduction processes, growth functions, and community composition. Ecologists participate with foresters in many planning activities. And some foresters develop advanced skills in ecology so they can preform this planning and evaluation as well. Some ecologists also work with government regulation agencies helping to formulate environmental assessments related to potential land and resource use activities. Forest ecologists also participate in many types of educational activities, including public interpretation programs. They work in state-level cooperative extension programs helping to educate landowners and others. In addition, ecologists and others schooled in that discipline may find employment teaching in high schools, community and junior colleges, and at the university level. Most applications of ecology, however, remain with professionals of many disciplines who apply that understanding in deciding how to manage and wisely use our forest resources. In essence, forest ecology has become the concern of many. It permeates the work of foresters, wildlife managers, watershed specialists, entomologists, pathologists, silviculturists, and many other people who work in and with forestry. As a result, colleges with programs in forestry, wildlife and recreation management, forest influences, watershed management, and allied fields usually require their students to take introductory ecology courses. Or they incorporate studies of ecology into their teaching of silvics and the fundamentals of silviculture. Later the basic concepts of ecology will serve as a steppingstone to studies in silviculture, animal population dynamics, habitat management, range management, watershed management, and others of the applied disciplines. Interested students may also pursue advanced course work in ecologic principles as a means of strengthening their interest and capabilities in that field. Through graduate studies, students can develop more highly specialized skills in ecology.

Those interested in focusing their careers on working with plants and animals should start by taking courses in basic natural and biologic sciences. Elective courses in plant and animal physiology, botany, pathology, entomology, organic chemistry, plant identification, geology, soil science, hydrology, genetics, population dynamics, computer applications, ecologic energetics, and others that deal with components of natural ecosystems and how they function. All will contribute understanding that strengthens a person's capability in ecology. These same courses have application in silviculture and wildlife management as well. At the graduate level, studies in ecology can take several different directions. These can range from work with community composition, to chemical ecology, mathematical modeling, and more. Potentials for specialization and diversity in ecology appear almost infinite for those persons who elect to pursue them through advanced degrees.

In many ways the foundation of ecology influences the ways in which professional foresters grow, harvest, and use plants and animals. It affects their management of water and recreation resources as well, and tempers the way in which they conduct the business and administration of government and industry. In the long run, effective management depends upon how well these people can carefully integrate demands upon resources with the capacities of forests and grasslands to support the use in perpetuity. Ecology serves as a critical foundation in accomplishing this important mission. It provides the base that supports the many other aspects of forestry.

REFERENCES

1. K. A. Armson, *Forest Soils: Properties and Processes,* University of Toronto Press, Toronto, 1977.

2. J. A. Bassham, "Increasing Crop Production Through More Controlled Photosynthesis," *Science,* 197:630–638 (1971).

3. F. H. Bormann, "An Inseparable Linkage: Conservation of Natural Ecosystems and Conservation of Fossil Energy, *Bioscience,* 26:754–760 (1976).

4. T. W. Daniel, J. A. Helms, and F. S. Baker, *Principles of Silviculture,* 2d ed., McGraw-Hill, New York, 1979.

5. T. Eisner, "Chemical Defense Against Predation in Arthropods," in E. Sondheimer and J. B. Simeone (eds.), *Chemical Ecology,* Academic Press, New York, 1970, pp. 157–217.

6. F. C. Ford-Robertson (ed.), *Terminology of Forest Science, Technology, Practice, and Products,* Multilingual Forestry Terminology Series No. 1, Society of American Foresters, Washington, 1971.

7. J. F. Franklin and R. H. Waring, "Distinctive Features of the Northwestern Coniferous Forest: Development, Structure, and Function," in R. H. Waring (ed.), *Forests: Fresh Perspectives from Ecosystem Analysis, Proceedings of the Annual Biological Colloquium,* Oregon State University Press, Corvallis, 1979.

8. A. D. Hasler, "Chemical Ecology in Fish," in E. Sondheimer and J. B. Simeone (eds.), *Chemical Ecology,* Academic Press. New York, New York, 1970.

9. J. E. Hardh, "Trials with Carbon Dioxide, Light, and Growth Substances on Forest Tree Plants," *Acta Forestalia Fennica* **81**:1–10 (1966).

10. N. E. Johnson, "Biological Opportunities and Risks Associated with Fast Growing Plantations in the Tropics," *Journal of Forestry* **74**(4):206–211 (1976).

11. E. J. Kormondy, *Concepts of Ecology,* 2d ed., Prentice-Hall, Englewood Cliffs, N.J., 1976.

12. T. T. Kozlowski, *Tree Growth,* Ronald Press, New York, 1962.

13. R. M. Krinard and R. L. Johnson, "Fifteen Years of Cottonwood Plantation Growth and Yield, *Southern Journal of Applied Forestry* **4**(4):180–185 (1980).

14. H. Lieth and R. H. Whittaker (eds.), *The Primary Production of the Biosphere,* Springer Verlag, New York, 1975.

15. H. W. Lull and K. G. Reinhart, *Increasing Water Yield in the Northeast by Management of Forested Watersheds,* U.S. Forest Service, Northeastern Forest Experimental Station, Upper Darby, Pa., 1967.

16. D. H. Marx, "The Influence of Ectotrophic Mycorrhizal Fungi on the Resistance of Pine Roots to Pathogenic Infections: I. Antagonism of Mycorrhizal to Root Parasitic Fungi and Soil Bacteria; II. Production, Identification and Biological Activity of Antibiotics Produced by *Leucopaxillus cerealis var. piceina,*" *Phytopathology* **59**:133–163, 411–417, 549–558, 559–565, 614–619 (1969).

17. P. Mikola, "Comparative Observations on the Nursery Technique in Different Parts of the World," *Acta Forestalia Fennica* **98**:1–24, 1969.

18. E. P. Odum, *Fundamentals of Ecology,* W. B. Saunders, 3d ed., Philadelphia, 1973.

19. J. D. Ovington, "Quantitative Ecology and the Woodland Ecosystem Concept," in J. C. Cragg (ed.), *Advances in Ecological Research,* Academic Press, New York, 1962, pp. 103–192.

20. J. D. Ovington, *Woodlands,* English University Press, London, 1965.

21. S. Petterssen, *Introduction to Meterology,* 2d ed., McGraw-Hill, New York, 1958.

22. W. E. Reifsnyder and H. W. Lull, *Radiant Energy in Relation to Forests,* U.S. Forest Service Technical Bulletin 1344, 1965.

23. D. L. Reukema and D. Bruce, *Effects of Thinning on Yield of Douglas-fir: Concepts and Some Estimates Obtained by Simulation,* U.S. Forest Service General Technical Report PNW-58, 1977.

24. F. E. Smith, "Ecological Demand and Environmental Response," *Journal of Forestry* **68**:752–755 (1970).

25. E. Sondheimer and J. B. Simeone (eds.), *Chemical Ecology,* Academic Press, New York, 1973.

26. S. H. Spurr and B. V. Barns, *Forest Ecology,* 3d ed., Wiley, New York, 1980.

27. D. H. VanLear, N. B. Goebel, and J. G. Williams, Jr., "Performance of Unthinned Loblolly and Slash Pine Plantations on Three Sites in South Carolina," *Southern Journal of Applied Forestry* **1**(4):8–10 (1977).

28. R. H. Whittaker and P. P. Feeny, "Allelochemics: Chemical Interaction Between Species. Chemical Agents Are of Major Significance in the Adaptation of Species and Organisms of Communities," *Science,* 171:577–770 (1971).

29. R. H. Whittaker and G. E. Likens, "The Biosphere and Man," in H. Lieth and R. H. Whittaker (eds.), *Primary Productivity of the Biosphere,* Springer-Verlag, New York, 1975.

30. S. A. Wilde, *Forest Soils, Their Properties and Relation to Silviculture,* Ronald Press, New York, 1959.

31. E. O. Wilson, "Chemical Communications Within the Animal World," in E. Sondheimer and J. B. Simeone (eds.), *Chemical Ecology,* Academic Press, New York, 1970.

CHAPTER **5**

FORESTS OF THE WORLD

Forests are both places and natural systems. As places they form administrative units or ownerships. In an ecologic sense, they make up one of the 12 major types of ecosystems that occur on the earth.[14] Consequently forest resource managers must think in both an administrative and an ecologic way. Administratively they regard forests as units of land for which they must plan management operations and uses. But actually to develop a plan for managing the vegetation present, a forester must know what forests amount to ecologically, how they differ in different areas, and what factors influence the way in which a tree community develops and responds to treatment.

TREES AND STANDS

Ecologically the plant communities called forests primarily have trees and other types of woody vegetation that grow close together.[6] By definition, a tree is any perennial plant with a large woody stem and a definite leafy crown. At maturity it also will have a height of at least 6 m. Shrubs usually branch close to the ground into several erect or prostrate stems. This gives the shrub a bushy appearance.[9] In contrast, trees generally have a single stem that extends upright to a considerable height, especially if it is growing in a suitable environment.

In describing forests, foresters frequently refer to stands of trees. This word also has both ecologic and administrative meanings. Ecologically a stand means a community of trees that has a species composition, age arrangement, spacing, and overall character sufficiently unique to distinguish it from adjacent forest communities.[6] Stands of trees may grow next to each other, but each one will differ from the stands that surround it. These stands also serve as the smallest administrative units for managing forest vegetation, and that gives them managerial importance.[4] They must cover sufficient area to operate efficiently during harvesting and other management treatments, and are considered the primary unit of land for record keeping and planning. Thus in forests the trees form distinguishable communities called stands. And these stands serve as the principal unit of area for managing the forest to answer the needs of the landowner.

FOREST COMPOSITION

The species of trees, associated shrubs, and herbaceous plants found in forests will differ from place to place; so will the density of tree cover. As a result, foresters find it useful to know what kind of forest or other vegetation system should develop at a site. Such understanding helps them in planning for management and use. Experience has shown, however, that the exact species composition cannot always be predicted easily. Rather the forester or ecologist must often take into consideration many environmental factors that influence plant growth and survival, and assess the interaction of these in order to reach a reasonable judgment.

Often the abundance or surplus of one environmental factor will compensate for shortages of another. These compensating effects may modify the environment sufficiently for a species to grow in areas where it otherwise would seem out of place. To illustrate, limited amounts of precipitation fall on the forests of the western United States. That would seem to preclude development of stands containing all but drought-resistant species. Yet, because of reduced temperatures at high elevations in the Rocky Mountains, summertime losses of available moisture to evapotranspiration remain relatively low there. Generally the higher the elevation, the greater is the compensation of temperature for limitations of growing season precipitation. As a result, many dense stands of conifer species survive and grow well with as little as 250–500 mm of annual precipitation. Species that use the greatest amounts of water usually occur at the higher elevations. Lowland areas support less dense forests with trees that tolerate drier conditions. Through experience and research, ecologists have learned to appreciate these compensating effects as well as other complex ways in which elements of the physical environment interact. In response, they will evaluate more than a single factor in reaching a judgment about the type of vegetation that may develop at a site.[9]

To facilitate this process, ecologists have developed different schemes for vegetation classification. These systems use features of the physical environment as an index in delineating vegetative regions where specific kinds of plant communities commonly occur. They consider such factors as ranges and patterns of precipitation, light, temperature, soil nutrients, soil moisture-holding capacity, and influences of air movement upon local climate in assessing characteristics of the physical environment.[2,7] All directly influence the growth and survival of vegetation and control the type of plant community that might occur. By ascertaining how these environmental factors balance each other, the ecologist can determine which species will grow at the place, and reach a reasonable judgment concerning the probable composition of the ecosystem.

One early vegetation classification by Thornthwaite recognized five climatic regions for the world.[18] Based on the difference between water gain from precipitation and its loss by evapotranspiration, he subdivided the earth into superhumid, humid, subhumid, semiarid, and arid zones. Each of these zones has a specific environment conducive to the development of a particular kind of vegetation. For example, forests develop where the supply of moisture exceeds evaporative losses. Grasslands or desert communities dominate the landscape in other regions. Later classifications, such as that of Holdridge, improved upon this system and permitted the development of a finer classification that foresters and others could utilize for land-use planning.[10] Even more recent advances permit the application of ecologic classifications to more localized situations and aid forest managers in answering questions about wildland management for a particular property.[4]

WORLD FORESTS

Forests, woodlands, and shrublands cover about 38 percent of the earth's continental area, or about 57 million square kilometers.[25] Forests and woodlands occupy 28 percent, and of this broad leaved types make up approximately two thirds, with tropical hardwoods accounting for 50 percent, and temperate hardwoods for 15 percent. Softwood species dominate the remainder and supply nearly three fourths of all wood consumed for industrial purposes. The Food and Agriculture Organization (FAO) of the United Nations has grouped these forests into six major vegetation formations, each encompassing a wide variety of forest conditions.[20] Figure 5–1 depicts the distributions.

Cool Coniferous Forests

Cool coniferous forests of pines, spruces, firs, or larches extend in a broad belt around the earth at the high northern latitudes. They cover most of forested Alaska and Canada, and extend into the northern fringes of the continental United States (Figure 5–1). They also occur in Scotland, Scandinavia, Finland, the U.S.S.R., northern Japan, and the Himalayan region.[20] These forests provide important amounts of excellent pulpwood and good sawlogs of moderate sizes. Despite the relatively slow growth rates in the most northern latitudes, the sheer expanse of cool coniferous forests makes them important as a source of wood products. Few alternative uses compete successfully for the land there. Because of its short growing season and the condition of the soil, the zone is unsuitable for agriculture, and major population centers are found primarily to the south in more moderate

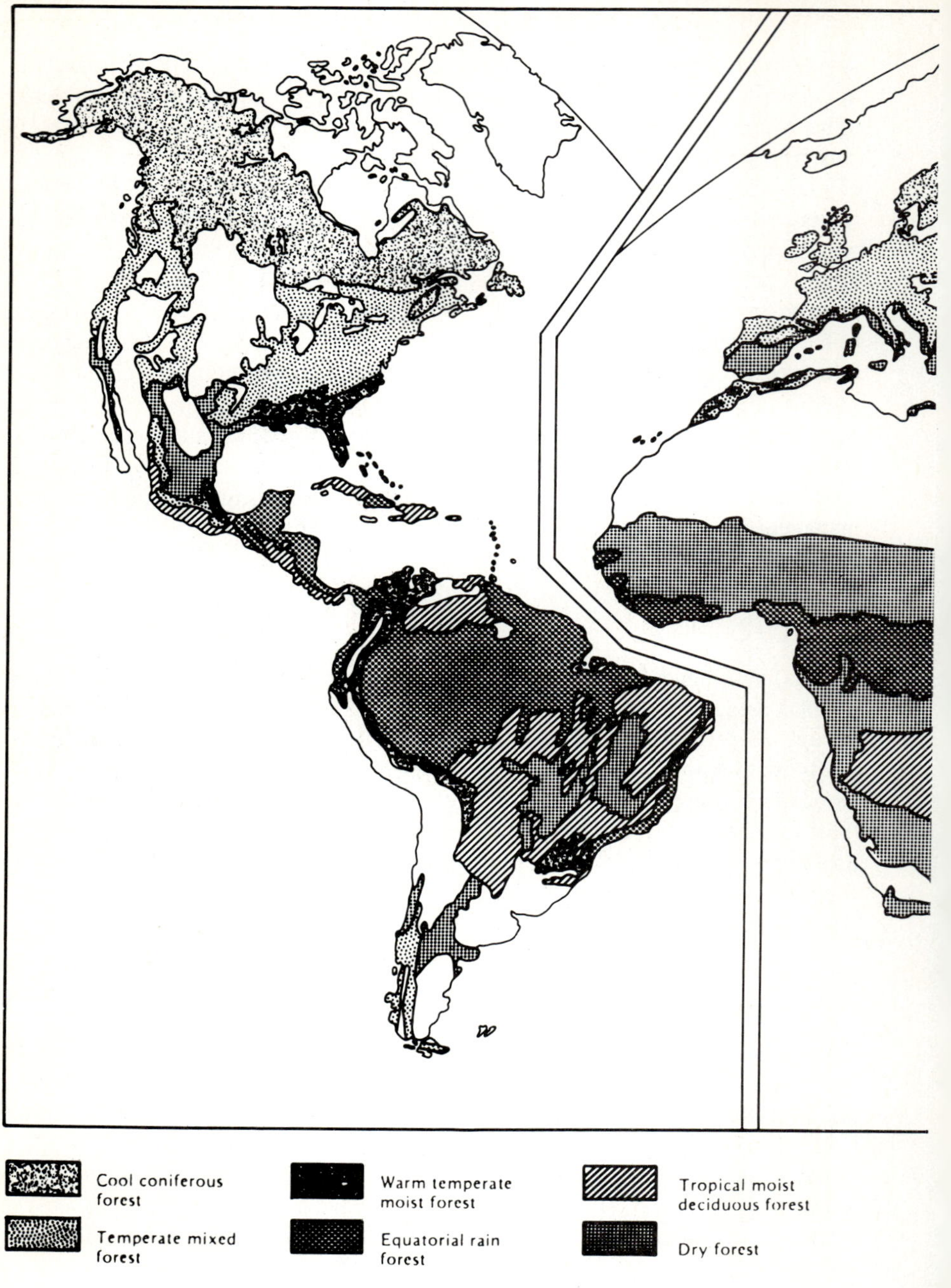
Cool coniferous forest
Temperate mixed forest
Warm temperate moist forest
Equatorial rain forest
Tropical moist deciduous forest
Dry forest

THE WORLD'S FORESTS

(The map does not show the actual extent of the world's forests, but only the main vegetational regions.)

Figure 5–1.
Forests of the world by major biome groups. *(Adapted by UN Food and Agriculture Organization, Forestry Department, from Oxford Atlas of the World, Oxford University Press, 1962.)*

climates. However, the northern regions do contain a wealth of mineral resources and oil as well as forestland.

Temperate Mixed Forests

Most well-settled industrialized nations have developed in the temperate mixed-forest regions of the world. These forests occupy large areas in western and eastern North America, Europe, southern U.S.S.R., northern China, and Japan. Some also grow in the Himalayas and in the Andes Mountains in southern Chile and Argentina.[20] Major commercial species include Douglas-fir, most of the pines, hemlocks, oaks, beech, birches, maples, ashes, walnut, and yellow-poplar. They provide diverse kinds of high-quality products, as well as pulpwood. Temperate mixed forests also serve as important watersheds for the world's population, house much wildlife, and provide people with many amenities and recreational opportunities. Though much of the land area within these regions now supports agriculture or development, the temperate mixed forests in conjunction with cool coniferous forests furnished about 85 percent of the industrial timber of the world in 1968.[21]

Warm, Moist Temperate Forests

Warm, moist temperate forests are found along the southeastern United States coastal plain, on the east coast of Mexico, and in Japan, China, Taiwan, the Andes Cordillera, the uplands of southern Brazil, southern Australia, and New Zealand.[20,21] These forests include the pines of southern United States and Mexico, the Eucalyptus of Australia, and the Auracarias of Brazil and Chile. All have desirable utility and fast growth. Forests in southeastern United States also contain many valuable hardwoods. Species of warm temperate forests transplant easily, and many have been planted extensively outside their natural ranges. Examples include large commercial pine plantations in South Africa, New Zealand, and Australia. Because of the hospitable climate, soil conditions, and relative freedom from natural pests, these exotic plantations are highly productive. Also eucalyptus from Australia now grows as an introduced tree in Brazil, Argentina, Chile, and southeastern United States. Rapid growth and high productivity at these places make the plantations desirable for many commercial uses.

Equatorial Rain Forests

Equatorial rain forests once formed an almost continuous belt extending 10 degrees north and south of the equator. As a result of clearing, they now cover extensive areas only in the Amazon and Congo river basins, and on the Malay archipelago (Figure 5–2). Smaller remnants of rain forest also grow in Central America, along the Gulf of Guinea in West Africa, and in humid areas of India, Sri

Figure 5–2.
Equatorial rain forests in Bislig, Surigao del Sur, Philippines, contain valuable mahogony and other dipterocarp species prized throughout the world. *(Martin Hurwitz.)*

Lanka (Ceylon), and Australia.[17,20] These warm well-watered forests may outproduce those found elsewhere, and contain large volumes of old-growth timber. Commercially valuable species include mahogany, green heart, Spanish cedar, balsa, oloume, obeche, sipo, and limba. The equatorial rain forest also contains hundreds of other species, but many have little commercial value at the present time. Paradoxically these forests occupy soils of inherently low fertility, and heavy rainfall leaches out nutrients quickly following removal of forest cover. The remote locations of many remaining equatorial rain forests make access difficult, limiting the commercial potential. Heavy demand for local-use wood and clearing to expand agriculture threaten much of the forests in Africa, in parts of Central and South America, and in some other regions.

Tropical Moist Deciduous Forests

Tropical moist deciduous forests grow in areas with a distinct dry season, when the leaves drop off. The largest areas are in south central Africa, in South America south of the Amazon basin and in the Orinici basin, and in India and southeast Asia.[20] Less extensive areas remain along the west coast of Mexico and Central America, in the Caribbean islands, and in central New Guinea. Their best known commercial species include Asia's teak and sal. Lack of moisture during the dry season limits the density of tree cover, but forage grows under and between the widely spaced trees. That makes part of these forests suitable for grazing. The forage potential, timber for local use and export, and their suitability for management give these forests worldwide importance. However, much area has been cleared for agriculture and degraded by shifting cultivation, and the extent of tree cover continues to decline throughout these regions.

Dry Forests

Dry forests are found in Mexico, east central Brazil, Argentina, much of Africa south of the Sahara, western Madagascar, the Mediterranean coastal region, Spain, West Pakistan, and India, and bordering the central desert areas of Australia.[20] These forests receive little rainfall and produce no timber for export. They do furnish some local-use materials, much firewood, fence posts, and other small-tree products. While dry forests are relatively unimportant as worldwide timber resources, indigenous peoples depend upon them for local wood and fuel needs. Unfortunately ill treatment through overgrazing, fire, and steep-slope cultivation has led to extensive soil degradation and a decline of forest cover in many areas. Especially in drier regions, the loss of forest cover poses important problems in satisfying local needs, even for fuel.

FORESTS OF THE UNITED STATES

The Society of American Foresters recognizes 145 different forest types for the United States. By definition, a forest type comprises a plant community with trees as a dominant component, and consisting of a particular combination of tree species that occur together over a fairly large area. Some of these aggregations have occupied the same area for long periods, whereas others only dominate a site for relatively short periods of ecologic time following disturbance or succession.[5] Ninety different forest types are found east of the Great Plains, and 55 to the west.

Forest Regions

Figure 5–3 depicts a generalized classification of forest types, and their geographic distributions in the United States. These follow overall patterns of climate across the continent. For the most part, areas east of the plains get abundant precipitation well distributed during the year. As a result, forests occur through-

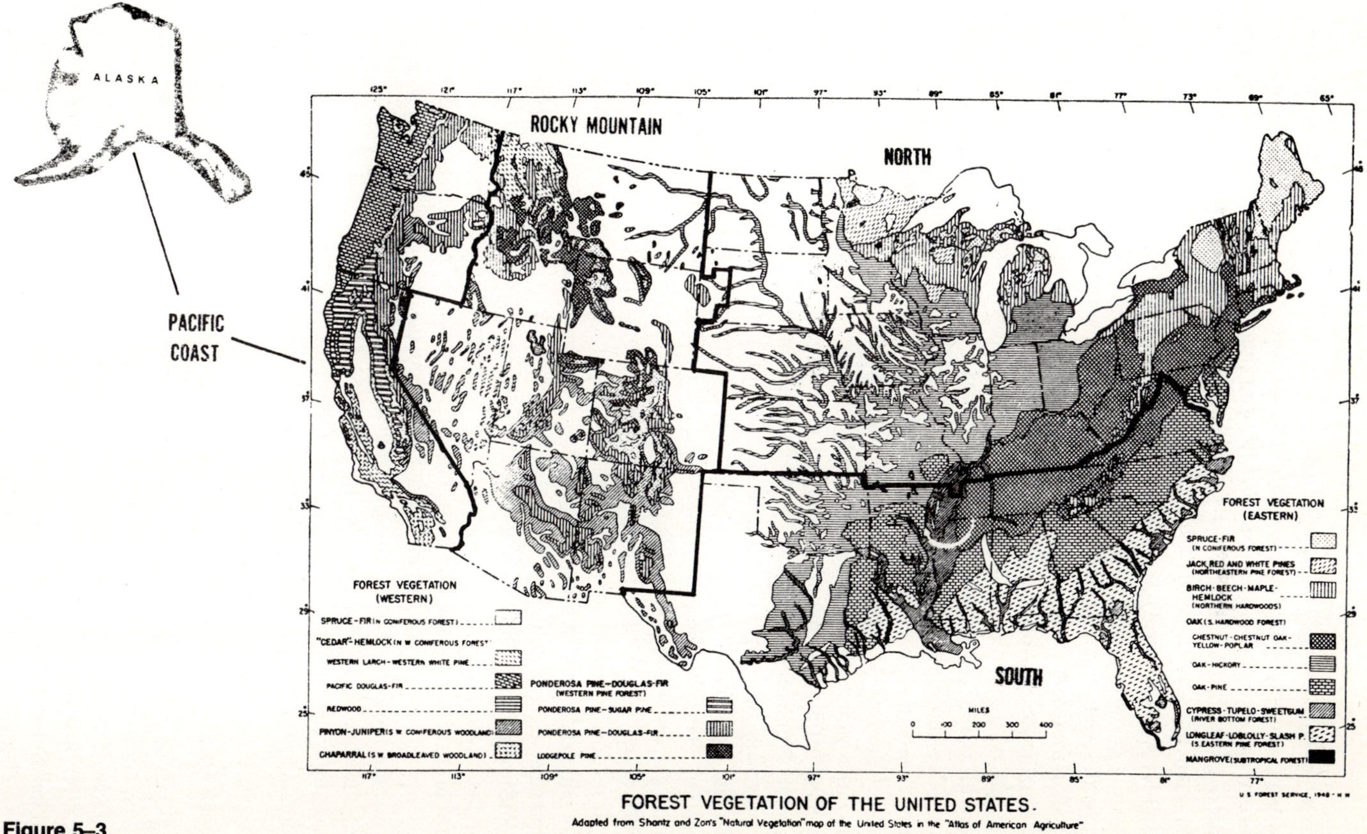

Figure 5–3.
Natural forest types of the United States and the geographical regions used by the U.S. Forest Service for amassing statistics for its assessment of the nation's forests and rangelands. *(Adapted from U.S. Forest Service map.)*

out, except where people have cleared the land for other uses. Because of the effect of climate and precipitation, plus influences of soil and other features of physical environment, deciduous forests cover most of the eastern United States. Eastern conifers grow in abundance mostly at the more northern latitudes, highest elevations, and along the southeastern coastal plain. At the western edge of the deciduous forests, the vegetation changes to prairie. There and throughout the Great Plains, precipitation comes mostly as summer rains, and much less falls during the year than farther east. Available moisture does not support forest growth in the Great Plains, except in mountainous areas and along stream bottoms.

From the Rocky Mountains west, precipitation occurs mostly in winter months. Amounts of available moisture sustain forest growth only at upper elevations where the cooler temperatures result in reduced evapotranspiration, and along the coast. The same pattern holds true for the semidesert region of the southwest where forest growth has developed in the mountains and along streams. Also, many of the forests there contain only shrublike trees that do not grow to large sizes. In contrast, the upper Pacific Coast receives abundant winter precipitation and has sufficient moisture to support dense forests of large trees. Along the Olympic Peninsula, the stands even have a rain forest appearance with dense undergrowth of many nontree species. Conifers dominate most commercially valuable western forests and species composition changes with latitude and elevation. These reflect conditions of available moisture.[1] In addition, aspen covers extensive areas in the Rockies.

For some needs, statisticians find it more convenient to group the nation's forest resources into economic regions. This treatment recognizes political rather than natural boundaries and has mostly economic utility. It overlooks several important vegetal features. Political boundaries, for example, do not include the tongues of northern forests that extend southward along upper elevations of the Appalachians, and the southern types that reach into Maryland, Pennsylvania, and as far north as Long Island in New York. Classification by economic regions also does not fit the broad global forest formations described earlier. It misses the representation of cool coniferous forest in Alaska by grouping that state regionally with the Pacific Northwest. Yet these regional schemes prove convenient for amassing forest statistics. They serve as the framework for national assessments under the Forest and Rangeland Resources Planning Act of 1971.

Forest Area

Regional data show that forests cover one third of the total land area in the United States, and about 22 percent of the land supports forests of commercial quality.[23] By U.S. Forest Service standards, these areas can produce at least 1.42 m^3 of wood per hectare per year.[24] In contrast, slightly over 11 percent of the nation's land has forest cover of little commercial promise. It will not produce continuous crops of sawlogs that landowners make available for harvesting. These include areas reserved for nontimber uses such as National Parks, wilderness areas, and state or local preserves. However, forests of all kinds may serve as watersheds, habitats for wildlife, grazing areas, and places for recreation.

About three fourths of the commercial forest area lies in the eastern United States, almost equally divided between the Northern and Southern states.[23] Conversely 89 percent of the noncommercial forest area is found west of the Great Plains, and much of it in Alaska. Yet despite the more limited area of commercial stands in the West, those regions contain 57 percent of the nation's sawtimber reserves suitable for manufacture into lumber. Many western forests have large trees, such as giant redwoods and Douglas-fir. They also contain many high-volume stands of old-growth trees never exploited commercially. On the other hand, the total growing stock of wood in trees at least 12.7 cm in diameter at breast height (1.37 m from the ground) is about equally distributed between eastern and western forests. Trees of the East generally have smaller diameters and heights and many do not meet the sawtimber standards for manufacture into lumber. They would have value only for pulpwood, fuel wood, and other products where tree size is less important.

Northeastern Forests

Forests of northeastern United States cover 65.73 million hectares and represent both the cool coniferous and temperate mixed-forest formations (Figure 5–3).[24] Stands in the more northern areas of Minnesota, Wisconsin, Michigan, and Maine, and at the higher elevations in New York, Vermont, and New Hampshire, contain spruce, balsam fir, aspen, and birch. These conifer-dominated stands form a transition between the vast area of boreal forest that covers much of Canada, and the mixed temperate formation to the south. At the lower elevations and more southern latitudes, the species composition changes to the northern hardwoods association. There sugar maple, American beech, basswood, red maple, yellow birch, and paper birch occur in abundance. Drier sites may have red, white, or jack pine. The moister sites may support stands of hemlock. Throughout this area and to the north, black spruce, northern white cedar, and tamarack grow in swampy places of the glaciated terrain.[1]

South of a line from southern Wisconsin, through lower Michigan, southern New York, and southern New England, the forest changes to the temperate mixed formation with a different species composition. Stands may contain all of the hardwoods that grow farther north. In addition, the species include sweet birch, red oak, elms, ashes, black cherry, white pine, hemlock, and several other trees. Further south and west and on drier sites, these disappear and one or more species of oak and hickory plus Virginia pine and shortleaf pine predominate. Silver maple, river birch, sycamore, and the elm-ash-cottonwood forest type grow in the wetter places. Richer soils support a wide mix of such species as black walnut, butternut, yellow-poplar, and red oak (Figure 5–4). Perhaps no other part of the

United States contains a greater number of tree species having such high technical use than do the middle, southern, and western parts of the northeastern forest region.

The area of commercial forest occupied by the major forest types includes the following:[24]

Type	Area, million hectares
Maple-beech-birch	14.95
Aspen-birch	8.23
Spruce-fir	8.58
White-red-jack pine	4.78
Oak-hickory	17.63
Elm-ash-cottonwood	7.22

Despite the long, cold winters and short growing seasons, these forests have high potential for productivity. Currently stands of the northern region grow an average of about 2.5 m^3/ha/yr. They potentially can yield an average of 4.7 m^3/ha/yr, if well stocked and under proper management.[24] The better soils in warmer portions of the region may grow an average of 5.4 m^3/ha/yr, if occupied by well-stocked stands.

Conifers of the Northeast are valuable for pulpwood since the spruces, balsam fir, red pine, and jack pine have long fibers, with high yields of pulp for papermaking. Hardwoods also have value for pulpwood. Conifers of sawlog sizes yield good structural lumber as well as poles and veneer logs. The large high-quality hardwoods yield beautiful veneers and provide lumber for furniture, interior trim, cabinets, flooring, and specialty uses. Many species are in high demand for export to other countries, as well as to manufacturing plants in the South. As of 1972, forests of this broad northern region yielded about 13 percent of the nation's total timber products. Altogether they supplied about 10 percent of the volume of sawlogs, 2 percent of that used for veneers, and 18 percent of the pulpwood. The combined value of these products accounted for about 9 percent of the total generated from U.S. forests.[15]

The northern region supports extensive forest cover, especially where rugged topography dominates the landscape. It also has several of the nation's largest cities and a fairly dense population. Corporations and public agencies hold large tracts of land for timber production, but largely at the northernmost part of the region and in the more mountainous areas. For the most part, limited-acreage ownerships have fragmented the land and many people own property for nontim-

Figure 5–4.
Oak-hickory forests on good Midwestern sites contain a rich mixture of hardwood species, including shelbark hickory such as this one.

ber purposes. Because of their proximity to large population centers, these forests are important for recreation. The glaciated and mountainous terrain, lakes, streams, and open areas lend amazing beauty to the region. Yet the many industrial holdings and an active commercial lumber and paper industry give the timber resources great economic importance as well. As a result, resource managers must often accommodate both amenity and commodity values in planning forestry operations.

Southern Forests

The southern forest region (Figure 5–3) has 88.67 million hectares of forest spread across Virginia, Tennessee, Arkansas, Oklahoma, and states south of these, plus portions of southern New Jersey.[24] Along the coastal plain, forests of loblolly, slash, longleaf, and shortleaf pines predominate. Other less common species include Virginia, pond, and sand pines. Bottomlands that intersect the coastal plain have mostly southern oaks, sweetgum, green ash, cottonwood, black willow, soft maples, and several other valuable hardwoods. Highly productive cottonwood plantations have become increasingly abundant between levees of the lower Mississippi River. True swamps with standing water contain bald cypress and water and black tupelo. Several other bottomland species grow where the land rises slightly and has better drainage.

The Piedmont's rolling to hilly topography along the coastal side of the Appalachians has oaks, hickories, yellow-poplar, and southern pines. Fingers of northern forests interrupt them along the higher elevations, and rich mixtures of oaks, yellow-poplar, maple, cherry, ash, and several other valuable species form stands on moister soils of coves that cut into the slopes. Toward the western reaches, the forests merge into the oak-hickory type that interfaces with the Great Plains as far north as Wisconsin. Regionwide the major forest types cover the following areas:[24]

Type	Area, million hectares
Loblolly-shortleaf pine	19.33
Longleaf-slash pine	6.90
Oak-pine	12.85
Oak-gum-cypress	11.60
Oak-hickory	28.06

Uses of southern hardwoods parallel those described earlier, including sawtimber, pulpwood, and railroad ties. In the pine regions, loblolly pine has gained greatest popularity. Foresters can regenerate it easily. It grows rapidly, responds to silvicultural manipulations, and has wood suitable for papermaking and lumber. Longleaf pine develops somewhat more slowly but also has strong wood of good fibers that is prized for pulpwood, construction uses, poles, and pilings. Slash pine has intermediate growth, thrives on drier sandy soils, and regenerates well naturally or artificially (Figure 5–5). Its wood compares to any of the southern pines. Along with longleaf, it also yields resins, turpentine, and similar products called naval stores. Shortleaf pine grows on upland sites and produces excellent lumber, but its growth is slower. These four species provide an important resource for the major timber industry that thrives throughout the southern and eastern portions of the region. Overall the South produced about 47 percent of the total timber volume harvested in the United States in 1972. That included 34 percent of the sawlog

Figure 5–5.
Longleaf pine, such as this stand in Alabama, has value for lumber, plywood, and poles, as well as many noncommodity uses. *(Allan P. Drew.)*

volume, 37 percent of the veneers, and 72 percent of the pulpwood. In combination these accounted for about 35 percent of the total value of products harvested from the nation's forests.[15]

Across the region, stands currently yield an average of about 3.92 m^3/ha/yr as growth, with a potential for about 5.47 under adequate management.[25] Excellent markets for pines in the pulp and paper and veneer industries make intensive cultural practices profitable and possible. Plantations on carefully prepared sites planted with improved genetic stock yield harvestable crops in 20 years for pulpwood. It takes 40 years for sawlogs. With methods such as these, many landowners have converted extensive areas of less valuable hardwoods to pine, especially along the coastal plain and the Piedmont. However, interest in hardwoods has increased throughout the southern region, especially for fast growing species suitable for culture in plantations for fiber and small sawlogs. Traditional markets for hardwoods in the western parts of the area remain strong.

Commodity uses have great importance in the southern region. The active forest industry provides strong markets for both hardwoods and the southern pines. Private ownerships predominate. And while several large industrial firms own extensive areas within the region, the bulk of timber volume remains in limited-acreage ownerships. In general, the southern forests receive less pressure for recreation than do ones farther north. Yet amenity values are important and do influence management programs. Hunters make extensive use of southern forests and portions of the region support active tourism. Mountainous areas serve as valuable watersheds. Grazing has become fairly widespread throughout much of the South and competes with forest uses in some areas. Overall the southern region seems destined to play a vital role in the supply and manufacture of wood products in the future.

Rocky Mountain Forests

The Rocky Mountain region includes eight states with nearly 55.21 million hectares of forestland[24] (Figure 5–3). It extends from the Great Plains to the Sierras and Cascades, and from Mexico to Canada. These represent the temperate mixed-forest formation, but predominantly coniferous species. Major ones include pines, spruces, true firs, hemlock, cedars, larch, and Douglas-fir singly or in mixtures. Hardwoods attain no large sizes and have only minor economic importance, except for quaking aspen in the central and northern Rockies. Principal types that dominate the commercial forestland area include[24]:

Type	Area, million hectares
Ponderosa pine	6.68
Douglas-fir	7.06
Fir-spruce	6.66
Lodgepole pine	6.78
Western hardwoods	3.10
Larch	0.84

Rocky Mountain forests of commercial importance occur almost exclusively in mountainous terrain. Further, different species mixes grow separately in fairly distinct altitudinal zones.[1] For example, from central Colorado northward, vast areas of lodgepole pine or aspen grow at elevations between 1850 and 3500 m. These shift to Douglas-fir and then to true firs and spruces at higher elevations. Timberline occurs at 3350–3650 m in the central Rockies, and as low as 2400–3000 m near the Canadian border. At these elevations, the tree cover stops and a tundra vegetation occupies the sites. In Idaho and Montana, western white pine, ponderosa pine, and western hemlock form excellent stands with trees exceeding 40 m in height. From Colorado southward, the trees do not grow to such large sizes and have less value for lumber. In the southwestern part of the region, stands at lower elevations of 2100–3000 m contain mostly pinyon pine and juniper of no

Figure 5–6.
In many areas of the Rocky Mountains ponderosa pine grows in open stands, intermixed with grasslands at the lower elevations.

commercial value at the present time. Species such as ponderosa pine, some true firs, Douglas-fir, and other species grow in limited areas at upper elevations of the mountain ranges. In other parts of the southwest, and in the Black Hills of South Dakota, forests of ponderosa pine intermix with grasslands. While the trees grow closely together on the better and moister sites, they are widely spaced at places of more limited soil moisture (Figure 5–6).

The diversity of conditions across the region results in great variability in potentials for timber production. Fewer than one half of the forests have commercial potential and about 60 percent of this area lies in Montana, Idaho, Wyoming, and the Black Hills of South Dakota.[23,24] On a broad scale, forests of the Rocky Mountains currently grow an average of 2.03 m^3/ha/yr, with a regional potential for about 4.21 m^3/ha/yr of growth among well-stocked stands under management.[24] As a whole, the region contributed about 7 percent of the timber volume harvested in the United States in 1972. This included 11 percent of the sawlog volume, 6 percent of that used for veneer, and 1 percent of the pulpwood.[15]

Recreation, tourism, timber products, mining, and grazing are the backbone of the region's economy; population densities are relatively low. Vast expanses of treeless land characterize the southern portions of the region, while in the northern Rockies the forests provide almost continuous cover across the landscape. The three northern states alone contributed 77 percent of the roundwood products harvested from Rocky Mountain forests in 1972.[15] Land farther south has the greatest value for watersheds, grazing, recreation, and limited timber production. Federal, state, and municipal agencies control 75 percent of the forest area, and virtually all of that classified as noncommercial.[23] These public owners generally have a commitment to multiple-use, sustained-yield management and the enhancement of environmental quality. This region also has several of the largest National Parks and many extensive wilderness areas that draw people from across the country for recreation.

Pacific Coast Forests

Southern and central California generally lacks large areas of commercial forest, but supports extensive coverage by a shrub association called chaparral. The Pacific Coast temperate mixed-forest formation extends from northern California up through Oregon and Washington (Figure 5–3). These forests contain the nation's largest and most valuable conifer species. East of the Cascades, the vegetation changes to treeless and open shrub associations, with forests generally limited to the higher elevations. But costal stands contain Douglas-fir, ponderosa pine, sugar pine, western white pine, western red cedar, Port-Orford cedar, Sitka spruce, redwood, and giant Sequoias (Figure 5–7). The region's 37.71 million hectares of forestland support the following major types[24]:

Type	Area, million hectares
Douglas-fir	8.52
Ponderosa pine	6.41
Fir-spruce	5.48
Lodgepole pine	1.69
Hemlock-Sitka spruce	2.31
Larch	0.29
White pine	0.09
Redwood	0.31
Western hardwoods	3.77
Pinyon-juniper	2.07

The Pacific Coast has the highest potential for timber growth of all the continental United States. Under management, well-stocked stands across the region can average 6.87 m^3/ha/yr. They currently grow an average of 3.49 m^3/ha/yr.[24] The western hemlock-Sitka spruce type group has the highest potential, with mean annual growth reported to be as much as 25–30 m^3/ha/yr. Along the coast in Washington and Oregon, stands may support 1130 m^3 or more of

Figure 5–7.
Temperate mixed forests of the Pacific Northwest contain several valuable conifers, such as those in the Coos River drainage in Oregon. *(U.S. Bureau of Land Mangement.)*

sawtimber per hectare, with Douglas-fir 2.0–2.5 m in diameter and 35–60 m tall for usable height. A limited area of redwoods along the northern coast of California has even larger trees. Farther inland, forests cover slopes of the Cascade and Sierra mountains, with timberline at 3200–4800 m. These stands have giant Sequoia, ponderosa pine, sugar pine, western white pine, whitebark pine, Jeffrey pine, lodgepole pine, Douglas-fir, white fir, red fir, and incense-cedar. Though less heavily stocked than stands along the coast, the trees grow to large sizes and have important commercial value.

Forests of this region account for less than 14 percent of the nation's commercial forest area, but have one half of the sawtimber volume. In 1972, they yielded about one third of the total national output. This included 45 percent of the sawlog volume, 55 percent of that used for veneer, and 8 percent of the pulpwood. Their combined value accounted for 47 percent of the total generated by roundwood products harvested from the nation's forests in 1972.[15] However, limited old-growth timber remains and future harvests will come increasingly from younger stands regenerated following the cutting of old ones. Logging and lumber manufacture contribute importantly to the economy of the region and receive great local attention. Pulp and paper and plywood manufacture also are major enterprises of the upper Pacific Coast.[8] The region contains several National Parks, wilderness areas, and other private reservations. Recreation and tourism contribute importantly to the economy, and noncommodity uses receive considerable attention. Upward of 10 million hectares in southern California have critical value as watersheds.

Alaskan Forests

Forests cover 48.22 million hectares of Alaska, spreading across about one third of the state.[22,24] They represent 16 percent of the total forest area of the United States. Much of Alaska, however, supports unproductive tree cover and tundra, with commercial forests on less than 4.51 million hectares. Further over three quarters of these remain remote and economically inaccessible for commercial use. Altogether the major forest types that cover most of Alaska include the following:

Type	Area, million hectares
Fir-spruce	33.77
Hemlock-Sitka spruce	5.13
Western hardwoods	9.20
Lodgepole pine	0.11

Coastal lands comprise 13.33 million hectares, including hundreds of islands and a narrow strip of mainland broken by numerous fiords and inlets. Commercial-quality forests cover 2.85 million hectares there, and represent the state's most productive ones (Figure 5–4). They contain the same species as farther south along the coast, except for Douglas-fir. Western hemlock and Sitka spruce stands alone occupy 95 percent of the coastal commercial forests.[23] Close to the ocean, steep and rugged topography rises 600–900 m from sea level, with the timberline at 600–760 m. Forest development diminishes in quality with elevation. The old-growth coastal forests actually contain 85 percent of Alaska's total sawtimber volume. Stocking varies from 50-580 m^3/ha. These stands supply logs for about 2 million cubic meters of lumber annually. On a national scale, however, the forests of Alaska contributed less than 1 percent of the total volume harvested in 1972, about equally divided between sawlogs and pulpwood. These also had a combined value of just less than 1 percent of the total for roundwood products harvested during that same year.[15]

Alaska's interior of 133.44 million hectares has wide variations in topography, weather, and vegetative cover. Severe winters, continuous and discontinuous permafrost, low annual precipitation, and other adversities limit the composition of the vegetation and its growth. Typical cool coniferous formations called boreal forests cover 42.83 million hectares, mostly along major drainages. Commercial-quality stands include only 1.66 million hectares,[23] but growth averages less than 3.5 m^3/ha on about 98 percent of the total land area.[11] Forests consist mostly of extensive white spruce stands, with inclusions of paper birch, aspen, and balsam poplar (Figure 5–8). On the best sites, spruce trees reach 30 m in height and stands support 58 m^3/ha.[26] However, on north-facing slopes and poorly drained wetlands, much shorter black spruce forms open woodlands with tamarack. Tall shrubs, willow, and alder stands intermix between the forests, creating a mosaic of vegetation types across the landscape.[18] Fire historically has played a major role in shaping these ecosystems.[13]

Alaska has extensive fish and wildlife resources as well as much forest. Major species include caribou, wolf, moose, Sitka deer, brown bear, Dall sheep, and mountain goats. The forests also have many other species of smaller animals and

Figure 5–8.
Interior Alaska's best forest growth occurs along major water courses, such as these stands of white spruce and poplar bordering the Kuskokwim River. *(U.S. Forest Service.)*

birds. These, plus the state's vast wilderness, appeal to persons interested in hunting, fishing, outdoor recreation, and tourism. Inhabitants of the interior region use the game and timber for domestic needs. However, remoteness, severe climate, and scattered distribution of commercial-quality stands make more intensive use of interior Alaska's forest unlikely. Coastal areas will continue to offer greater economic benefits from timber and related industries.

Most of Alaska remains in public domain, through provisions of the Alaska Statehood Act and the Alaska Native Claim Settlement Act of 1971 will eventually alter ownership patterns. The first authorizes transfer of about 42.5 million hectares to state use. The other grants about 18.2 million hectares to Alaska's natives, and authorizes 32.4 million for new National Parks, wildlife refuges, wild and scenic river areas, and National Forests. Such distributions will alter potentials for use and development of timber and mineral resources and influence the state's future economic development.

Tropical Forests

Tropical forest formations occur at the southern tip of Florida, and in Hawaii, Puerto Rico, and the Virgin Islands. Those of Florida consist largely of subtropical swampy thickets with mangrove and other saltwater shrubs, and of coconut palm on better drained areas along the coast. These forests have little commercial value for timber but do protect soil from erosion and provide important habitats for many species of birds and animals, including alligators.

Hawaii's eight main islands and several islets originally supported extensive forests on the better soils. Important species included sandalwoods and koa. However, clearing or cultivation of the level, deep soils for crops of sugarcane and pineapple has left only 48 percent of the state in forest cover, mostly on steep slopes and thin-soil lava flows. Much of the residual area at low elevations has only desert scrub or mesquite. Taller forests occur on slopes between 600 and 2400 m above sea level. Commercial-quality stands include about 402.0 thousand hectares.[23] Overcutting and grazing by wild goats, pigs, and cattle have left many stands badly abused. They serve mainly for watershed protection and recreation with limited amounts of timber harvested for local use. Reforestation of burned-over and denuded areas serves primarily to protect the soil and enhance watershed qualities.[12]

True tropical forests grow in Puerto Rico and the Virgin Islands, though the limited area of forest on the Virgin Islands has little commercial significance. Puerto Rico has a great diversity of vegetation, and dense tropical rain forests clothe the mountains from east to west. In these moister areas, species include giant tabanucos, guaraguaos, ausubos, and the valuable *Magnolia splendens* (Figure 5–9). Many of these hard, heavy, beautifully grained woods are suitable for cabinet work. Drier soils of the island grow mahogany, silk, cottonwood, Spanish cedar, teak, and other valuable trees, while tabanuco occurs on moderately dry lands. Only cactus and lignum vitae survive in the driest areas. The forested areas serve important recreational uses as well as supplying timbers.

REGIONAL OPPORTUNITIES FOR FORESTRY CAREERS

The diverse and extensive cover of U.S. forests provides many and varied opportunities for careers in forestry and related disciplines. These differ somewhat between regions, depending upon the types and qualities of tree cover, who owns the land, and the extent of industrial activity that draws upon the forest resources. While concerns discussed here focus attention upon future opportunities within the United States, careers await individuals who might find service in other countries appealing as well.

Figure 5–9.
The rain forests of Puerto Rico contain a dense growth with several desirable species, including this giant tabanuco tree. *(U.S. Forest Service.)*

Conifer resources of western forests, the southern coastal plain, upper New England, and the northern Lake States have traditionally proved attractive for industrial development. These regions currently support well-developed commercial activity for sawtimber and veneer logs, pulpwood, fiberboard, pilings, and poles. Deciduous forests in other eastern areas have also attracted lumber and pulpwood industries, though these tend to remain smaller and more dispersed than in the major conifer areas. Greater fragmentation of the land into smaller properties, higher population densities, and less governmental or industrial ownership give much of these eastern deciduous forests a different character.

The bulk of forests and rangelands in western states remain in governmental control. Industries draw upon these resources for raw materials, and control considerable area along the west coast. Forest industry also has acquired much property in the eastern pine forests and the spruce-fir regions to the north. These support growing pulp and paper companies. Depletion of old-growth stands west of the Great Plains will likely place increased importance on the eastern forests as a supply of greater amounts of wood for a wide variety of uses.

Most states with appreciable areas of tree cover maintain a school of forestry or natural resources offering forest-related curricula. These have courses of instruction leading to careers in forestry, wildlife, water resources management, recreation, wood products engineering and utilization, and many supporting sciences and disciplines. Schools tend to emphasize management and use for the forest types and conditions common to their environs. However, the nature of forestry as a science-based profession allows individuals schooled in one region to adjust easily to employment and conditions elsewhere.

The employment and resource management opportunities available in a region tend to reflect the extent of industrial and government ownership, the type of forests, and the demand placed upon the timber relative to other associated resource values. While each region has a rather diverse forestry profession, some types of activity seem to characterize each area. For example, governmental agencies employ large numbers of people in the western conifer areas, with industrial employment more prevalent along the Pacific Coast. Industrial employment also may dominate the south and the conifer forests far to the north. National Forests or other federal properties represent a relatively small portion of these areas, and government employs fewer people than in the West. Governmental employees in the East tend to work for states and counties instead. Of course, in all regions foresters find opportunities for consulting work. The fragmentation of forests of the eastern United States into predominantly small, nonindustrial private holdings make the potential there great.

Not all foresters work in positions related to timber production. Several of the governmental regulatory agencies hire forestry professionals and many serve in recreation, wildlife, water resources, and similar programs. Many of the larger forest industries maintain active programs in these noncommodity areas and employ professional specialists to staff them. Also, especially near the population centers of the eastern forest regions, foresters assume important roles in managing vegetation in and around urban areas to enhance its values for a host of noncommodity benefits.

Overall the diversity of forests, their uses, and the demands we place on the associated resources create an almost endless variety of forestry-related employment opportunities for people who graduate from the schools of forestry and forest technology throughout the United States. As the pressures of resource use intensify, the role of professional forest resource managers will grow. Increased demand will pressure foresters to find new and better ways to draw upon the nation's forests for essential goods, products, and services.

REFERENCES

1. J. W. Barrett (ed.), *Regional Silviculture of the United States,* 2d ed., Wiley, New York, 1980.
2. W. T. Daniel, J. A. Helms, and F. S. Baker, *Principles of Silviculture,* 2d ed., McGraw-Hill, New York, 1979.
3. K. P. Davis, *Forest Management,* 2d ed., McGraw-Hill, New York, 1966.
4. L. S. Davis and J. A. Henderson, *Ecosym—The Conceptual Framework,* Department of Forestry and Outdoor Recreation, Utah State University, Progress Report no. 1, 1976.
5. F. H. Eyre (ed.), *Forest Cover Types of the United States and Canada,* Society of American Foresters, Washington, 1980.
6. F. C. Ford-Robertson (ed.), *Terminology of Forest Science, Technology, Practice and Products,* Society of American Foresters, The Multilingual Forestry Terminology Series no. 1, 1971.
7. H. A. Fowells (ed.), *Silvics of Forest Trees of the United States,* U.S. Department of Agriculture, Forest Service, Agricultural Handbook no. 271, 1965.
8. J. A. Guthrie and G. R. Armstrong, *Western Forest Industry: An Economic Outlook,* Johns Hopkins Press, Baltimore, 1961.
9. W. H. Harlow, E. S. Harrar, and F. M. White, *Textbook of Dendrology,* 6th ed., McGraw-Hill, New York, 1978.
10. L. R. Holdridge, W. C. Grenke, W. H. Hathaway, T. Liang, and J. A. Tosi, Jr., *Forest Environments in Tropical Life Zones—A Pilot Study,* Pergamon Press, New York, 1971.
11. K. O. Hutchinson, and D. R. Schumann, "Alaska's Interior Forests: Timber Resources and Utilization," *Journal of Forestry* **74**(6):338–341 (1976).
12. C. S. Judd, "Forestry in Hawaii," *Journal of Forestry* **33**(12):1005–1006 (1935).
13. H. J. Lutz, "Fire as an Ecological Factor in the Boreal Forests of Alaska," *Journal of Forestry* **58**(6):454–460 (1960).
14. E. P. Odum, *Fundamentals of Ecology,* 3d ed., W. B. Saunders, Philadelphia, 1963.
15. R. B. Phelps, *Timber in the United States Economy 1963, 1967, and 1972.* U.S. Forest Service General Technical Report WO–21, 1980.
16. B. Rhett, "Sugarwater Forests," *American Forests* **43**(1):8–10, 47–48 (1947).
17. P. W. Richards, *The Tropical Rain Forest,* Cambridge University Press, London, 1952.

18. H. L. Shantz and R. Zon, "Natural Vegetation," in *Atlas of American Agriculture,* U.S. Government Printing Office, Washington, 1924.

19. C. W. Thornthwaite, "The Climate of North America According to a New Classification," *Geographical Review* **21**:633–655 (1931).

20. UN FAO, *Wood: World Trends and Prospects,* United Nations FAO, Rome, 1967.

21. Ibid., "World Consultation on the Use of Wood in Housing," *Unasylva* **25**:1–168 (1971).

22. U.S. Forest Service, "Areas Characterized by Major Forest Types in the United States," *National Survey of Forestry Resources,* U.S. Forest Service, Washington, 1949.

23. Ibid., *Forest Statistics of the U.S. 1977,* U.S. Department of Agriculture, Forest Service, Washington, 1978.

24. Ibid., *An Assessment of the Forest and Range Land Situation in the United States,* U.S. Department of Agriculture, Forest Service Report FS–345, 1980.

25. R. H. Whittaker and G. E. Likens, "The Biosphere and Man," in H. Lieth and R. H. Whittaker (eds.), *Primary Productivity of the Biosphere,* Springer-Verlag, New York, 1975.

26. J. C. Zazada, "Alaska's Interior Forests: Ecological and Silvicultural Considerations," *Journal of Forestry* **74**(6):333–337 (1976).

CHAPTER **6**

TIMBER

Altogether the continental United States plus Hawaii have about 298,201,781 ha of forestland, with 66 percent classified as commercial.[37] Timber inventories totaled 22,437 million cubic meters in 1977. This included 339.7 million cubic meters of salvagable dead trees killed by insects, fire, diseases, competition from other trees, and similar causes. About 64 percent of the total volume was in sawtimber-sized trees. These have a diameter at breast height (1.37 m above the ground) of at least 23 cm for softwoods, or 28 cm for hardwoods. They also contain at least one harvestable log. Another 26 percent of the total volume is attributable to smaller trees called poletimber. These have a diameter of at least 13 cm, a main stem of good form, and good vigor. Pole trees will become the future sawtimber after they increase in size and mature. Poorer trees of undesirable quality or that are at least one half rotten comprise about 8.5 percent of the total. Thus, overall, about 90 percent of the national timber inventory comes from living trees of potential commercial value.

Total withdrawals for wood products of all types included about 339.5 million cubic meters in 1976, with amounts projected to reach as high as 555.8 million by 2000. Removals could run as high as 704.8 million cubic meters by 2030 (Table 6–1).[39] Forecasts indicate that actual demand may even rise above the levels projected, with the difference met by wood imported from other countries.

In 1952, timber removals across the nation exceeded the growth of standing trees. But from 1962 on, growth has exceeded harvests by 10 percent.[37] This generally has resulted from considerable net increases in hardwood resources. In fact, as of 1976 the hardwood species added 1.7 times more volume than was

TABLE 6–1
USE OF ROUNDWOOD PRODUCTS FROM FORESTS OF THE UNITED STATES, INCLUDING PREDICTIONS FOR FUTURE DEMAND.

	Demand for roundwood products, by years (millions of cubic meters)		
Species	1976	2000	2030
Hardwoods	82.1	169.9	266.2
Softwoods	257.4	385.4	438.6
Total	339.5	555.3	704.8

Source: Adapted from U.S. Forest Service [38]. Used by permission.

harvested from them. For conifers, removals about equaled growth. Thus the eastern hardwoods by far offer the greatest potential for sustaining heavier cutting to satisfy greater demands for wood products.

The forests of the United States have greater potential to supply the nation with wood products than current inventory suggests.[37] In 1977, the total commercial forestland supported 20,132.4 million cubic meters of growing stock trees. Others classified as rough and rotten contained an additional 1905.4 million cubic meters.[36] Under better management, the space and resources used by these undesirable trees might be given over to trees of better quality and a higher potential for use. Another 4 percent of the commercial area does not support sufficient numbers of trees per hectare to keep the stand adequately stocked. Improved regeneration of these areas could increase the productive potential as well. In addition insects, diseases, fire, and storms kill about 113.3 million cubic meters annually. If recovered before it deteriorates, or if more intensive management and protection programs reduced these losses, the potential national timber output could rise substantially. All together, for these and other reasons, forests of the United States now produce only at about 60 percent of their potential. Private lands rank higher than public lands.[37] Those owned by forest industries currently are in the best condition, and produce at about 68 percent of capacity. Some differences occur between regions, but generally the application of more intensive management would result in a higher growth rate for most ownerships. That would add to the nation's potential to satisfy its ever-increasing demand for wood products.

GROWING TIMBER TO MEET THE NEEDS

In past years, the United States enjoyed abundant timber supplies. Untapped reserves remained sufficient to allow significant area for preserves for purposes other than timber production. Past management mostly focused upon keeping removals in balance with growth, or at least not cutting more wood than grew in the vast area of unmanaged stands. Recent assessments, however, indicate that demand by a growing population will put much greater pressure on the nation's forests for wood products. This will come even as the total area of producing forest may decline by as much as 2 percent.[37]

Projections for the future suggest a need to change from casual approaches of the past. To satisfy the demand, the nation must implement more intensive culture of its timberlands to capitalize fully upon their productive potential. This will include steps to improve stocking on poorly occupied sites. For other stands, managers will attempt to recover greater proportions of the wood volume now lost to disease, insect attack, fire, and natural mortality resulting from intertree competition. Available information indicates that the nation would gain appreciably by applying improved management and by better utilizing more of the materials in trees cut as a part of harvesting operations.[37]

In these activities, foresters operate according to an important concept of conservation that separates timber management from some other kinds of natural resources use. It deals with their renewability. To illustrate, oil, coal, and other mineral resources appear in fixed quantities across the earth. Once used, they are not replenished. Water resources are different, since precipitation renews the supplies regularly and people can use the water as it flows by in streams or rests temporarily in ponds and lakes. But renewable biologic resources such as tree crops are replenished naturally as long as nothing causes a major disruption of their reproductive potential.[3] Hence forest management strives to use tree resources at a rate consistent with their renewal, and to harvest them in ways that ensure timely regeneration of future crops.

This concept of conservation provides for use of renewable resources as needed. But foresters control the way and intensity of use to assure regeneration and regrowth of another crop. Foresters apply this concept to timber management, and call it sustained yield.[3,11] It implies that care, tending, and use will maintain the productive capacity of the land. At the same time, management will provide a perpetual yield of goods and services at predictable intervals. Realization of such a goal separates forest management from exploitation. It permits maintenance of a healthy ecosystem offering the full benefits to be derived from forested lands.

SILVICULTURE FOR NATURAL STANDS

A knowledge of how trees grow and respond to the environment around them is the basis for the field of silvics, or forest ecology as it is also known. Silviculture encompasses processes of regenerating, tending, and eventually harvesting forest crops to obtain yields of goods and services. It gives attention to the ways trees grow in response to their environment, plus economic factors.[8,11,28] Silviculture works to alter the composition, growth, and condition of forest stands to better serve human needs. It involves ways to influence the species present, the number of trees per unit area, and regeneration of new tree crops. In some cases, the forester may even replace natural forests with artificially established ones in order to change the stand to fit the requirements of a landowner.

The Silvicultural System

Foresters blend several of these practices into a long-term plan of management called the silvicultural system (Figure 6–1). It provides for an orderly sequence of treatments over the long run to influence conditions and make the crop more useful. Further, the silvicultural system provides a plan of management for an individual stand. For the entire property composed of several stands, the forester will prepare several different schemes and merge them into an overall management program.

Foresters manage tree crops by one of two broad approaches: even-aged management or uneven-aged management.[22] Even-aged means that all the trees in

Figure 6–1.
Components of silvicultural systems for sustained yield management.

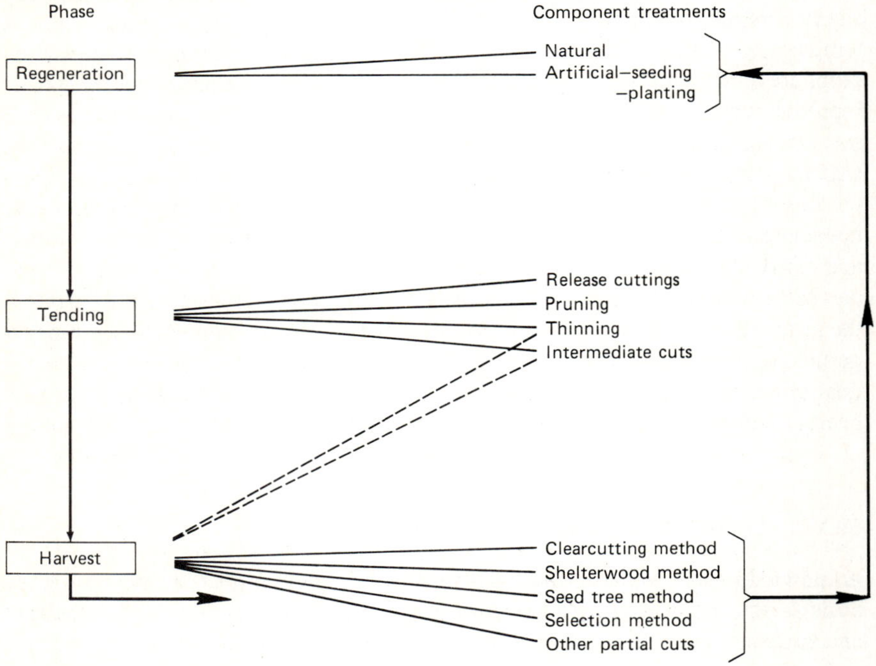

a stand are about the same age and will mature at the same time. Uneven-aged implies that trees of several different ages grow intermixed in the stand. They will mature at different times. The silvicultural systems used for both kinds involve three types of treatments—measures to regenerate the stand at maturity, to tend it during intermediate ages, and to harvest tree crops from it at appropriate times. The harvesting may serve as one means of tending immature trees. And by cutting trees in given ways, a forester can create conditions to promote regeneration of a new crop once the existing one matures. All silvicultural systems will incorporate some program to regenerate, to tend, and to harvest. The combinations and timing of these options will depend upon the nature and condition of the stand, the forest type, and the goals of the owner (Figure 6–1).

For convenience, foresters name the silvicultural system for the cutting method used to regenerate new trees from time to time.[28] Silviculturists call it the reproduction method to emphasize its purpose. The even-aged approaches include clearcutting, shelterwood, and seed tree systems. Uneven-aged ones are individual tree and group selection systems. With even-aged reproduction methods, a final cutting eventually removes all of the standing timber at one time. This leads to the regeneration of a new even-aged stand over the entire stand area. With uneven-aged methods, the periodic cuttings remove the mature trees, but leave the younger ones. This sets up conditions to regenerate a new age class to replace only the mature trees. The new age class covers only a portion of the stand area. In conjunction with the harvest, the forester will also tend the immature residual trees to enhance their growth for a longer time until they reach maturity (Figure 6–2).

Foresters do not arbitrarily pick the treatments to include in a silvicultural system for a given stand. They evaluate the characteristics of the species present or desired, the existing age arrangement among the trees, the nature of the physical environment at the site, the interests of the landowner, and an array of economic considerations. From these evaluations, a forester can identify appropriate alternatives and discuss them with the landowner. In concert, they can select the particular approach that would seem best to satisfy the objectives for management consistent with the biologic potentials of the site.[10] Here the forest manager draws upon an understanding of silvics and forest ecology to judge how stands of different kinds might respond. The forester must also decide whether the available species will grow well and reproduce, and what effects various treatments might have upon the physical and biotic components of the ecosystem. The forester then must reconcile plans for individual stand treatments with those for the property as a whole. This means scheduling cuttings to provide regular income for the owner, to capitalize upon existing and potential markets, and to facilitate other uses of the property. In short, the forest manager must plan ahead for long periods to allow a continuing yield of forest goods and services. The management must also maintain and enhance the overall productive capacity of the ownership.[9]

Figure 6–2.
These uneven-aged northern hardwoods are managed by an individual tree selection system that harvests mature trees periodically, thins the immature ones, and regenerates a new age class to replace the mature one.

Regenerating Tree Crops

Unlike many other plants, trees do not begin to flower and produce seed for many years. The age when flowering commences differs between species. Jack pine, for example, may have cones and viable seeds at three years of age, while many other pines take 15–30 years. As trees mature to middle age, the quantities of seed produced will usually increase. Then seed production among vigorous trees may continue for centuries. However, large crops of seed do not develop annually. Instead many species have bumper crops only periodically and produce smaller quantities during intervening years. Such factors as weather, insects, animals, and diseases affect seed and fruit production from year to year.[14]

A species' shade tolerance seems related to its seeding characteristics and growth. Shade tolerance refers to the species' photosynthetic efficiency at various levels of light.[30] The shade-tolerant species can grow and survive for long periods even in shade. In them, photosynthesis occurs over a wide range of light levels from low to high. These include grand fir of the Rockies, dogwood of the Southeast, and red spruce of the Northeast. Other species photosynthesize efficiently only at higher light levels and will not survive long periods of shading. The lodgepole pine of the West, longleaf pine of the South, and black cherry of the

Northeast have this characteristic. Foresters call these the shade-intolerant species. Additional trees have shade tolerance between these extremes and survive under medium light levels that occur under partial shading, or at full sunlight in the open. Douglas-fir of the West, southern red oak, and yellow birch of the North have such intermediate shade tolerance. Within each region, foresters will find species that represent a continuum of shade tolerances. Such mixtures even occur in a single forest type, or in any one stand.[35]

In general, shade-intolerant species tend to grow most rapidly and have more frequent good seed years than do the shade-tolerant ones.[30] As seeds fall to the ground, many will germinate once the proper combination of temperature and moisture triggers chemical reactions within them. But for the shade-intolerant species, the new seedlings do not generally survive if overtopped by larger trees in the main crown canopy. With the more shade-tolerant species, the seedlings survive under the shading. This advance regeneration may become dense and grow quite well. If the forester makes a cutting to harvest the tall trees, the advance regeneration will begin to grow fairly rapidly to replace the trees removed from the stand.

To get abundant quantities of the seedlings, foresters must often cut down some trees. This will reduce the density of the crown canopy and let light filter through to the forest floor. For some species, such as white pine, the forester will try to schedule the reproduction method cutting to anticipate or coincide with a good seed year. Such timing proves especially important with shade-intolerant species or if the seeds remain viable in the forest litter for only a short time. Hence, such reproduction methods as clearcutting, shelterwood, seed tree, and selection create their effect by creating openings in the forest. Through these cuttings, foresters can influence environmental conditions at ground level and make light, temperature, and moisture more conducive to the germination and subsequent survival and growth of seedlings.[8,27]

Managing Stands for Long-term Production

Silviculturally, the regenerating of a new tree crop by one of the reproduction methods begins a long series of events designed to influence the character of the new stand and its development.[8,28] For even-aged crops, early treatments during the sapling stages might include efforts to favor the selected species by removing others from around them. Then later, as the trees grow into pole sizes, the silvicultural system may call for a series of thinnings. These operations attempt to reduce the density of stands and to foster more rapid diameter growth of the best trees. When properly applied, thinnings reduce the number of trees present to concentrate the growth potential of a site on fewer trees than would develop naturally. Each residual tree will get more light, moisture, and nutrients. As a result, they grow more rapidly and reach sizes required for large-diameter products sooner. These treatments also remove trees that otherwise would die from competition. In this way, a forester can utilize such trees for pulpwood, firewood, and other products for which small trees are suitable. As the even-aged stands mature and the trees get larger, the thinnings yield sawlogs as well. Finally, following such tending, the stand will reach a degree of financial maturity and the forester will once more regenerate it to begin to grow another crop for another cycle of time. Foresters have a name for this interval of time from the regeneration of a new even-aged stand until its final harvest at maturity. They call it the rotation. It represents the period covered by the silvicultural system for an even-aged stand.

Most silvicultural treatments applied to established stands of trees involve cuttings of various types. Whenever possible, foresters also use these occasions to recover and sell products and to generate income for the landowner. If the cutting removes mature trees, foresters call it a reproduction method and use it to regenerate a new age class to replace the mature one. If they design a treatment to influence the development of immature trees, they call it an intermediate cutting. They usually try to influence stand composition by removing the poorest trees to favor development of better ones. The choices may involve tree quality as well as species.

Thinnings comprise one group of intermediate cuttings applied to even-aged stands. These reduce crowding between trees, which then reduces the withdrawal of moisture and nutrients from the soil, and makes more of these available for the residual trees. The cutting of some trees also creates temporary openings in the main crown canopy. As a result, the lower portions of the residual tree crowns are bathed in sunlight. The added light triggers a higher level of photosynthetic output in the residual trees. Then gradually, as the branches grow longer, the crowns become larger. They fill in the spaces created by the cutting. Also, because of the higher light levels at the lower parts of the tree crowns, the bottom branches stay alive longer. As the tree height increases through annual growth, the crown length gradually increases. This also results in a greater capacity for food production in the tree. And with the increase in food production, diameter growth of the individual trees is more rapid. The trees become larger sooner.

Controlling Stocking

Foresters can reduce the stocking of forest stands by a considerable margin and still realize full volume production per hectare.[28] This happens because, as they reduce the number present, the residual trees grow faster. Each tree increases in volume at a greater rate than before the silvicultural treatment. Collectively the fewer faster growing trees increase in volume sufficiently to make up for the growth lost by having cut away some of the poor trees. In fact, foresters have considerable latitude in reaching this goal. To achieve full-site utilization, they need only know how many and what kinds of trees to leave behind. This well-tended stand will yield full net volume production per hectare.

As an aid in designing cuttings either to thin even-aged stands or to treat uneven-aged ones, foresters rely upon marking guides. Researchers have developed such guides for many forest types. They usually include both a method for evaluating the conditions in a stand before cutting and a technique for determining what kind of residual stand to leave. The guides take several forms but generally indicate the combinations of residual basal area, average stand diameter, number of trees, and number per diameter class. Foresters determine these conditions by various inventory techniques whereby they count trees per hectare and measure the diameters. Basal area expresses stand density. It represents the sum of cross-sectional area of tree stems at 1.37 m from the ground for all trees growing on a hectare.

One type of stocking guide provides foresters with a tool for designing thinnings of even-aged stands (Figure 6–3). These use numbers of trees and basal area to determine the conditions of stocking present. They define the maximum basal area that stands are likely to have at various stages of development. Foresters can compare conditions present with those described by the guide and evaluate the need for thinning. The guide also suggests a B-level of stocking for the residual

Figure 6–3.
A stocking guide for designing thinnings in oak stands uses basal area and numbers of trees per hectare to determine the level of stand stocking. (Note that oak stands with combinations of basal area and numbers of trees between the A and B lines represent stand conditions that fully use the available growing space and site resources.) *(R. Rogers, Evaluating stocking in upland hardwood forests using metric measurements. U.S. Forest Service Research Paper NC-187, 1980.)*

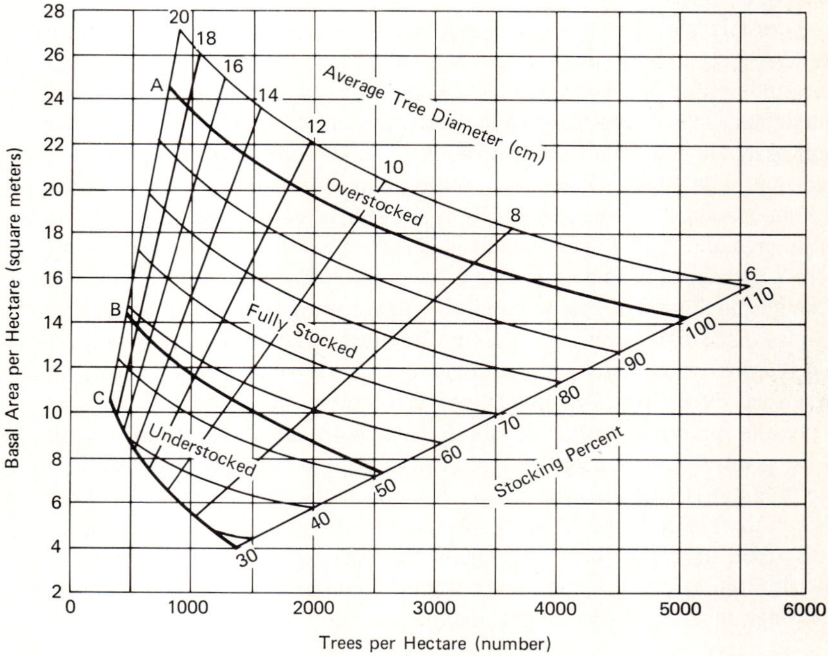

after thinning. As long as the stand has this minimum level of stocking, it should produce the full volume of growth possible for the amount of incoming radiation and other site resources. To achieve this result, the forester cuts the excess numbers of trees, usually removing the poorer ones and those of the lowest vigor.[23,24]

With uneven-aged stands, the silvicultural treatments operate differently. In these cases, the forester also returns to a stand every few years, but each time both harvests mature trees and tends the younger ones. Removing the mature trees makes space for regeneration of a new age class. The thinning among immature trees helps to improve their diameter growth just as with the tending of even-aged stands.[12] However, treatment of the uneven-aged stands differs in at least two regards. First, trees of several ages grow intermixed with each other so the thinning must treat several age classes rather than just one. Second, by cutting the mature trees, the forester makes a lasting opening in the crown canopy. The opening must be large enough to let in adequate light to sustain the growth of the new seedlings over a long time period. Also, to use uneven-aged systems, the forester must work with species that will survive and grow under partial shading.[5,21] Engleman spruce, sugar maple, American beech, and western hemlock have that characteristic. Less shade-tolerant species such as Douglas-fir, black cherry, white pine, and the oaks will not grow well under the shaded conditions and do not regenerate successfully with uneven-aged reproduction methods.[35] Foresters generally manage these more shade-intolerant species by even-aged systems.

Foresters also use stocking guides for planning treatments of uneven-aged stands, but their form is quite different from those for even-aged thinnings. Here the guides take into account the fact that trees of different ages have different diameters. The young trees have small diameters; the large trees are older. In managing such stands, the forester strives to allocate equal amounts of space to each age class, and to keep the site fully occupied to ensure complete utilization of the site resources. To achieve this, the forester sets a goal for the level of residual basal area and volume per hectare. Also, the marking guide will call for cutting across the size classes and leaving a fixed number of trees per diameter from the youngest to those that will mature by the next cutting. But because small trees take up less space than larger ones, the forester must leave more younger ones. Usually the progression of numbers of residual trees per diameter class follows a pattern described by an inverse-J-shaped curve (Figure 6–4).

At the time of each cutting, the forester will remove all the trees larger than some given threshold size that represents a state of maturity. In addition, to tend the immature trees, the forester cuts from each diameter class any excess beyond the numbers called for by the J-shaped curve. After the cutting, if the stand has this prescribed structure, the forester calls it a balanced stand.[8,28] This means that it should support sustained periodic cuttings of comparable volume per hectare over the long run. And at each cutting, the forester will find it possible to restructure the

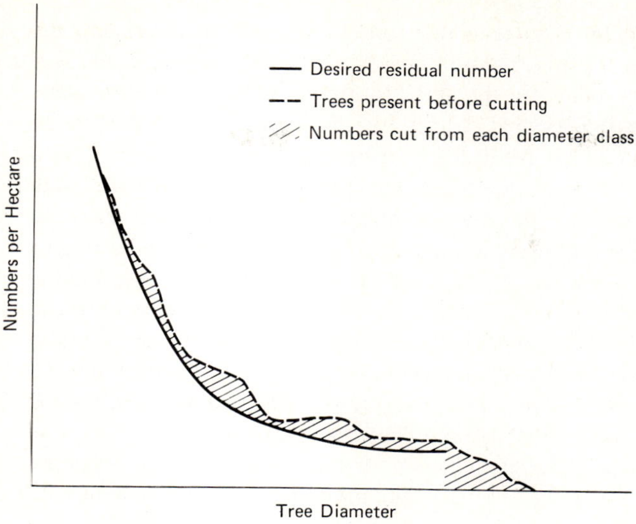

Figure 6–4.
Marking guides for uneven-aged stands call for leaving a residual crop with trees of many diameters, as described by an inverse-J-shaped distribution, and cutting the excess numbers from each diameter class.

diameter distribution to a condition like that left after the previous harvest. When the trees are treated in this manner, the forester applies all elements of the silvicultural system each time the management plan calls for a cutting in an uneven-aged stand.

Effects of Management

The thinning of immature trees in uneven- or even-aged stands does not increase the total production of organic matter per hectare.[28] Ultimately the amount of incoming solar energy and the availability of moisture and nutrients from the soil determine the quantity of biomass produced. The thinning will concentrate that productive capacity onto fewer trees per hectare, which speeds their growth and gets them to minimum product sizes sooner.[8,28] The forester can also use thinning to remove poor-quality trees and favor the growth of better ones. It helps to rid the stand of diseased and damaged trees, or of ones that might provide a good habitat for the buildup of insect pests and diseases. In addition, through thinning the forester can harvest trees that might die as a result of crowding by neighboring trees, and market them to generate income for the landowner. In these ways, a forester can control the development of trees in forest stands and affect their growth to better serve the interests of the landowner.

With even-aged silvicultural systems, the forester will thin in a stand several times before the trees eventually mature. The interval between treatments will depend upon how long it takes for the stand to grow enough volume to make another harvesting operation feasible. During the first thinnings in a stand, the trees may be so small that they have little commercial value. Foresters call this a precommercial thinning. Later, as the trees get larger and larger, they will have value for pulpwood, sawlogs, and other products. In uneven-aged stands, the forester will couple a thinning of the immature trees with the periodic harvest of the mature ones. The tending will speed the growth of the young trees, and the cutting of mature ones will serve as a method for regenerating a new age class of seedlings.[28] In both cases, the thinning will result in increased diameter growth of the residual trees without a loss of overall volume production from the stand.[8]

Through various silvicultural practices, the forester will work to influence the composition, character, and growth of trees in forest stands. Thinnings are one way to achieve these results. In the end, the trees mature and the forester will apply a reproduction method of cutting to remove the old trees and replace them with new seedlings to grow into another crop. Here again, the forester will remove the mature trees from the woods and sell them for sawlogs, pulpwood, timbers, wood chips, firewood, and many other types of wood products. These provide income to the landowner to pay off the costs of management, plus some for profit. In addition, many owners couple their timber management with treatments to enhance wildlife habitat, improve water yields, create conditions more conducive to recreation, or to derive other benefits. In all cases, the forester will design the treatments to facilitate the uses the owner wants, and to provide for continuing crops and benefits.

The reproduction method cuttings serve as the means to make space and resources available for new seedlings to regenerate and grow. This involves cutting down trees to open the main crown canopy so that sunlight can penetrate to the forest floor. By cutting the mature trees, the forester also reduces withdrawals of moisture and nutrients needed to sustain the growth of the new trees. Frequently these methods depend upon natural regeneration from seeds that fall to the ground at the time of the cutting. With some forest types, the new crop may develop from seedlings that started just prior to the cutting. Foresters call this advance regeneration.

With the even-aged systems, the reproduction method must remove all the trees present and replace the entire stand with a new one. For uneven-aged stands, the reproduction method will establish a new age class only in the spaces that previously had mature trees. These new seedlings will grow up between the remaining older ones, maintaining the uneven-aged character of the stand. In both cases, the silvicultural plan will provide for regular subsequent tending until the crop eventually matures and the forester can remove it and regenerate still another new crop.

ARTIFICIAL REGENERATION

Under some schemes of management, the forester may not use natural regeneration. Instead the plan may require stricter control over the species composition, genetic constitution, spacing, or arrangement of the new stand. Then the reproduction method may involve planting trees or artificially sowing the seed to create the new tree crop. In some cases, the owners may also wish to convert open land to forest cover. Then they will use one of the artificial regeneration methods to accomplish the task.

Seed Quality and Genetics

In many ways, the decisions a forester makes well in advance of any tree planting will affect the future yields, character, and quality of a forest crop. These include the choices about where to collect the seed, and what trees to gather it from. By those decisions, the forester can influence the genetic composition of the forest stand, and ultimately how the new crop will grow.[15,19] This includes the growth rates, the form of the trees, and the qualities of the wood.

Ideally the forest tree improvement specialist or nursery manager will gather seeds only from stands of good quality and growing conditions comparable to the sites scheduled for reforestation. Within these stands collection is limited to individual trees with good external characteristics (called the phenotype) and satisfactory growth. The manager or specialist hopes that these good-looking trees will also have excellent genetic qualities (called the genotype) and will produce offspring of outstanding form, quality, and growth.[19] Thus, by exercising control over the kinds of seed they use, the nursery manager and tree improvement specialist try to provide the landowner with seedlings that satisfy the needs of the ownership.

Individual trees of the same species may differ in growth, form, wood quality, or other characteristics. Some of this results from influences of the microsite and competition. But the variability also reflects inherent genetic differences among trees in any single stand, and between individuals of the same species growing at different places.[15,42] By selecting individual trees that exhibit the most desirable traits, the tree improvement specialist can limit production to trees of the most desirable economic growth form.[13,19] This may pertain to their straightness, size and condition of branches, rate of growth, density of foliage, bole form, internal wood qualities, or other features that potential users deem desirable.

Forest geneticists and tree improvement specialists have made important progress in helping forest managers to appreciate the importance of these differences in seed source, and of limiting their tree planting to seedlings and seeds of proven quality. For convenience, these specialists may gather branch sections, or scions, from selected parent trees. They graft them onto small trees or seedlings called rootstocks, or root them by special techniques. Once artificially propagated in this way, the scion develops into another tree. And it will have a genetic character identical to that of the parent. The geneticists then plant these propagates in orchards and manage them for seed production. The resultant seed should have genetic qualities reflective of the carefully selected parents, and develop into trees of good phenotypes. Further, the seed orchard can contain trees from many different, widely separated geographical areas. In this way, the geneticist can cross-fertilize trees that otherwise would not pollinate each other. In some cases, geneticists even pollinate the trees artificially to obtain particular crosses of interest and to test different combinations.[15,25,42]

After pollination, the trees produce cones or fruits that contain new seed. Tree improvement and nursery specialists gather the seed as it matures, clean and otherwise process it, and store it until ready for use. They take great care to keep good records so that at some future time they can check on the actual performance of different seed sources and crosses after outplanting.[25] In some cases, they even make specially designed test plantings to compare the offspring of different trees. With results from these trials, they can cull out individuals that prove undesirable and concentrate upon the better trees to produce future supplies of seed.[42]

Geneticists call these tests provenance trials. Provenance means a population, so a provenance trial compares different populations of trees. These may come from seed collected in different stands at geographically separated locations. For the trial, the geneticists gather the seed and grow it into seedlings. Then they can plant seedlings from different populations together in a common environment. After several years of growth, they look for differences that show up between the populations and see if one has better offspring than others. By making several replications of these plantings under varying environmental conditions, they can determine whether any one population grows better over a variety of sites. If it does, they may attempt to infer whether the differences observed have a genetic control.[30] In tests where they compare the performance of trees from selected crosses with that of natural populations, the geneticists can also determine how much improvement they gained by their programs of selecting, crossing, and propagating trees to make them genetically better.

Nurseries and Seedling Production

Seed of many tree species will remain viable up to 10 times longer than in nature, if properly prepared and stored under suitable conditions.[31] Preparation involves extracting seed from the cones or fruits, removing the wings or pulp, drying it to relatively low moisture content, and sealing the cleaned seed in containers. After this preparation, the nursery manager or seed specialist will usually store the seed under refrigeration at a low temperature. Seed can be drawn from this storage periodically and used for growing crops of seedlings as needed. By periodic testing, the potential for germination is monitored and the results used to determine sowing rates in the nursery.[6] Larger nurseries usually maintain facilities for extracting, cleaning, and storing the seed for future use. By gathering large amounts during years of bumper crops and storing it, the nursery manager or seed specialist can assure a steady supply even during years when the trees produce little seed.

Traditional methods for tree seedling production parallel those for agricultural crops. For convenience, the managers usually divide the nursery fields into long narrow plots called seedbeds. They cultivate the surface, smooth it, and then sow the seed in rows lengthwise along the beds. They may add organic matter during the initial cultivation, mix in fertilizers, and apply fungicides to kill off harmful soil organisms.[32] After the seeds germinate, the managers will irrigate the beds to supplement natural precipitation. They also will fertilize periodically to ensure an ample supply of nutrients essential for adequate plant growth. And they will occasionally weed the seedbeds and thin the crop to prevent crowding and promote rapid growth of the seedlings.[1,32] With these methods, a manager can normally get a crop ready for digging and outplanting in two or three years in northern latitudes, and in a single year in the South.

Recently many nurseries have adopted a technique called containerization. Instead of open seedbeds, they grow the seedlings in small containers inside greenhouses.[33] The containers may be molded from such materials as styrofoam, paper, plastic, and compressed peat moss. They are filled with special mixtures of soil and organic matter, and one seed is sown in each small container. The manager can control the temperature and day length, alter the quality of light in the greenhouse, irrigate and fertilize as needed, and otherwise provide an environment optimum for seedling development (Figure 6–5). This process speeds seedling growth and shortens the time needed to get the plants ready for field use.[8,31]

Container methods have proved especially useful for growing seed of carefully selected genetic makeup. The system essentially eliminates the natural environmental hazards that might threaten the seedlings or limit their growth in open nursery beds. Thus the geneticists can get high survival from the seed and make sure that it develops into thrifty seedlings. With the growing environment maintained at carefully controlled levels, seedling growth accelerates and the plants develop into sizes suitable for outplanting in as little as six months. Field trials have generally given excellent results. They also have shown that container seedlings will survive well even if planted fairly late in the summer. However, due to the elaborate greenhouse facilities and the greater problems of handling the stock during transport, costs of container-grown seedlings can run about 1½ times more than for those grown by conventional methods.[8]

Tree Planting

Tree planting gives the forest manager considerable flexibility in controlling the character and composition of a forest crop. Foresters call this type of stand a plantation, meaning that they used an artificial method to regenerate the land. In most cases, this involves tree planting. It gives the foresters strict control over species, spacing, and the arrangement of trees in the new crop. They even can use

Figure 6–5.
Growing trees in containers within climate-controlled greenhouses speeds their development and yields a high-quality seedling in six months or less. *(Weyerhaeuser Co.)*

specially selected genotypes of a given species in the hope of influencing the phenotypic character of the trees. They also can put the trees into rows and arrange them in a fashion that will facilitate future harvests and thinning operations. By using proper spacing between the trees, the managers reduce the crowding during early years of the rotation. This will result in more rapid tree growth and shorten the time it takes to get the trees to a size suitable for harvest.

To prepare for planting, foresters must carefully evaluate the site conditions to determine what species and genotype will grow well in that physical environment. They will consider the soil, climate, existing plant cover, indigenous pests, available moisture, and other factors critical to tree growth. They also must take cognizance of the types of products the landowner wants to grow, and what species the local markets demand. These decisions become especially critical when the forester wants to grow a species that does not normally live in the locality. In addition, foresters must determine how many seedlings to purchase, and arrange for the equipment and work crews needed to complete the planting on schedule.

At the time of planting, a forester obtains seedlings from a nursery, transports them to the planting site, and supervises the crew in setting them out according to a prearranged spacing and arrangement. Where site conditions permit, the forester may use special tree-planting machines. These devices plow a narrow slit in the soil and open the ground so a worker can insert the tree roots into the hole. Then a set of packing wheels pushes the soil back in place around the tree roots. Where rockiness, poor drainage, or other site features prevent use of these machines, people are hired to plant the trees by more traditional methods. The work crews use shovels, mattocks, planting bars, and other hand tools to dig the holes. Generally these methods take longer and cost more than machine planting.

Site Preparation

Often foresters must get a site ready for planting before the crews put the seedlings into the soil. Silviculturists call these practices site preparation. In some instances, they may need to break apart and flatten the logging slash to facilitate the planting operation or eliminate the habitat for pests that might harm the trees. They can do this by burning the debris, or by crushing it and breaking it apart. At some sites, they may need to clear the land of other vegetation by cultivating the ground, applying herbicides, or using fire to kill the plants. At other times, they may find it advantageous to till the soil to improve drainage and structure. Or they may want to mix up the organic matter on the surface in order to make the seedbed more conducive to the germination of a particular species. All of these practices help to improve chances for seedling survival and subsequent development.[8,28]

Forest engineers have developed special machinery for the site preparation work. These implements are rugged in construction and often look like gigantic versions of those used in agriculture. They have built large, heavy rollers to help crush the logging slash. They use bulldozers outfitted with special blades to help clear vegetation and push debris out of the way. They have implements called scarifiers to drag across the surface to mix the organic litter and upper soil layers. And they have heavy-duty plows and disks to cultivate the soil. In some cases, foresters will even cultivate the soil, form it into mounded beds before planting, and apply fertilizer. All of these practices add cost to the tree planting program, but pay off in better seedling survival and more rapid tree growth. They usually have a role in intensive management programs where the owner seeks to maximize production in a manner similar to that for agricultural crops.

Subsequent Care

Any regeneration program really serves as the beginning of a long period of management to nurture the crop until it becomes ready for harvest. During the first few years, the foresters may do little more than watch the development and protect the plantation from fire, insects, and diseases, although in some cases they will need to control unwanted vegetation that might interfere with the trees. As the trees get larger and more crowded, they will begin thinning. Just as with natural stands, the thinning will serve as a means of removing intermediate products from the stand, and of concentrating the growth potential of the site on fewer trees. As a result, each gets more sunlight, moisture, and nutrients than before the thinning, and the diameter growth will increase. Throughout the rotation, foresters must continue to protect the crop from harm, and repeatedly thin it until the trees reach a stage of maturity. Then they harvest the final crop and once more prepare to replant the land and grow still another crop. However, tree crops take a long time to grow, and a forester who plants one may never see it reach maturity. Each forester will likely tend and harvest the crops planted by someone else, and in turn regenerate new stands that future generations of foresters can harvest to provide for a continuing yield of products from the land.

TIMBER HARVESTING

The objective of silviculture is to make forest stands more useful. It produces trees the owner can benefit from. Thus, when the time arrives, forest managers employ logging to harvest the mature trees, and to regenerate a new crop to replace the old one. They also use logging to implement many other cultural operations. These include thinning, improvement cutting, salvage and sanitation, wildlife habitat manipulation, and watershed management. All necessitate the cutting of trees to bring changes in the forest. In the process, the managers also recover the usable portions of the felled trees and sell them to produce income for the ownership.

Logging means harvesting and transporting the usable portions of trees from forest stands.[7] It includes:

1. Felling or cutting down the trees
2. Delimbing the main stem
3. Bucking or cutting the stem up into logs and bolts
4. Skidding or yarding to move the logs to a roadside landing
5. Sorting and loading the pieces onto trucks, railcars, or other means of transportation
6. Transporting or hauling the products to a mill or processing plant
7. Unloading at the mill yard

These basic steps have always characterized logging. However, different logging operations can vary considerably in the degree of mechanization used to accomplish the basic tasks.

Harvesting Equipment

Modern logging operations increasingly employ sophisticated machines for phases of the work once done by people and animals in order to make the job easier, to increase productivity, and to reduce costs. It all begins when a logger cuts down the tree, delimbs it, and bucks the trunk into logs. In most cases, loggers use hand-held chain saws, axes, and a variety of accessory equipment. In some regions, contractors have purchased machines called feller-bunchers to replace the people and saws for these operations. These machines have large hydraulic shears, auger devices, or machine-mounted saws attached to a tracked or wheeled vehicle. The operator can drive the machine from tree to tree and cut them down without leaving the driver's seat. The feller-buncher will also have hydraulic holding devices that the operator can use to pick up the trees and move them into piles for skidding or forwarding.[27]

Machines also have replaced animals for moving the logs and bolts from the woods to a roadside concentration point called a landing. These vary in construction and character depending upon the terrain and types of products handled. In pulpwood operations, forwarding machines may carry the logs or bolts completely suspended off the ground. Other pulpwood and most sawlog harvesting operations use articulated, four-wheel-drive machines called skidders.[7,27] Most of these powerful tractors have a winch mounted over the rear wheels (Figure 6–6). An attached cable runs off the winch and over an elevated armlike structure called an arch. The logger can pull out the cable, hook it onto the logs using chains called chokers, and then pull the logs up to the skidder and raise the front end of the pieces off the ground by reeling in the winch cable. Other skidders have hydraulic grapples instead of a winch. From the cab of the tractor, the operator can lower these, open the tongs, grab hold of the logs, and lift their front ends off the ground.[27] This reduces the friction and the work required to pull them along the skid trails to the landing.

In many areas of the West Coast with large trees and/or steep topography, and more recently in difficult topography in the East, logging engineers substitute cable yarding systems for the skidders. These employ a polelike tower called a spar. At the base, it has powerful winches. The workers can unwind the cable off one of the winches, then pull the end over pulleys on the spar and out across the logging site. The loggers can then string the cable through pulley blocks attached to trees in the stand, and make a long loop back to the second winch. By releasing one winch and reeling in the other, they pull the cable around the loop and out into the stand. They also attach chokers to the cable and hook the chokers to logs. Then, by engaging the return winch, they pull the chokers back to the spar, with the logs attached to them. For some systems, the engineers suspend the cables off the ground by using spars at both ends of the cable.[27,41] In other places, the engineers use large helium-filled balloons to lift the logs, and they pull the balloons back and forth using a cable system such as the one described. They also have used helicopters to lift and carry out the logs. Balloon and helicopter logging systems cost a great deal to operate, however, and consequently have received only limited use.

In many parts of the United States, logging contractors will skid or yard the stems in tree-length pieces rather than as short logs. This adds efficiency to the skidding operations. They can then load the logs onto tractor-trailer trucks for hauling in lengths as long as 10–15 m. In other instances, they buck the tree-length pieces into short logs at the landing and haul them to a mill on trucks with a shorter wheelbase. But since both short and long logs weigh a great deal, heavy-duty forklift loaders or cranelike ones must be used to handle the pieces. The type of

Figure 6–6.
Large four-wheel-drive, frame steering, rubber-tired tractors have replaced animals and crawler tractors for skidding logs throughout most of the United States.

equipment will have to be fitted to the size and weight of the logs in order to operate efficiently and at the least cost. Once loaded, the logs go directly to a primary processing plant for manufacture into lumber, sheets of veneer, or other products.[6,27]

Improvement in hydraulic systems in machines has brought new efficiency to logging; it has given the contractors increased productivity at a lower cost. They can use mechanical devices to fell trees, for example. They can have grapple skidders to pick up the logs and move them to a landing without the operator having to leave the cab. They can use other machines to delimb a tree and cut it into logs or short bolts. With multifunction harvesters, they can even combine several of these operations. Also, the growing market for chips for paper, pulp, and fuel has encouraged in-woods chipping rather than having this operation carried out at the primary processing plant.[27] With these chip-harvesting operations, the contractor will have machines that receive the whole tree, delimb it, remove the bark, cut the wood into chips, and blow the chips into trailers for hauling to a remote location for use (Figure 6–7). In these ways, logging has become increasingly automated, or at least employs more and better machines to do much of the work once performed by people and animals.

Figure 6–7.
Multifunction processors can debark, delimb, and cut up trees and long logs into chips for shipment to a pulp mill or other manufacturing plant for use in a variety of products.

Safeguarding the Environment

New machinery and other advances have revolutionalized many aspects of timber harvesting and affected other forestry operations as well. They have precipitated new approaches to silviculture and other management operations in order to accommodate the greater mechanization. At the same time, logging contractors have had to become more sensitive to the ways their activities affect the forest environment. Formerly loggers had to get out the timber expeditiously and cheaply. They paid most attention to cutting costs and producing at maximum rates. But present-day environmental mandates for protecting water quality and safeguarding the productivity of our natural ecosystems make such approaches intolerable. Fortunately improved technology, better understanding of the options available, and new methods to log efficiently give today's contractors the potential to keep impacts within tolerable limits.[18,20]

Recently the federal and state governments adopted new and widespread policies to protect water resources. These involve a nationwide federal program that affects logging as well as many other types of forest and open land activity. In fact, provisions of the Federal Water Pollution Control Act of 1972 require use of what are called "best management practices." These represent the best known ways to conduct silvicultural and other management activities, so as to have minimal effects upon water quality.[35] The program's objective is to control siltation, addition of foreign substances, increases in water temperature, inflow of nutrients, and deposition of organic matter. The best management practices (BMPs) also should allow efficient logging and prove cost effective for the contractor.

The forester and logging engineer have the responsibility to plan for and lay out timber-harvesting operations in a way that permits efficient and orderly logging with minimum risk to workers, vegetation, and the environment. Above all, the forester must keep the land productive for future timber crops and other forest values. Advance planning must include steps to ensure minimum impacts upon the soil and water, and to secure prompt regeneration to replace the harvested crop.[29] Logging must become one phase of a long-term system for the permanent management of land. This requires careful linking of silviculture with logging to bring about the sought-after result.

Logging as a Business

Logging contractors really run a business enterprise. As with any other business, they must operate at a profit and generate sufficient revenues to pay the cost of keeping equipment, purchasing timber, and paying workers' wages and benefits. To succeed, they must deal with many more matters than simply cutting down the trees and hauling them to a mill. They also must locate the harvestable timber, arrange for its purchase, lay out and develop the skid trails and roads, schedule the cutting to match with demands of the primary processing plant, comply with the forester's regeneration plans, and oversee several other activities needed to keep the operation running smoothly and efficiently. The logging supervisor must control costs of these activities, schedule the use and maintenance of expensive machinery, and employ the skilled workers to do the logging and associated tasks.[7,41] This calls for special knowledge, much practical know-how, and good administrative skill. Persons in these positions must serve both as long-term planners and imaginative improvisors to keep operations running smoothly and still meet the many unexpected emergencies that may arise.

Above all, the skillful use of timber-harvesting operations provides a means to implement silvicultural programs. It requires good coordination between the forest manager and the logging contractor or supervisor. Silviculturally timber harvesting provides a way to affect the culture and regeneration of forest stands to satisfy the interests of owners in sustaining the land's yield of goods and services. At the same time, the forest products industry looks to logging to procure raw materials to satisfy society's demand for manufactured goods and paper products. The two go hand in hand. Silviculture programs ensure the continued production of marketable trees over the long run. Timber harvesting and processing offer landowners the payments needed to cover costs of landownership, plus a reasonable profit to make forestry an attractive business enterprise.

FOREST MEASUREMENTS

To a large measure, foresters can succeed in their silviculture and logging only if they can first determine what kinds and amounts of resources they have to work with, and how these change from time to time as a result of the management. They require information on which to base decisions, and need various statistics to use in reporting the results of their programs and advertising the crops they have available for harvesting. And in many situations, they need to know not just about the timber, but about other resources of the forest as well. As a result, they must frequently integrate several data-gathering needs into multipurpose inventories. These will provide a picture of the various resources available, including the timber, water, rangeland, and wildlife.[17]

The inventory of forest resources falls within the responsibility of two technical specialties. Forest mensuration encompasses the measurement of timber, area, and other features of the forest. From these, the mensurationist can prepare estimates of wood volume, land area, tree sizes, and other statistics useful to the forest manager. Biometrics deals more with mathematical models, statistical analyses and probabilities, and computer simulation to represent forest conditions and suggest how these change over time. Both specialties are important to forestry operations.

Many techniques have become available to forestry as a result of advances in mensuration, biometrics, measurement methods, and equipment development.[17] The array of possibilities available in designing and implementing such an up-to-date forest resources inventory requires that the forest manager first consider the use to which the data will be put. This means determining the need for a forest inventory and the ways in which it can become an integral part of management decision making. It also requires decisions about cost, and how to obtain adequately precise information most efficiently.[2] Such concerns fall under the purvue of the forest mensurationists, and become part of the technical routine of the general forest manager.

The Use of Inventories

Basically timber inventories involve sampling the volume, kinds, sizes, and condition of trees. Foresters also want information about the accessibility, operability, and location of different forest areas (Figure 6–8). In addition, they must know the species' composition, quality, degree of development, and condition of stocking among the various stands on a property.[26] Silviculturally these provide a picture of needs for stand treatment and the kinds of cultural measures that might prove appropriate. Forest managers use basic statistical information such as stand density, age, species composition, and growth rates to schedule the timing of intermediate treatments for immature stands, and to determine appropriate kinds and intensities of cutting needed to bring a stand to some prescribed level of stocking. For sawtimber stands, they use similar data to ascertain the degree of maturity, the appropriate time for harvest cutting, and the levels of residual stocking that will result in effective regeneration. The measurements taken through basic forest inventories provide the managers with information needed to design and schedule a wide array of programs for creating, maintaining, and regenerating forest stands.

Timber inventories also indicate the amounts of volume available for sale, and its quality. They portray changes resulting from cutting, mortality, and growth.[2] These data allow calculation of timber values and the negotiation of sales agreements between landowners and timber buyers. Wood-using industries also use comparable inventory statistics for keeping track of the general condition of their forest operations, for scheduling the harvests of materials off their lands, and for planning a regular flow of wood volume to sustain production in their mills.[26] This usually involves assessing the amounts of timber added annually by growth. They balance that against losses to mortality, and program harvests to remove from the forest an amount equivalent to the net volume added periodically. Foresters call these plans cutting budgets. They form a critical element of forest yield regulation needed to keep harvests in balance with capabilities of the land to produce recoverable products.[9]

As a standard practice, foresters measure tree diameters at 1.37 m above the ground. They call this the diameter at breast height (dbh). It avoids the swelling that occurs closer to the ground above the root collar and gives a better indication of the tree trunk circumference. In addition, foresters measure the merchantable height of the usable portion of the trunk. For hardwood sawlogs, they normally select an upper diameter of about 30 cm, and for conifers about 20–25 cm.

Figure 6–8.
By using measurements of tree diameter, height, and condition, foresters can determine the volume of boards in a tree, calculate its value, and make determinations of silvicultural importance. *(U.S. Forest Service.)*

Pulpwood pieces may run as small as 8–10 cm at the smaller end. These represent the smallest diameters traditionally used for commercial roundwood. However, in catering to developing markets taking wood chips for fuel and similar purposes, the contractors may utilize smaller branches. In some cases, they utilize the entire above-ground portion of the tree.

From measurements such as these, foresters can determine a tree's volume as a function of the cross-sectional area of the trunk, and its length.[2] A commonly used equation to express this relationship is $V = a + bD^2H$ where a and b represent coefficients determined by regression analyses, and D and H the tree diameter and height respectively. Regression analysis amounts to a mathematical way of determining how to predict values of one variable (such as volume) by using other, more easily measured variables, such as dbh and height. The result will provide an estimate of the cubic volume of solid wood. Foresters also may predict the volume of boards that can be sawn from logs in the tree. This measure of lumber results in less volume than the total cubic content, since some of the log ends up as sawdust, slabs, and other nonlumber materials that the sawmill owner will chip for fuel or sell to a pulp mill.

Monitoring Forest Conditions

Given the means to estimate the volume of individual trees, foresters can take sample measurements at randomly selected places in forest stands and across a forest property. They measure characteristics of trees on these sample plots, and combine the data to provide reliable estimates of the total timber inventory on the entire property.[2,9] Such forest inventory procedures are even applied across broad geographical areas such as entire states, and the nation as a whole. The U.S. Forest Service has maintained a "Forest Survey" of timber inventories and forest area within states for many decades. Periodically it combines the data to portray conditions across the United States.[37] This inventory system has served as a primary source of information for the national assessment mandated by the Forest and Rangeland Renewable Resources Planning Act of 1974. The Department of Agriculture published the first results in 1980.[37,38] Like earlier inventories, it gives a picture of the amounts and condition of timber and forestland available. The assessment also provides information about the number, sizes, and species of trees. It details estimates of growth, mortality, and removals. And it presents many other facts needed in properly planning for the wise management and use of the nation's timber resources.

Many times, an agency or corporation will wish to assess the change of forest conditions by periodic remeasurement of the same stands. To enhance this comparison, the mensurationist will often install permanently marked sample points. For these, the forester will lay out an area of fixed size, often about ⅒ ha. To identify trees on the plot, foresters paint numbers on them, or use aluminum tags for this purpose. Usually they mark the tree at dbh so they can remeasure the diameter at the same place each time. Measurements may also include species, tree height, and condition. The mensurationist then compares data gathered on two occasions to determine how much the volume increased or how the sizes and number of trees changed. Through this process, called continuous forest inventory, foresters can get a good estimate of growth and change in forests on an ownership.

Many broad-based inventories that cover vast areas of land, even of entire states, require unique approaches. Costs prohibit detailed measurement of a large segment of the forest area, and so mensurationists design sampling systems whereby they gather data from a few selected areas. They use these samples to estimate the magnitude of resources over the larger area. Such approaches have application to large industrial or public holdings, many of the bigger nonindustrial private properties, and even to stands that cover a sizable area. In these cases, mensurationists may employ aerial photography or other types of remotely sensed imagery. From these, they classify the forest into groupings of similar kinds, and evaluate many conditions of the land and the resources on it.[16]

Through specialized techniques, persons skilled in photo interpretation can easily measure crown diameters, estimate tree heights from shadows, count numbers of trees in sample areas of the photo, and combine these with other measurements to provide an estimate of forest area, timber volume, stand condition, and the like.[2] As such information needs verification, they also field-check some of the photo-interpretation plots and develop adjustment factors for improving the accuracy of the photo data. Permanent sample plots serve well for this purpose. When undertaken properly by skilled persons, these approaches yield amazingly useful information. They help to give the foresters a better understanding of the nature of the resources in question. They show how conditions change over time. And they help the forester to develop strategies for managing the resources.

Increased availability of high-speed data-processing equipment at relatively low cost has facilitated greatly the storage, analysis, and retrieval of forest resources information. These machines and the companion programs for controlling the processing operations give forest managers great flexibility, and allow the timely and efficient handling of even large amounts of data. The modern data management systems also provide for many types of specialized uses, including the preparation of harvesting plans, tabulation of volumes by management units, forecasting of growth and yield, listing of the stands and areas needing silvicultural treatment, determination of costs and yields from various operations, and even computer mapping of various sets of information the manager can easily retrieve and display in graphic or tabular form.[40] Aside from these complex analyses, many types of resource inventory solutions require only simple approaches, such as short single-purpose programs that the forest manager can quickly write from time to time to use on small computers in the office. In some cases, the programs may be fed into a terminal and through a telephone line linkage to large processing systems at remote locations. Many of these operations also lend themselves to in-field solution through the use of programmable hand-held electronic calculators.[4] All reduce the work and speed up the processing of information needed for wise decison making concerning forest resources management. Their increased availability has given new dimensions to information processing and use at all levels of timber management, harvesting, processing, and planning.

MANAGEMENT FOR SUSTAINED YIELD OF GOODS AND SERVICES

From a management viewpoint, individual forest stands are the basic land units for implementing a silvicultural treatment. But a forest property will often support a mixture of stands with a variety of species composition, age arrangement, degree of development, and other features that make the units different from each other. As a result, the forest manager must develop a plan for each stand and integrate those many plans into a well-orchestrated program for the entire forest property. It will include an assessment of the volumes and other resources on the property, of how much these grow each year, and of what amounts to harvest annually to keep the removals in balance with growth. When implemented, such a plan will ensure the continuous production of goods and services, or what foresters call sustained yield.

For ownerships that have several properties or large holdings, foresters will need fairly complex schemes for planning and information management. With more limited ownerships, these become less complicated, though not necessarily less demanding of skill and imagination. Foresters must devise ways to integrate the silvicultural treatments, timber harvesting, inventory, property and road maintenance, and other activities into a coordinated package so that each part complements the others. This means providing for the regular and timely culture of the timber, harvesting the products when ready, and regenerating new crops to replace the old ones (Figure 6–9). It means finding ways to accomplish these tasks in a manner that fully capitalizes upon the productive potential of the land, and protects the resources from harm and loss. It means anticipating what timber and other resources might die or be lost due to competition and other causes, and harvesting them while they still have value. It means controlling the composition and development of individual stands so that the trees in them eventually will provide the types of products the owner wants. And it takes skillful marketing of the resources to generate income for the landowner. Of course, this all depends upon having good markets for the products and selling the goods at a price that makes the investment in management economically attractive.[9]

This role of management applies not just to timber, but to other values of the

Figure 6–9.
With skillful integration of silvicultural practice, timber harvesting, inventory control, and program planning, the forester can sustain yields of products and services from a property and make management more attractive economically.

forest as well. When developing plans for managing timber crops, foresters should take cognizance of the impact upon soil and water resources, wildlife habitat, recreational potential, and the ways that the management activity might facilitate or impose upon the many values people derive from forested lands. Ultimately the purposes of the ownership determine the balance between alternative uses and the dominance placed upon any one of them. Schemes appropriate to one owner or type of ownership will not necessarily apply in other cases. Thus the forest manager must understand why the employer owns and manages the forest. And the forester must develop the skills and capability to devise creative plans for satisfying these purposes.

CAREER OPPORTUNITIES IN TIMBER MANAGEMENT

Growing and selling of timber crops are major elements of most forestry programs, and provide the principal revenues to pay the costs of operations. Probably most people engaged in professional forestry become involved in some way with timber activities. These include many support services such as mensuration, biometrics, forest engineering, management planning, harvesting supervision, silviculture, forest tree improvement, and nursery practice. Through the coordination of these disciplines, foresters bring about the creation, culture, and harvest of continuous crops of timber from forested lands.

The degree to which individuals work in one of these areas of forestry depends in large measure upon the type of organization that employs them. Large corporations and public agencies often have staff personnel who provide specialized services fitted to the types of resources and operations at hand. Consulting firms also may focus upon limited aspects of forestry, and individual landowners or smaller firms and agencies may hire these consultants to assist with different land and timber management operations. In many small land management and forest products companies, however, the forester must serve in many diverse capacities. Commonly the local professional will function part-time as silviculturist, part-time as mensurationist, part-time as logging supervisor, and part-time in a host of other activities needed to grow and harvest timber. Such diversity, in fact, generally characterizes the early professional experience of most foresters.

Many sawmills, pulp mills, and other primary wood-using plants depend heavily upon timber from lands owned by private individuals. In such cases, the landowner may obtain the services of consultant foresters for growing the timber and scheduling its harvest. Then the firm's staff foresters will purchase the timber and supervise the logging. Many companies also provide timber-marking services and may offer other types of technical assistance as part of their continuing procurement programs. These foresters, too, must function in many diverse ways to provide a range of professional services to their employers and the landowners whose timber they purchase.

As foresters gain experience, they win promotion within the firms and organizations. With this advancement, they find new professional challenges, and usually experience a change in the types of work performed. It may mean less direct involvement with in-woods activities, and more planning and supervision. Here skills in management planning, business and personnel management, decision making, and other related areas begin to dominate professional activity. Consequently forestry education must provide a broad background of courses, founded in the basic sciences. But it must also prepare the forester in management, economics, business, and the social sciences. With this body of theory, the individual can move successfully through the series of different responsibilities that make up the practice of forestry.

Forestry curricula begin students with courses in basic sciences—mathematics, chemistry, botany, zoology, physics. In addition, the students take economics, English, social and political sciences, and similar basic courses. With these as a foundation, they move into ecology and silvics, silviculture, forest management, forestry economics, entomology, pathology, wildlife management, policy, administration, computer science, mensuration, biometrics, and other specialized technical courses needed for the effective management of timber resources. They also will have time for electives to strengthen their understanding in some particular aspect of forestry, or to broaden their skills. Because of the many options available, students should discuss the opportunities with a faculty adviser.

Many colleges offer advanced degrees beyond the bachelor of science, including master of forestry, master of science, master of forest science, and doctor of philosophy. Through these, individuals can strengthen their basic preparation, or develop a specialization in some area of forestry. Most advanced degrees demand independent study or research leading to the preparation of a thesis to partially fulfil the requirements, or in some cases, a special project instead. And some schools allow a student to substitute additional course work and take a comprehensive examination in lieu of a thesis. The exact requirements differ among institutions. Students will need to work closely with a major professor to determine the best approach to take.

Some schools of forestry also maintain technical training programs that offer an associate of science degree in forest technology following two years of study. Graduates will usually find employment in field-oriented positions. They supervise timber and land management activities, and often do much of the timber marking, cruising, overseeing of harvesting, and related activities. Graduates with the B.S. degree will find opportunities at the professional level with a wide range of employers. What they do depends upon who hires them, so these graduates have many options from which to choose. Individuals with the master of science, master of forestry, or master of forest science degree may also work within forestry organizations and become involved directly in timber management. But these individuals often provide specialized services related to one of the important subdisciplines of forestry. Along with persons holding Ph.D., the master of science graduates will also find opportunities in research and in teaching, in forestry extension work, or on planning staffs for companies and agencies.

Clearly one must exercise judgment in planning elective courses, especially to strengthen a particular area of interest. The kinds of electives taken determine the types of skills a person has, and that often influences the types of employment opportunities available. However, as a result of its foundations in the basic sciences and the breadth of subject areas normally required as a core for a forestry curriculum, most forestry professionals develop a capacity to function well in carrying out the many diverse activities that make up forestry practice today.

REFERENCES

1. K. A. Armson and V. Sedreika, *Forest Tree Nursery Soil Management and Related Practices,* Ontario Ministry of Natural Resources, Division of Forestry, Forest Management Branch, Toronto, 1974.

2. T. E. Avery, *Natural Resource Measurement,* 2d ed., McGraw-Hill, New York, 1957.

3. R. Barlow, *Land Resource Economics,* Prentice-Hall, Englewood Cliffs, N.J., 1958.

4. J. P. Barrett and D. S. Linden, "Using Short Programs in Natural Resource Inventories," in H. G. Lund, V. J. LaBau, P. F. Ffilliott, and D. W. Robinson (eds.), *Integrated Inventories of Renewable Natural Resources: Proceedings of the Workshop,* U.S. Forest Service General Technical Report RM-55, 1978.

5. B. M. Blum, "Regeneration and Uneven-Aged Silviculture—The State of the Art," in *Uneven-aged Silviculture and Management in the United States, Proceedings of In-service Workshop,* Timber Management Research, U.S. Forest Service, Washington, 1978.

6. F. T. Bonner, "Seed Testing," chap. VIII, in C. S. Schopmeyer (ed.), *Seeds of Woody Plants in the United States,* U.S. Department of Agriculture, Forest Service, Agricultural Handbook no. 450, 1974.

7. W. E. Bromley (ed.), *Pulpwood Production,* 2nd ed., Interstate Printers and Publishers, Danville, Ill., 1969.

8. T. W. Daniel, J. A. Helms, and F. S. Baker, *Principles of Silviculture,* 2d ed., McGraw-Hill, New York, 1979.

9. K. P. Davis, *Forest Management: Regulation and Valuation,* 2d ed., McGraw-Hill, New York, 1966.

10. R. G. Florence, "The Silvicultural Decision," *Forest Ecology and Management* 1:293–306 (1978).

11. F. C. Ford-Robertson, *Terminology of Forest Science, Technology Practice and Products,* Multilingual Forestry Terminology Series no 1, Society of American Foresters, Washington, 1971.

12. C. B. Gibbs, "Uneven-Aged Silviculture and Management? Even-Aged Silviculture and Management? Definitions and Differences," in *Uneven-Aged Silviculture and Management in the United States, Proceedings of In-service Workshop,* Timber Management Research, U.S. Forest Service, Washington, 1978.

13. H. W. Hocker, Jr., *Introduction to Forest Biology,* Wiley, New York, 1979.

14. S. L. Krugman, W. I. Stein, and D. M. Schmitt, "Seed Biology," chap I, in C. S. Schopmeyer (ed.), *Seeds of Woody Plants in the United States,* U.S. Department of Agriculture, Forest Service, Agricultural Handbook no. 450, 1974.

15. C. S. Larsen, *Genetics in Silviculture* (translated by M. L. Anderson), Oliver and Boyde, Edinburgh, 1955.

16. P. G. Langley, "Remote Sensing in Multi-stage, Multi-resource Inventories," in H. G.

Lund, V. J. LaBau, P. F. Ffilliott, and D. W. Robinson (eds.), *Integrated Inventories of Renewable Natural Resources: Proceedings of the Workshop,* U.S. Forest Service General Technical Report RM-55, 1978.

17. G. H. Lund, V. J. LaBau, P. F. Ffilliott, and D. W. Robinson (eds.), *Integrated Inventories of Renewable Natural Resources: Proceedings of the Workshop,* U.S. Forest Service General Technical Report RM-55, 1978.

18. S. R. Miller (ed.), *Timber Harvesting Methods and Equipment of Today and Tomorrow—Needs, Goals, and Limitations in New York State,* SUNY College of Environmental Science and Forestry, Applied Forestry Research Institute, AFRI Miscellaneous Report no. 2, 1960.

19. H. Nienstaedt and E. B. Snyder, "Principles of Genetic Improvement of Seed," chap. II, in C. S. Schopmeyer (ed.), *Seeds of Woody Plants in the United States,* U.S. Department of Agriculture, Forest Service, Agricultural Handbook no. 450, 1974.

20. R. D. Nyland, P. J. Craul, D. F. Behrend, H. E. Echelberger, W. J. Gabriel, R. L. Nissen, Jr., R. Uebler, and J. Zarnetske, *Logging and Its Effects in Northern Hardwoods,* SUNY College of Environmental Science and Forestry, Applied Forestry Research Institute, AFRI Research Report no. 31, 1976.

21. R. D. Nyland, D. M. Marquis, and D. W. Whittemore, "Northern Hardwoods," in *More Than One Way . . . Choices in Silviculture for American Forests,* Society of American Foresters, Washington, 1981.

22. B. A. Roach, *Selection Cutting and Group Selection,* SUNY College of Environmental Science and Forestry, Applied Forestry Research Institute, AFRI Miscellaneous Report no. 5, 1974.

23. B. A. Roach and S. F. Gingrich, *Even-Aged Silviculture for Upland Central Hardwoods,* U.S. Department of Agriculture, Forest Service, Agricultural Handbook 355, 1968.

24. R. Rogers, *Evaluating Stocking in Upland Hardwood Forests Using Metric Measurements,* U.S. Forest Service Research Paper NC-187, 1980.

25. P. O. Rudolf, K. W. Dormann, R. G. Hitt, and A. P. Plummer, "Production of Genetically Improved Seed," chap. III, in C. S. Schopmeyer (ed.), *Seeds of Woody Plants in the United States,* U.S. Department of Agriculture, Forest Service, Agricultural Handbook no. 450, 1974.

26. P. R. Russell, "Information Requirements for Timber Management: Just Part of an Integrated Information System," in H. G. Lund, V. J. LaBau, P. F. Ffilliott, and D. W. Robinson (eds.), *Integrated Inventories of Renewable Natural Resources: Proceedings of the Workshop,* U.S. Forest Service General Technical Report RM-55, 1978.

27. F. C. Simmons, *Handbook for Eastern Timber Harvesting,* U.S. Department of Agriculture, Forest Service. Northeastern Area State and Private Forestry, Broomall, Pa., 1979.

28. D. M. Smith, *The Practice of Silviculture,* 7th ed., Wiley, New York, 1962.

29. Society of American Foresters. *Timber Harvesting Guidelines for New York State,* Pamphlet, Society of American Foresters, New York Section.

30. S. H. Spurr and B. V. Barnes, *Forest Ecology,* 3d ed., Wiley, New York, 1980.

31. W. I. Stein, P. E. Slabaugh, and A. P. Plummer, "Harvesting, Processing, and Storage of Fruits and Seeds," chap. V, in C. S. Schropmeyer (ed.), *Seeds of Woody Plants in the United States,* U.S. Department of Agriculture, Forest Service, Agricultural Handbook no. 450, 1974.

32. J. H. Stoeckler and G. W. Jones, *Forest Nursery Practice in the Lake States,* U.S. Department of Agriculture, Forest Service, Agricultural Handbook no. 110, 1957.

33. R. W. Tinus and S. E. McDonald, *How to Grow Tree Seedlings in Containers in Greenhouses,* U.S. Forest Service General Technical Report RM-60, 1979.

34. U.S. Environmental Protection Agency, *Processes, Procedures, and Methods to Control Pollution Resulting from Silvicultural Activities,* U.S. Environmental Protection Agency, Office of Air and Water Programs, EPA 430/9-73-DIO, 1973.

35. U.S. Forest Service. *Silvics of Forest Trees of the United States,* U.S. Department of Agriculture, Forest Service, Agricultural Handbook no. 271, 1965.

36. Ibid., *Forest Statistics for the U.S., 1977.* U.S. Department of Agriculture, Forest Service, Washington, 1978.

37. Ibid., *An Assessment of the Forest and Range Situation in the United States,* U.S. Department of Agriculture, Forest Service, FS-345, 1980.

38. Ibid., *A Recommended Renewable Resources Program—1980 Update,* U.S. Department of Agriculture, Forest Service, FS-346, 1980.

39. Ibid., *The 1980 Report to Congress on the Nation's Renewable Resources,* U.S. Department of Agriculture, Forest Service, FS-347, 1980.

40. J. V. van Roessel, P. G. Langley, and T. D. Smith. "Timber-Pack—A Second Generation Forest Management Information System," in H. G. Lund, V. J. LaBau, P. F. Ffilliotti, and D. W. Robinson (eds.), *Integrated Inventories of Renewable Natural Resources: Proceedings of the Workshop,* U.S. Forest Service General Technical Report RM-55, 1978.

41. A. E. Wackerman, W. D. Hagenstein, and A. S. Michell, *Harvesting Timber Crops,* 2d ed., McGraw-Hill, New York, 1966.

42. J. W. Wright, *Introduction to Forest Tree Genetics,* Academic Press, New York, 1976.

CHAPTER 7

SOILS

Soil is one of the most critical of the earth's natural resources. It serves as the medium to support plant growth, and its properties profoundly affect the species that will grow in a particular place and how well they develop. Together with climate, soil determines the productivity of the land for growing food and fiber, and the uses people can make of an area. This historically has influenced the ways in which societies have developed, and where populations have thrived.

By definition, soil includes only the upper portion of natural material that covers the earth. In this context, it differs from the regolith. The regolith includes all the unconsolidated weathered rock and the soil material, whereas soil means only the layers suitable for supporting plant growth.[25] Thus soil characteristically differs from the unconsolidated material below it by virtue of the organic matter, the roots of vegetation, the degree to which the inorganic materials have weathered, and how well it has formed into a series of layers with distinct colors and characteristics.[8,22,25,26]

Soil varies in depth and character from place to place. It may range from a few centimeters to more than 3–8 meters in thickness. It contains not just solid materials, but also air, gases, water, and a host of living organisms—the abundance of which changes throughout the year. Further, as the different edaphic factors interact, the soil undergoes a constant and gradual evolution over time. This makes it a dynamic body. Thus soil scientists must learn to recognize both its physical and chemical properties, and decide how these will change. In this way, they can judge the potentials for different types of use, and the limitations.[1,8,27]

SOIL DEVELOPMENT

Soils develop by the weathering of underlying rock material through the action of physical and chemical agents. This involves freezing and thawing, the dissolving of minerals through various chemical reactions, abrasion of one rock against another, and many other forces. In the process, rock breaks apart into fragments and these disintegrate into fine particles. Chemical action also releases the component minerals, soluble salts, bases, and acids. Such weathering even takes place well below the surface. It constitutes the initial stages of soil formation, and continues within the soil mantle as an ever-present process.[5,7,33]

Vegetation and animals also modify the soil. They add organic litter and by-products. Further, the animals help to carry organic particles from the surface downward into contact with mineral soil, and they add nitrogen.[7,8,22] Additional mixing comes with freezing and thawing, movement of water through and within the soil, vibrations from earth tremors, stirring of roots when trees topple over, and other physical actions.[7] As the organic material decays, released humic acids percolate into the soil and further help to break down the minerals into chemicals used in the plants. The organic litter also protects the surface from erosion.[8,22]

Past action of glaciers, water, volcanic eruptions, wind, and gravity have all profoundly affected the earth's soils. These helped to transport surface particles from place to place, and often moved them long distances from their point of origin.[5,7,8] Thus soil formed originally from one type of rock may lie over another of different geologic origin. In the process, the rock abraded and became ground into finer particles. The abrasion and mixing also tended to intersperse small and large soil fragments, giving the soil a different texture from what would have developed only from weathering of the underlying rock. In addition, the mixing of particles from different kinds of rock blended materials of various chemical constitution. This helped to introduce material containing additional essential plant nutrients. As a result, mixed and transported soils such as loess (wind-deposited soils), alluvium (water-deposited soils), and those in moraines (glacier-deposited soils) often rank among the most productive.

The sizes of mineral particles, amounts of organic matter incorporated in the soil, the soil depth, the nutrient status, the aeration, and the moisture content all influence plant growth. But the condition of one may compensate for shortages or an overabundance of some other edaphic factor. The way they balance one another to provide a particular soil environment determines how well the plants grow, and what species occur at a site.[22] And while foresters can alter some characteristics, such as nutrient status, they must learn to identify the limitations and potentials. Then they can judge what types of vegetation will grow successfully and plan their management to capitalize upon that potential.

Soil Texture and Horizons

Soil particles range from small to large. At any one place, the sizes may be reasonably uniform, giving the soil a characteristic texture. Soils with particle sizes larger than 0.05 mm are called sands. Very coarse sands may even have particles as large as 2.00 mm. Silt soils have particles of intermediate size, ranging from 0.002 to 0.05 mm. Clay soils have smaller ones. In some soils, a mixture of sizes makes the texture intermediate. Soil scientists call such soils loams.[25,26] Because of their characteristics, the loam soils are important for agricultural use.[2]

Generally sandy soils have greater infiltration capacity. That means water can move into and through them quite readily. Such soils hold limited amounts of water. Clays hold more water, but water infiltrates these soils slowly. Additions of organic matter alter the properties, giving the sandy soils greater moisture-holding capacity and improving the porosity of clays. Consequently the presence of a vegetation cover, accumulation of organic litter under it, and incorporation of decomposed materials into the mineral soil importantly alters the infiltration and water-retention capacity of a soil.[8] For these reasons, forests have a major influence on hydrologic conditions.[9]

The percolation of water through a soil, the action of soil organisms, and other forces move solutes and decomposed organic material from the surface. This mixing and leaching may result in the formation of definitive layers or horizons within the upper soil layers.[7,8] Soil scientists use the prominence and character of these horizons, plus texture, depth, and similar features to classify soils into different types, known as series.[27] Knowing the classification helps one to decide the best ways to use a soil to grow different types of plants, and to support other uses.

The differentiation of the soil into layers occurs in a characteristic sequence of horizons. This sequence of layers and their vertical arrangement make up what scientists call the soil profile. Soils subjected to similar processes of weathering, comprised of materials of like origin, and subjected to comparable climate tend to develop much the same profile. Consequently soils at different places may look alike, thus allowing classification into well-defined taxonomic categories. Such a classification usually relies upon the presence or absence of specific soil horizons. Soil scientists identify these in the field by digging a soil pit and examining the thickness, color, texture, and other easily observed features of the soil profile.[1,8]

Variation in Soils and Vegetation

Soils under forest cover differ from those in other types of ecosystems. The forest soils usually have thicker accumulations of organic litter on the surface (Figure 7–1). In addition, the trees shade the ground and modify the mircroclimate at the surface and in the upper soil layers. As a result, forest soils contain different species and diversity of microorganisms. These act upon the mineral soil and the fallen organic materials. As the organic debris decays, the released organic acids

Figure 7–1.
Forest soils have accumulated litter on the surface, a humus layer beneath that, vertical layering in a sequence of horizons, tree roots, and many microorganisms that aid decay and decomposition. *(U.S. Forest Service.)*

wash through the soil and move the bases downward by percolation. This leaching and other processes give forest soils a character quite distinct from soils under other types of plant cover. Further, deep penetrating tree roots leave organic channels through the soil as they die. These provide pathways for water movement into the lower horizons.

As a result of these influences, soils underneath a forest will differ from soils in the open. Compared with a cultivated field, the forest soil remains cooler, holds more moisture, contains more surface organic material, shows greater signs of leaching, and exhibits a more dynamic cycling of nutrients between the vegetation and the soil material.[22] These features make forest soils sufficiently different that a specialized discipline of forest soil science has developed. And since most forestry operations do not modify the soil characteristics to any extent, compared with the extensive cultivation in agriculture, foresters must learn to recognize how the natural features of soil influence plant growth and to adapt their management to that inherent potential.

At any specific place, the texture, temperature, acidity, nutrient status, moisture conditions, and related properties of the soil limit the type and growth of vegetation that develops. The plant community does help to modify some soil features, but only gradually over long periods of time. The effect of soil can be so dominant that even at nearby places with comparable climate the vegetation may change with the soil type. Areas with materials deposited by wind, water, ice, or other phenomena may support a plant community that differs considerably from the vegetation that develops in soil resulting only from the weathering of the underlying rock. Such effects may become manifest not just in the kinds of plants, but also in their rates of growth and the ultimate sizes a species may attain.[22,33]

In areas of extreme climate, the soil itself may have less influence over the characteristics of the vegetation that grows there. Under such circumstances, effects of climatologic factors often offset those of the soil.[33] For example, soils in arctic and subarctic regions have weathered less than ones in temperate regions, and tend to be more shallow. They may also have an underlying frozen layer called permafrost. This permanent frost impedes soil drainage regardless of soil texture, reducing leaching of nutrients from the upper layers, and keeping the soils moist during summer. Permafrost also prevents root penetration, limiting the amount of soil material the plants draw upon to replenish nutrients used in photosynthesis. All of these features control the type of vegetation and how well it grows.

In contrast, tropical soils never freeze. High temperatures prevail, with little seasonal variation. Humid tropical regions also receive great amounts of rainfall, though often interrupted by distinct dry seasons. As a result of such climatic factors, soil-forming processes are more rapid than in temperate regions, and so tropical soils show advanced stages of weathering. High temperatures also help accelerate organic decomposition and turnover, and the release of nutrients. However, these leach through the soil rapidly due to the heavy rainfall and the great amounts of water that infiltrate and move through the upper horizons.

Within tropical soils, the nutrients are found primarily in a thin layer within and just below the rapidly decomposing organic material at the surface. Plant uptake from these decomposing layers results in a rapid recycling of nutrients, and plants derive little nutrition from the inorganic soil material at greater depths.[3] Consequently, as in arctic regions, the nature of the parent rock material in the tropics has less influence over the nutrient status of the soil than do climatic factors. In both cases, the climate may more profoundly affect the type of vegetation growing there, and its productivity.

SOIL CLASSIFICATION AND SURVEY

Over long periods of recorded history, people who have worked with land and cultivated it for crops have categorized soils in an effort to describe their productivity and potentials for use. Some of these early classifications are recorded on maps that date back more than 42 centuries. Written history also tells of people using manures, ashes, and sulfur in ancient Greece and Italy to amend soil properties.[8] Even in the United States, the government organized a "Soil Survey" as early as 1899, and charged the staff with the responsibility to make and interpret a detailed soil survey of the country.[25] Early records of forest soil science date back to the mid-1800s.[1]

These classifications reflect human descriptions of soil characteristics. They group soils according to common and easily recognized features. Ideally this permits soil scientists to fit any particular soil into a previously described taxon based upon readily observed physical features, and to separate any soil of interest from others in the adjoining landscape. In this way, they can identify where a soil occurs and map its geographic extent so that others can use the information in planning their management and use of the land.[25] More broadly, these soil classifications help people to organize knowledge, and to clarify relationships among individual soils and the classes of different soils subjected to classification. They also serve as a convenient device to use in recalling the properties of a soil, for learning new information about them, and for grouping soils in practical ways to facilitate planning and management.[7]

The *United States Soil Survey* follows this approach. It describes soils, classifies them by common groups, and maps their location and distribution. It also provides information about the suitability of different soils to support growth of various types of vegetation, as well as a basis for determining appropriate management practices. The survey separates soils according to suitability for supporting other uses on the basis of physical properties, even though it provides information principally for agriculture, forestry, and grazing.[26] As with any system of classification, it must reflect actual conditions encountered. Further it must work for a wide variety of potential users, since soil scientists and foresters have many different purposes for classifying soils.[27] These depend upon who needs the information.

Schemes of soil classification employ somewhat different approaches in different countries. In the United States, soil classification uses the characteristics of diagnostic horizons. The soil series represents the smallest unit of classification, and differentiation between the soils depends upon the kind and arrangement of horizons in the profile. Soil scientists separate these by color, texture, structure, consistency, acidity, and the chemical and mineralogic properties. Within a series, the soils would have essentially similar profile characteristics, except for the texture of the surface. Scientists can readily recognize these differences and similarities in the field, though some supplementary laboratory measurements often prove useful.[1,7,8,21,27]

The *United States Soil Survey* maintained by the Soil Conservation Service (SCS) uses this system of classification to identify and map soils across the country. As of 1981, it had described over 10,000 different soil series.[27] Through field investigations, SCS scientists have also mapped these soils to delineate their geographic distribution. Such information serves as the basis for making predictions about effects of land use and its potentials and limitations, and in developing strategies for good soil management.[21] Additionally private corporations and other forest landowners may make independent soil surveys to provide specific information for their own use. They may want to judge land productivity, soil trafficability, engineering characteristics, erosion potentials, and the like. This information aids in planning the wise use and management of lands, and in evaluating the forest vegetation growing on them.[27]

INFLUENCE OF SOIL ON TREE GROWTH

Soil serves as a medium for the anchorage of tree roots, and as a storehouse for nutrients and water needed to sustain the physiologic processes in trees. The surface also provides a bed for seeds to land upon and germinate, and for the new seedling to develop in. In these ways, the soil influences the growth and productivity of the forest, consistent with limitations of the climate.

During recent years, there has been some question as to whether timber harvesting has any adverse effect upon soil nutrients. Some people argue that the wood and bark contain stored nutrients, and that removal of the logs reduces the total stockpile available at the site. They especially are concerned about clearcutting and whole tree harvesting. These have become more prevalent with the growing demand for fiber products and for wood as an alternative energy source. The importance of these concerns depends upon whether additions of new supplies from weathering and atmospheric input compensate for the quantities removed in the products.

Recent investigations indicate that, over the long run, the addition of nutrients through natural weathering of the soil and rock, precipitation, dust, and biologic fixation do seem sufficient to replace most losses from timber harvesting. Available information suggests that soils in temperate forest regions usually recover naturally and can sustain common management schemes with relatively long rotations between reproduction method cuttings in a stand.[6,22,31]

The presence of certain compounds in high concentrations in the soil may also adversely affect forest productivity, or even prove toxic to trees. Effects may show up rather abruptly, or may become apparent as a gradual deterioration of the vegetation. Though uncommon in most forest soils, toxicity may result from pollution by harmful substances, and from careless application of fertilizers. It may develop if evaporation leaves a concentrated residue of harmful salts on the surface, or from other types of contamination. Even excessive accumulations of animal wastes may prove harmful to vegetation.[1,33] Unfortunately foresters can do little to ameliorate these problems. They can only wait for the condition to change naturally, or try to introduce species that will grow in soils despite the high concentrations of the toxic substance.

Nutrient Status of Soils

Trees take in nutrients through movement of water from the soil into tree roots. The water contains dissolved minerals and enters the trees in rather large quantities each year. Through recycling, large amounts of elements also fall back to the surface, leach into the soil, and once more cycle back into the trees. By this process, forests survive and grow well, even in relatively infertile soils.[22,33]

The fertility of the soil, or the status of essential nutrients available from it, affects the amounts of biomass produced in the trees. When nutrients occur in too limited amounts in a soil, or when they remain unavailable for uptake in plants, deficiencies become manifest in the form of poor or abnormal growth. When critical in a tree, these nutrient shortages will show up as a discoloration of the foliage. Earlier detection in a less acute stage requires chemical analyses of the foliage in a laboratory.[1,33] For these tests, soil scientists gather leaf or needle samples from the ends of the branches. They grind the plant materials into a fine powder, and treat it with chemicals to extract the mineral elements. Later they can analyze the solutions using machines called spectrophotometers, and measure the amounts of each element present in the foliage. If the tissues have shortages of some nutrients, the soil scientists can apply chemical fertilizers to the stand to overcome the limitations on growth caused by the deficiency. They usually test the soil as well to determine whether it lacks adequate amounts of the element, or if something inhibits its release and uptake into the trees.

Forest Fertilization

Forest fertilization has gained increased use as a means of adjusting the nutrient status of soils. Its benefits have been widely appreciated in forest nurseries for many years. More recently, research has demonstrated positive effects on tree growth and stand yields in forested areas, especially in soils with nutrient deficiencies. As a result, several corporations now fertilize in conjunction with site preparation before tree planting. The practice has gained widespread use in the Southeast with pine species grown at short rotations for fiber crops. Aerial applications have been used to fertilize old-growth Douglas-fir in the West, and established stands in the Southeast. The techniques are not highly sophisticated as yet, but will improve with experience and practice.

Uncertainties about economics have slowed forest fertilization. Users have difficulty judging the prices they will get for wood in the future, and determining whether the investment will pay off. As they gain more experience and accumulate more data about effects on tree growth, they will develop better capabilities for making these determinations. Also, forest soil scientists still need to learn a great deal about the timing of applications, the appropriate means of applying fertilizer, the best stage of stand development to treat, and the rates of application that will give adequate results most economically.[22] These programs will also gain impetus as soil scientists develop better diagnostic techniques for evaluating more precisely the response of the trees to treatment,[18] and for assessing the economic benefits.[22]

While nitrogen occurs most abundantly in forest ecosystems, it often seems to limit tree growth. Potassium and phosphorus also have proved important in many forest ecosystems. Trees use calcium, magnesium, and sulfur in large quantities, though these normally appear in sufficient supply in most forest soils. Such micronutrients as boron, copper, iron, manganese, molybdenum, and zinc seem always to occur in adequate amounts in natural forests.[22] Among all of these, nitrogen, phosphorus, and potassium most commonly are utilized in commercial fertilizer applications.[1]

Exchange of Nutrients and Soil Acidity

Soil scientists cannot easily measure the actual nutrient status of soil without elaborate and time-consuming laboratory analyses. They must judge both the amounts of different elements held in the soil and the quantities that would go into solution in the soil water. Only this latter amount becomes available for transport into plants through the root systems. Consequently the soil scientist must render a judgment about the capacity of a soil to exchange nutrients, as well as determine whether the essential ones occur in sufficient amounts to satisfy needs in the growing plants that occupy a site.

Several mineral elements go into plants as cations. These become attached to the outer surfaces of small, insoluble, nondiffusable organic and mineral particles called colloids. The soil colloids have an electrical charge different from that of the cations and hold them through this bonding. Some cations detach more readily than others. Thus both their amounts in the soil and the degree to which they go into solution influence the nutrient status for plants. This characteristic, called the cation exchange capacity (CEC), differs with the amount of colloid surface area and the acidity of water held within the soil.

Generally colloidal clay and organic matter have the greatest surface area of all soil particles, and thus the highest cation exchange capacity. But even with large amounts of these in a soil, the more acidic the conditions, the lower is the CEC. Thus, in highly acid soils, such elements as calcium, magnesium, and potassium will be less likely to go into solution. Elements such as boron, copper, manganese, and iron become more readily available in acid soils and remain largely unavailable in neutral ones if present in limited supply. Some anions, including phosphates, sulfates, and nitrates, also attach to soil colloids, and detach more readily under acid conditions. As a result, these leach fairly readily from the acidic forest soils as incoming precipitation percolates through, moves underground off the site into streams and ponds, or goes into groundwater storage.[11,22,24]

Soil scientists judge soil acidity by testing for hydrogen (H^+) ion activity. For this, they use a measure called pH. It gives a logarithmic expression of the reciprocal of the H^+ ion concentration as follows:

$$pH = \frac{1}{A_{H^+}}$$

where A_{H^+} represents the H^+ ion activity in moles per liter. The more acid the soil solution, the lower is the pH value. Pure water or a neutral solution will have a pH of 7.0. On the other hand, most forest soils run about 3.5 to 6.5, and the pH changes slightly with the season of the year. The acidity will also differ by position within the soil profile. Generally the upper layers have the greatest acidity.

A low pH of the litter and decomposed humus often characterizes soils under coniferous forests in particular.[22] Species such as the firs, pines, and spruces grow well on quite acid soils and decrease in growth with a high soil pH. In contrast, the oaks, sycamore, and tulip-poplar grow best at higher levels of pH. Most species have a range of acidity where they will survive and grow well, and the requirements of different ones may overlap each other. This may have some effect upon the species mix that will develop on some soils. Even in forest nurseries, the managers may need to apply acid-forming forms of nitrogen to keep the pH low enough for some conifers. Under extremely acidic conditions, they can add limestone to make the soil more neutral.[22]

Soil scientists obtain the most accurate index to soil acidity by measuring pH in the laboratory using a special electronic meter. They first make a mixture of soil and distilled water and then place the instrument's electrodes into the solution. From the dial, they get a direct reading of pH. For expediency in the field, foresters and soil scientists often use another approach. They place small samples of soil in a small dish, moisten the soil, and sprinkle a chemical powder on it. The powder changes color, and the hue designates the relative acidity. They get an indication of the general range of pH, but even that helps. For detailed assessments, soil scientists will often take samples from different layers of the profile. However, since most feeding roots grow within the upper 20 cm, the conditions of those layers have the greatest importance.

Soil Characteristics and Site Assessment

Survival and growth of trees also depend upon several physical aspects of soil. These include the effective rooting depth and the volume of soil from which the trees draw moisture and nutrients. Both affect tree growth, root development, and anchorage against winds. In this sense, such features as underlying rock, the presence of impervious soil layers, rock content, and a high water table may limit soil depth and volume.[10,22,33]

Soil texture and structure also influence the water-holding and drainage characteristics, the penetrability of roots, and the environment for soil microorganisms. These affect tree growth, especially for species poorly adapted to the particular site. They also influence how well a species will regenerate on it.[22,30] Especially in planning for tree planting, foresters must assess the soil characteristics to determine if the site will support tree growth of sufficient vigor to repay the investments. They use the assessments in deciding what species to use as well. Both survival and growth concern the forester. Physical characteristics of the soil can affect either one.

These needs have fostered schemes of forest site evaluation to help managers predict likely future productivity and to evaluate potentials for management investments. Where trees already grow on the site, direct assessments can often give an index of the potential productivity. Usually this involves measuring tree heights or other characteristics relative to their age. Then the forester compares these with index values derived from older stands that have grown well.[11] In other approaches, the forester or soil scientist might assess an array of soil features, plus characteristics of existing vegetation. Site quality is estimated from these.[4]

In areas lacking tree cover, such as old fields subject to reforestation, the evaluation must deal entirely with soil features. For these, foresters will use known requirements for a species in order to judge whether the soil seems appropriate for planting.[30] They often will look at soil depth, texture, moisture-holding capacity, drainage, and acidity as primary features. Slope steepness and position and the kinds and vigor of vegetation already growing on the site often give useful clues as well. All provide the forester with an indication of how well trees will grow and what productivity to expect as a result of management.

Soil scientists may supplement these field methods with more elaborate laboratory analyses of the soil for both physical characteristics and chemical status. For some applications, they also collect foliage samples in order to assess the nutrient content in the vegetation. These techniques historically have received greatest use in agricultural applications. Yet they also have value in forest nursery management. The analyses give managers vital information for planning fertilizer applications and other programs for improving and amending soils to enhance productivity.[2,21,22] In cases of reforestation, they indicate whether the site has the requisite nutrients and physical properties to sustain vigorous tree growth of the preferred species.

Soil Testing

Land grant colleges in most states maintain elaborate soil testing laboratories. These mostly serve agricultural needs within the state and operate as a service to landowners at a fairly nominal fee. In most heavily forested states, the laboratory will have access to a forest soil scientist who can make interpretations for forestry needs. Generally both agriculture and forestry use the same basic types of physical and chemical analyses. But trees and agricultural plants have different nutrient requirements and grow under different physical conditions. Thus the skills of forestry become important in deciding the implications of test results and their meaning to the economic interests of landowners.

Soil testing begins when a forester or soil scientist gathers samples to bring to a laboratory. As with other types of inventory procedures, several samples must be obtained from scattered places in a stand or field. From these, inferences are made about the general nutrient status and physical properties of the soil throughout.

To get a sample, a soil scientist will often dig a small hole and remove material from different horizons of the profile. These go into sterile bags to prevent contamination. At the same time, the soil scientist will examine the profile for such features as presence or absence of specific soil layers, their thickness, the overall depth and color of the soil, the amounts of rock, and the texture. Since most of the small feeder roots of trees grow within the upper 20 cm of the soil, the examinations focus upon these layers. Although used to some degree with established stands, this soil testing has great value in analyzing for soil management needs in tree nurseries, seed orchards, and seed production stands, and in association with reforestation.

SURFACE EROSION

The texture of the soil and the slope of land influence greatly the potential for erosion. Surficial erosion involves the detachment of particles from the soil surface, and transportation of them away from the site. In the process, surface layers gradually become lost, including the finer materials of highest fertility.[7,8,22]

Water most commonly acts as the primary agent of erosion; winds may detach and carry away soil particles, too. Water erosion occurs as a common, universal geologic phenomenon that has shaped the face of the earth over the eons of time. Even in undisturbed ecosystems, some soil erosion takes place, but slowly and at low levels.[5,7] When people upset natural surfaces in a way that intensifies the rate of soil loss, the rate of erosion accelerates.[8,25] When accelerated erosion becomes severe, water quality and soil productivity suffer.

Erosion by Water

Uncovered, exposed soils lacking an organic litter layer are most prone to erosion, especially if the topography slopes to any degree.[8,22] The greater the slope, the faster water may run across the surface, detaching and carrying away soil particles. Sandy and high-porosity soils generally erode the least. Precipitation and runoff water infiltrate them readily and percolate rapidly to the subsoil layers. This diminishes overland flow and the chance for erosion. Clays and silty soils erode most readily. They have finer particles. The water infiltrates them more slowly, increasing the likelihood of surface runoff. Wind, however, has greater erosive effects on sandy soil and generally causes much less erosion of the finer loams and clays.[8]

Soil scientists recognize three types of surface erosion—sheet, rill, and gully. Sheet erosion detaches a fairly thin and uniform layer of soil from across the surface over a wide area. That makes sheet erosion difficult to detect. In rill erosion, the water runs in tiny streamlets that form small shallow gullies irregularly dispersed across the surface. These can be seen quite readily. If the water flow becomes concentrated into distinct channels, the cutting action may intensify. Then, at scattered places, the water carves small, or even large, ravines (Figure 7–2). This is gully erosion and has the most conspicuous effect upon the landscape. However, all types of erosion may result in considerable loss of soil from the surface.

Landslides also result from movement of soil, though normally over shorter distances. They usually occur after long periods of heavy rain or snow melt. The

Figure 7–2.
Water running over the surface in distinct channels can cause gully erosion that may form deep ravines in extreme conditions. *(U.S. Department of the Interior.)*

water acts as a lubricant to reduce cohesion between soil particles, and along the surface of underlying rock. The soil becomes saturated and so liquid that it may flow like water over the bedrock or an impermeable subsoil. On slopes the soil, boulders, trees, and other vegetation may rush downhill at high velocity carrying everything in the pathway. Sometimes these landslides form semipermanent dams that block the flow of water through stream channels. This impounds water on the upstream side. Such catastrophes have occurred periodically in the United States, as with the devastation to northern California by landslides during January 1982. These caused substantial loss of property, and of several lives.

Erosion has at least two major effects. First, upper soil layers usually have the highest fertility and supply the greatest amounts of nutrients necessary for plants. Topsoils also have the better physical properties for plant growth. Their loss to erosion may adversely affect vegetation in a variety of ways.[8] Additionally the runoff waters may transport loosened soil and organic material down the slope into stream channels, rivers, and ponds. These particles become suspended in the water, adversely affecting its clarity and quality. In quiet, still waters the particles settle into the bottom, forming a layer of fine materials.[20] This process of siltation may eventually form broad, fertile deltas or flat flood plains of the type that occur along the bottomlands of the Mississippi Valley. However, siltation will also take place in reservoirs and lakes, gradually filling them. With accelerated erosion of soil, resultant siltation may even rapidly reduce the water-holding capacity of a large reservoir.[9]

Flooding and the erosion of soils in the upstream parts of watersheds do have a soil-building function. The runoff moves soil from exposed surfaces by the process described earlier. But that material goes someplace. In some cases, the soil moves only short distances before it settles onto the surface again. This usually happens as the water slows in velocity at the change of a slope, or at the bottom of a hill. The process shows up in sloped farm fields. Soil up the hill may appear light in color, like the lower subsoil layers. Toward the bottom of the slope, the soils will appear darker, like the topsoil. In these cases, the erosion has usually moved the upper dark-colored soil down the hill and exposed the lower layers at the higher elevation. If the soil moves into a stream or gets picked up by floodwaters, the particles may be transported over long distances. As the waters spread out on the floodplain, their velocity slows. Then the soil and other debris will settle out and form a layer on top of the old soil. This can enrich the land, and over many subsequent floods slowly will raise the height of the surface. The large deltas at the end of the Mississippi and other rivers exemplify this soil-building process. It all occurred at the expense of topsoil loss farther up the rivers.

Reducing Potentials for Erosion

Vegetation plays an important role in reducing soil loss to erosion.[20,22] It adds organic matter and protects the surface from direct solar radiation and the splash of raindrops. The vegetation also holds the soil in place, because the plant roots spread throughout the upper soil layers and hold it physically. The covering of organic material largely protects the surface from erosion by water and wind. The roots also help retard mud slides and mass movement of the soil mantle. Also, the incorporated organic matter increases the soil's moisture-holding capacity. Old root channels aid in infiltration and percolation through the soil. These effects help to reduce overland flow that causes surface erosion.

Activities that encourage a thrifty vegetative cover usually create good watersheds that yield the best-quality water. But wildfire, overgrazing, careless road building, poor cultivation practices, or other uses can destroy the plant cover. These may expose and compact the soil, or increase and channelize surface runoff. Then erosion increases.[20] At the same time, reduced infiltration results in more rapid runoff and increases the peak flows in streams, even causing flooding in the most severe circumstances. Programs of soil management, therefore, usually include major efforts to establish strong vegetation cover on exposed sites. In extreme cases, the watershed manager may also plan to develop structures and engineer the surfaces to reduce or direct overland flow.

Cutting of trees in itself causes little or no erosion. Rather the use of machinery during logging and the road building associated with it most likely are the cause of any damage to soil and water that results. This happens only if the logging contractors do the work improperly. On the other hand, when planned and carried out with an eye toward possible impacts upon the soil and water resources, logging becomes an acceptable activity in our forest. It can serve as a vegetation management tool essential to creating those conditions of forest cover deemed useful in influencing snow accumulation, water yield, and losses to transpiration.[1,20]

The classic experiments in the Coweeta Hydrologic Laboratory in North Carolina have provided considerable insight into planning for safe logging.[19,20] In these studies, when researchers cut all the trees on a watershed without removing the products, infiltration remained unchanged and no added surface runoff occurred. But when they installed poorly designed roadways and carelessly used heavy equipment to skid the logs, water became channelized in the skidways. Overland flow in these channels dislodged soil particles and carried them into small streams that drained the watersheds. Other work showed that with carefully engineered roadways plus measures for careful skidding, the contractors could avert potential problems.

Development of best management practices (BMPs) for logging under the Federal Water Pollution Control Act of 1972 has given logging contractors an effective means for safeguarding water quality from soil erosion. The exact measures may differ from state to state to reflect local needs and conditions, but they have a common goal—to reduce the potential for soil erosion during silvicultural operations. This includes the possible effects of logging. Frequently rather simple means suffice. For example, contractors can leave buffer strips along streams and keep machines back from the banks. This reduces soil disturbance near the water and minimizes chances of soil erosion into it. After logging, they can smooth the road and trail surfaces to reduce the channelization of water. And they can install water bars to divert the surface flow from the trails and onto the undisturbed forest litter adjacent to it. These water bars consist of shallow ditches that run from side to side across the road or trail, and usually at an angle to divert the surface water to the downhill side so that it will not flow along the road surface. Loggers can also reduce the erosion potential by removing debris from streams where it might impede the flow and cause undercutting of the stream bank. In some cases, they also will sow grasses on landings and road surfaces to establish a protective ground cover to help stabilize the soil.[23] These measures, coupled with proper design of skid trails and logging roads, provide foresters with assurance that logging and other activities will not cause harm to the forest ecosystem.[17]

Runoff and Flooding

Erosion may become especially severe during heavy storms that lead to flooding. The rapidly running surface water carries soil and organic debris into stream channels, muddying the water. As the velocity of flow through the stream and river channels increases with the rush of water, it undercuts the banks, picking up more silt and some stones. If the velocity increases sufficiently, the floodwaters may move boulders and other materials laying on the surface. Erosion loss during flooding becomes most severe on watersheds that lack a good, dense cover of vegetation.

The flooding that occurred on New Year's Eve 1933, in LaCanada Valley in California illustrates the destruction of floods. Heavy rains fell for 2½ days in the steep San Gabriel Mountains. Floodwaters rushed out of the hills, picking up silt, sand, and boulders that weighed as much as 60 tons. The water and debris wrecked homes and blocked highways. Significantly, just a few months earlier, a wildfire had destroyed the chaparral-covered mountains, laying bare 19.4 km^2. Surveys made after the floods indicated that these burned-over lands contributed 50 times more water than did nearby unburned areas.[13]

Similar floods have caused substantial loss of life and property in other parts of the United States as well (Table 7–1). Losses of valuable topsoil remain incalculable. The value of property destruction has continued to rise over the years, despite large expenditures for flood-control structures along the rivers and their tributaries. The growing destruction results at least in part from the fact that people have built their homes on the floodplains and adjacent to streams. Especially, urbanization around many cities has prompted construction of homes and other structures in flood-prone areas. When periodic flooding takes place, these areas become inundated, damaging property and disrupting most daily activities.

During earlier times, these same areas also had flooded periodically, but the lack of development along the floodplain kept the economic losses small. The receding floodwaters usually left a fresh layer of silt over the surface. At times, the rushing waters also eroded away the stream and river bank, just as the waters had worked on the ecosystem for eons of time. In cases of severe flooding, some trees and vegetation became damaged, pushed down, or covered over. And if the area

TABLE 7–1
LOSS OF LIVES AND PROPERTY TO FLOODING IN THE UNITED STATES, 1931–1974

	Lives lost	Property damage, millions of dollars
1931–1935	368	$ 187
1936–1940	607	879
1941–1945	346	605
1946–1950	306	843
1951–1955	502	2507
1956–1960	228	877
1961–1965	329	1844
1966–1974	797	3606

Source: Adapted from data by U.S. Bureau of the Census [32]. Used by permission.

flooded frequently, a vegetation community tolerant to this phenomenon gradually formed along the river. When people began to build in the lowlands, the economic losses began to mount. They will continue to rise as long as flooding takes place.

Erosion by Wind

Wind causes another type of erosion. It takes place when rapidly moving air blows over an exposed soil surface. The wind picks up loose mineral and organic particles and moves them along the surface. Or during strong winds, these particles may blow up into the air and be transported some distance away. Regions with dry climate suffer most, but soils in humid ones may also erode during dry periods. Losses occur with all types of soil, including the highly organic ones in the prairie.[8] The effect tends to be most pronounced in areas lacking vegetation or having only a sparse cover. In contrast, forested land and other places with a good cover of plants have less wind erosion because the vegetation stabilizes the surface and impedes wind movement.[29]

Wind erosion occurs commonly along the seacoast and the shores of large lakes. In these places, waves continuously wash new sand onto shore. The wind picks up the grains from the dry beaches and moves them inland. As the wind slows, the sands fall to the ground and eventually form a dune. The dunes gradually become larger and larger, until they resemble small hills. Once a dune forms, it tends to migrate inland due to continued wind action. In this case, the winds pick up particles from the upwind slope of the dune. As the velocit subsides on the leeward side, the soil falls out. Gradually the dune moves across the land, covering whatever lies ahead of it.

Sand dune migration can cause damage. In many places, it has ruined croplands, harmed recreation sites, and damaged residential land. Soil scientists have learned to stabilize the dunes by using a combination of structural devices and vegetation. The structures may include fences to slow wind movement, or surface coverings to stabilize the soil. Then, by planting grasses, shrubs, and trees, they create a more permanent cover that provides lasting protection. Since the water continues to wash up new sands, the winds move it inland. These will build up on the front side of the dunes, necessitating continuing efforts to stabilize the surface and to plant new vegetation to give permanent protection.

Foresters in Taiwan have used these dune stabilization procedures to build up new land out into the ocean. The results have been spectacular. First they laid lathing on the dunes and beaches to stabilize the surface. Then they planted grasses and Austrian pine to supplement the lathing. Within three to four years, they had the land stabilized. As the sea washed up new sand in front of the vegetation, they repeated the process until gradually the land moved farther and farther out to sea. Within a 10-year period, they had built up 8200 ha of new ground, extending about 0.5 km into the area that once had been open surf. This provides an unusual example of the way ingenuity and technical skills enabled foresters to address an important soil erosion problem and solve it to the benefit of the landowners.

Beach erosion does occur in the United States and poses important soil conservation problems in some areas. Still it does not represent even a fraction of the soil lost to wind erosion farther inland. For example, the Soil Conservation Service of the Department of Agriculture has estimated that during the period from November 1973 to May 1974, wind erosion damaged about 1.56 million hectares in the 10 Great Plains states.[14] Wind erosion became most severe in that region during the Dust Bowl days of the 1930s. Then great volumes of soil clouded the skies over much of the Great Plains, causing irreparable loss of valuable topsoil. To combat these losses, the government organized the Prairie States Forestry Project to establish shelterbelts and windbreaks from the Texas Panhandle to the Canadian border. During eight years 32,000 km of shelterbelts were planted, involving about 218,000,000 trees on nearly 31,000 farms.[15] Other efforts resulted in 9567 km in Wisconsin, 3220 in California, and 161 in the muck lands of Indiana.[29]

These belts and rows of trees and shrubs interrupt the flow of air across the surface. This slows wind erosion, improves soil water recharge, traps snow in drifts during winter, reduces surface evaporation off cultivated fields, and protects crops and livestock from hot drying winds in the summer (Figure 7–3). Shelterbelts have also proved valuable as wildlife habitats, for landscaping in the otherwise treeless plains, and in protecting homesteads. Unfortunately increased agricultural mechanization, pivot irrigation systems, crop sprays for weed control, and inadequate maintenance have resulted in a gradual deterioration or loss of many shelterbelts in recent years. Today landowners are removing as many windbreaks and shelterbelts as other people plant. This has raised new concern about wind erosion in the region.[12,13] Renewed efforts at shelterbelt planting and

Figure 7–3.
Shelterbelts of trees and shrubs on the Great Plains interrupt the winds, thus slowing surface erosion, trapping snow in drifts, and reducing surface evaporation. *(U.S. Forest Service.)*

work to maintain existing ones represent vital elements of forestry and soil conservation programs in these areas.[16]

CAREER OPPORTUNITIES IN SOILS

In many ways, the forest soil profoundly influences the ways trees grow. It also often limits how people use resources of the land. Consequently programs in silviculture depend upon an appreciation of how soil factors affect species suitability, stand productivity, and the likelihood of capitalizing upon investments in forest management. Physical soil conditions also influence machine trafficability, the suitability of a site to support structures and developments, the amount of human activity the land can take before signs of ecosystem deterioration become apparent, and many other operational matters. Overall, in both direct and indirect ways, the capabilities of a soil affect many types of decisions a forester faces in the course of routine business. And many times the potentials to realize goals of management depend upon how well the forester understands the soil as a critical element of the ecosystem, and adapts the scheme of management to the limitations of the soil and the possibilities it offers.

Soil becomes the business of foresters, not just the responsibility of a few soil scientists who have developed a specialized understanding of it. Attention to some aspect of soils influences most programs of forestry, wildlife management, watershed management, and recreation management. For this reason, curricula at forestry schools include required courses in basic soil science and the implications for management. In some schools, forestry students take the basic course together with agricultural students. They follow up with additional study to explore unique features of forest soils. Throughout they will learn about soil physical properties, their chemical status, and the ways these affect the growth of vegetation. Then they apply this knowledge in subsequent courses in silviculture, timber harvesting, forest engineering, and other areas that concentrate upon applying basic scientific information in managing resources or facilitating their use.

The federal government probably hires more soil scientists than does any other employer in the United States. The Soil Conservation Service actually does most of the soils-related work. But positions in forest soils exist within the U.S. Forest Service, the Bureau of Land Management, the Bureau of Reclamation, the Environmental Protection Agency, and some other departments. Jobs with private industry, consulting firms, university experiment stations, cooperative extension services, and some state departments of conservation also provide attractive opportunities. However, individuals wishing to work as soil scientists should plan an academic program that includes elective courses in soils beyond the basic core requirements for graduation. Faculty advisers can help them to identify ways to enhance the possibilities for a career in soil science.

Not all the preparation for work in soil science will necessarily involve soils courses. Students will need a good preparation in chemistry, physics, and laboratory technique. Other courses might include hydrology, soil physics, soil microorganisms, soil classification, and soil engineering. Persons interested in working in nutrient relations and their effects on plant growth should consider studies in plant physiology, ecology, silviculture, and other aspects of vegetation growth and management. They should place less emphasis on the engineering aspects of soils. By carefully selecting an array of courses, they can develop a sufficient concentration in soils to qualify for employment in many governmental agencies.

A real specialization in forest soil science will require study at the graduate level. In these programs, an individual would choose between the physical and chemical aspects of soils. The combination of courses would depend upon this choice. Many schools offer such advanced education leading to both the master of science and the Ph.D. degrees. Generally land grant colleges with a school of agriculture will have the broadest offerings. Students can combine studies in general soil science with specialization related to forest soils to prepare for employment with most of the organizations listed earlier. Also, they will qualify for teaching and research positions at universities, or for positions in corporate or governmental research stations.

Whether students decide to pursue careers in soil science or more generally in forestry, wildlife management, range management, or several other related disciplines, they will deal with soils. The basic course work will provide the foundations for communication with specialists. But it will also give them the basic appreciation of how conditions of soil influence the types of species that occur at a place, how well they grow, and what potentials the forest manager has for manipulating them. In cases where the owner strives to intensively manage the resources, the work may call for efforts at soil management as an important step in realizing the goal. Thus, both as a science supportive of other disciplines and in the opportunities it provides for specialization, the study of soils receives much attention in forestry. It has profound influence on the outcome of many forestry efforts.

REFERENCES

1. K. A. Armson, *Forest Soils: Properties and Processes,* University of Toronto Press, Toronto, 1977.
2. K. A. Armson and V. Sadreika, *Forest Tree Nursery Soil Management and Related Practices,* Ontario Ministry of Natural Resources, Division of Forests, Forest Management Branch, Toronto, 1974.
3. G. Aubert and R. Tavernier, "Forest Survey," in *Soils of the Humid Tropics,* Commission on Tropical Soils, National Academy of Science, Washington, 1972.
4. J. B. Baker and W. M. Broadfoot. *A Practical Field Method of Site Evaluation for Eight Important Southern Hardwoods,* U.S. Forest Service General Technical Report SO-14, 1977.
5. A. Beiser, *The Earth,* Life Nature Library, Time, Inc., New York, 1962.
6. F. H. Bormann and G. E. Likens, *Pattern and Process in a Forested Ecosystem,* Springer-Verlag, New York, 1979.
7. S. W. Boul, F. D. Hole, and R. J. McCracken, *Soil Genesis and Classification*. Iowa State University Press, Ames, Iowa, 1973.
8. N. C. Brady, *The Nature and Properties of Soils,* 8th ed., Macmillan, New York, 1974.
9. E. A. Coleman, *Vegetation and Watershed Management,* Ronald Press, New York, 1953.
10. J. D. Curtis, *Silvicultural Limitations of Shallow Soils*. U.S. Forest Service, Intermountain Forest and Range Experimental Station, Miscellaneous Publication 24, 1961.
11. T. W. Daniel, J. A. Helms, and F. S. Baker, *Principles of Silviculture,* 2d ed., McGraw-Hill, New York, 1979.
12. R. M. Davis, "Great Plains Windbreak History: An Overview," in R. W. Tinus (ed.), *Shelterbelts on the Great Plains, Proceedings of the Symposium,* Great Plains Agricultural Council, Publication no. 78, 1976.
13. B. Frank and A. Netboy, *Water, Land and People,* Knopf, New York, 1950.
14. L. Goldsmith, "Action Needed to Discourage Removal of Trees That Shelter Cropland on the Great Plains," in R. W. Tinus (ed.), *Shelterbelts on the Great Plains, Proceedings of the Symposium,* Great Plains Agricultural Council, Publication no. 78, 1976.

15. P. W. Griffiths, "Introduction of the Problem," in R. W. Tinus (ed.), *Shelterbelts on the Great Plains, Proceedings of the Symposium,* Great Plains Agricultural Council, Publication no. 78, 1976.

16. R. A. Hague, "SCS Technical Assistance in Windbreak Forestry," in R. W. Tinus (ed.), *Shelterbelts on the Great Plains, Proceedings of the Symposium,* Great Plains Agricultural Council, Publication no. 78, 1976.

17. J. N. Kochenderfer, *Erosion Control on Logging Roads in the Appalachians,* United States Forest Service Research Paper NE-158, 1970.

18. A. L. Leaf, "Where Are We in Forest Fertilization?" in E. Hegge (ed.), *Proceedings of the Workshop on Forest Fertilization in Canada,* Canadian Forest Service, Forest Technical Report 5, 1974.

19. H. W. Lull, "Ecological and Silvicultural Aspects," in V. T. Chow (ed.), *Handbook of Applied Hydrology,* McGraw-Hill, New York, 1965.

20. H. W. Lull and K. G. Reinhart, *Forests and Floods in the Eastern United States,* U.S. Forest Service Research Paper NE-226, 1972.

21. D. S. Pease, "Soil Surveys—The Basis for Management," in H. G. Lund, V. J. LaBau, P. Ffilliott, and D. W. Robinson (eds.), *Integrated Inventories of Renewable Natural Resources: Proceedings of the Workshop,* U.S. Forest Service General Technical Report RM-55, 1978.

22. W. L. Pritchett, *Properties and Management of Forest Soils*. Wiley, New York, 1979.

23. Society of American Foresters, *Timber Harvesting Guidelines for New York State,* Pamphlet, Society of American Foresters, New York Section,

24. S. H. Spurr and B. V. Barnes, *Forest Ecology,* 3d ed., Wiley, New York, 1980.

25. Soil Science Society of America, "Glossary of Soil Science Terms," *Proceedings of the Soil Science Society of America,* 29:330–351 (1965).

26. Soil Survey Staff, *Soil Survey Manual,* U.S. Department of Agriculture, Soil Conservation Service, Agriculture Handbook no. 18, 1951.

27. Ibid., *Soil Taxonomy. A Basic System of Soil Classification for Making and Interpreting Soil Surveys,* U.S. Department of Agriculture, Soil Conservation Service, Agriculture Handbook no. 463, 1975.

28. E. C. Steinbrenner, "The Techniques and Principles of Soil Surveys on Forest Lands," in H. G. Lund, V. J. LaBau, P. Ffilliott, and D. W. Robinsin (eds.), *Integrated Inventories of Renewable Natural Resources: Proceedings of the Workshop,* U.S. Forest Service General Technical Report RM-55, 1978.

29. J. H. Stoeckler and R. A. Williams, "Windbreaks and Shelterbelts," in *Trees, The Yearbook of Agriculture,* U.S. Department of Agriculture, Washington, 1949.

30. E. L. Stone, R. Feuer, and H. M. Wilson, *Judging Land for Forest Plantations in New York,* Cornell University, New York State College of Agriculture, Extension Bulletin 1075, 1970.

31. C. O. Tamm, "Nutrient Cycling and Productivity for Forest Ecosystems," in *Proceedings of the Symposium on the Impact of Intensive Harvesting on Forest Nutrient Cycling,* SUNY College Environmental Science and Forestry, Syracuse, N.Y., August, 1979, pp. 2–21.

32. U.S. Bureau of the Census, *Statistical Abstract of the United States: 1970 and 1976,* U.S. Department of Commerce, Bureau of the Census, Washington, 1976.

33. S. A. Wilde, *Forest Soils, Their Properties and Relation to Silviculture,* Ronald Press, New York, 1958.

CHAPTER **8**

WATER

The populations of industrial nations require large quantities of water. An adult may drink two to three liters daily, and use an additional 500–700 liters for other purposes. Irrigation, steam power generation, and industrial processes take more, pushing daily per capita amounts in the United States to over 6600 liters. This includes water for three types of needs: withdrawal use, consumptive use, and instream use.[2] Withdrawal use means taking the water, using it, and then returning it to a surface or ground source for reuse. Consumptive use includes losses to transpiration, natural and industrial evaporation, and discharge into locations where the water becomes irretrievable. Instream use includes on-site activities such as swimming, fishing, navigation, and hydroelectric power generation.[15] In these last cases, use does not alter the supply. It simply takes advantage of the water at some source. The other types of use actually remove water from reservoirs, lakes, ponds, streams, and natural underground storage. They reduce the amount available until replenished by precipitation, overland flow, underground recharge, or discharge following human use.

WATER USE IN THE UNITED STATES

People withdraw and use water to meet domestic needs and for mineral extraction and processing, crop irrigation, livestock watering, steam and electric generation, cooling, and many industrial purposes. Table 8–1 lists the amounts withdrawn and consumed. These data indicate that in 1975, total freshwater withdrawals in the United States averaged about 1281.3 gigaliters (GL) daily.[15] By far agricultural

TABLE 8–1
WATER WITHDRAWALS AND CONSUMPTIVE USE IN THE UNITED STATES IN 1975, AND PROJECTED THROUGH 2030.

	Gigaliters/day									
	1975		1985		2000		2030		Percent change 1975–2030	
Major use	Withdrawal	Consumptive	Withdrawal	Consumptive	Withdrawal	Consumptive	Withdrawal	Consumptive	Withdrawal	Consumptive
Domestic and commercial	108.96	27.92	122.46	31.38	140.25	35.74	178.22	44.95	64	55
Manufacturing	193.89	22.94	89.66	33.70	74.45	55.64	99.90	95.08	–48	314
Minerals	26.71	8.31	33.43	10.51	42.88	13.66	62.33	19.69	133	137
Irrigation	600.89	327.02	629.31	351.35	582.35	350.16	562.19	378.42	–6	16
Livestock	7.24	7.24	8.45	8.45	9.66	9.66	13.15	12.55	68	73
Steam electric	336.57	5.37	359.06	15.38	300.86	39.90	266.76	80.90	–21	1406
Other	7.06	4.68	8.18	5.53	9.32	6.55	12.11	8.80	86	88
Total	1281.33	403.48	1250.57	456.30	1159.81	511.32	1193.65	640.41	–7	59

Source: Adapted from U.S. Forest Service, 1980 [14]. Used by permission.

use for irrigation took the greatest volume, amounting to almost one half of the withdrawals. Electric generation and manufacturing accounted for 26 and 15 percent respectively. On the other hand, domestic use took only 7 percent of total water withdrawals. These data show that it takes a lot of water to support our lives each day, but the majority of these uses we never actually see.

Most uses do not result in water loss or consumptive use. Rather much goes back to the ground or a surface water source afterward. For example, manufacturing withdrew about 193.89 GL/day during 1975. The users later discharged about 88 percent of this. For all purposes combined, about 69 percent of the total amount withdrawn goes back to the earth. In 1975, the remaining 403.48 GL/day went to consumptive uses with no potential for later reuse. Crop irrigation took about 81 percent of that volume. Some did percolate into the soil and recharge groundwater supplies. However, most evaporated off the soil surface or transpired out of the leaves of plants directly back to the atmosphere.

Water demand for these various purposes has increased in the United States since 1900. At the beginning of the century, total daily per capita use averaged less than 2000 liters, or about one third of the amount needed in modern society. Projections of future demand (Table 8–1) suggest that withdrawals will actually decline about 9 percent by the year 2000, and thereafter will rise slightly to the year 2030. Improvements in the efficiency of water distribution, adoption of recycling to avoid pollution or to meet environmental standards, and other conservation measures will account for the reduction. On the other hand, consumptive use should increase by about 27 percent by 2000, and by 59 percent by 2030. This will come largely from increased demand for production of minerals, energy, and manufactured goods.[15] Overall the projections indicate a continually growing pressure upon available water supplies.

Fortunately the United States has generous supplies of water. Still it faces important problems with pollution, maldistribution, and inefficient use. Some parts of the country already must restrict water use to keep withdrawals in balance with supply. At other places, water occurs in excess, but its quality is threatened by careless discharge of waste, and by siltation. If these problems are satisfactorily resolved, the nation's water supplies seem sufficient to meet projected demands. Proper management of our natural watersheds likely will prove an important adjunct to the development of better systems for distributing, conserving, and safeguarding water.

Through the forseeable future, the eastern United States is likely to have few water supply problems. Water quality should also remain suitable and consumptive uses will not exceed 90 percent of the supply in any month of the year. In the north central area around the southwestern side of Lake Michigan, people probably will experience shortages in dry or droughty years by 2000. In the Southeast, only southern Florida will face significant dry-season water supply shortages. However, within many areas of the southern Great Plains and the Southwest, water supplies could become important or critical (Figure 8–1). In these regions, irrigation requirements already have depleted, or soon will deplete, groundwater

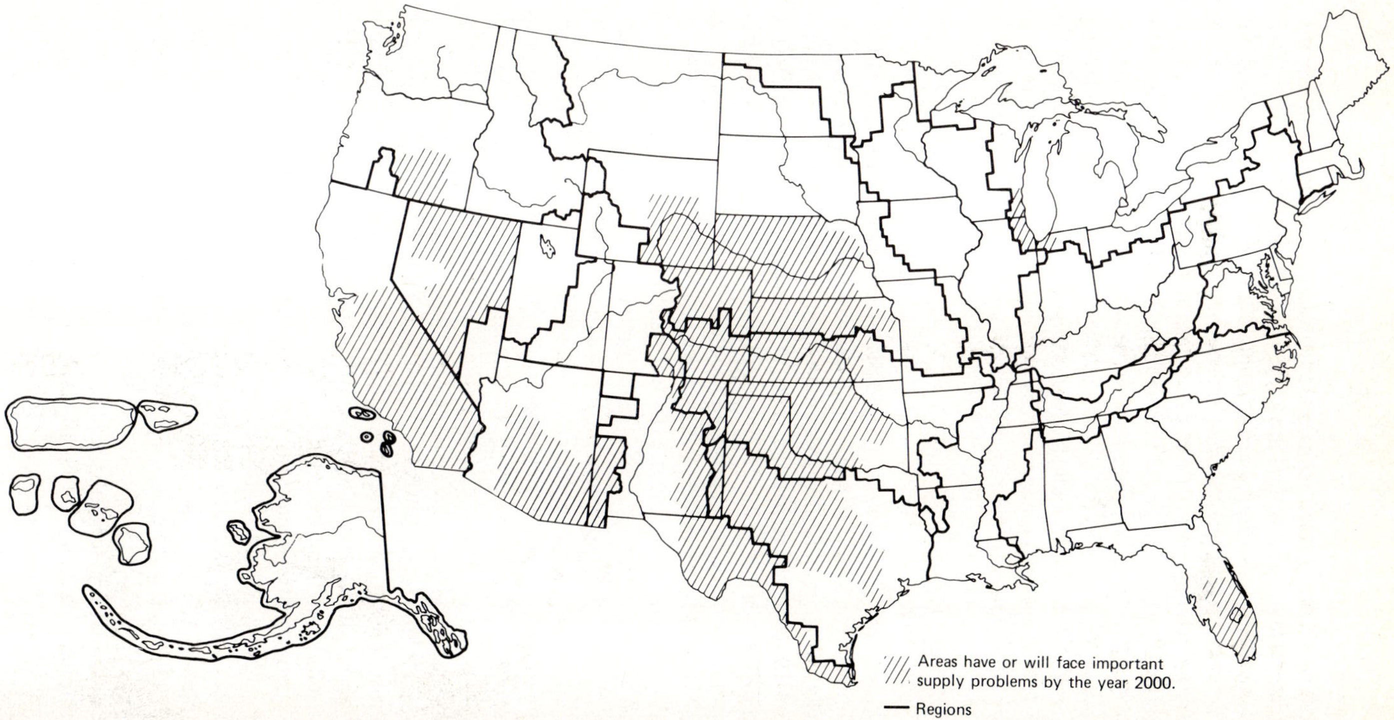

Figure 8–1.
Water resource regions of the United States, including areas with current or expected water supply problems. *(Adapted from U.S. Forest Service, An Assessment of the Forest- and Rangeland Situation in the United States. U.S. Department of Agriculture, Forest Service,* FS-345, *1980.*)

sources at a greater rate than natural recharge replenishes them during dry months. Consumptive use exceeds 90 percent of the supply, or will do so by the year 2000. In some places, withdrawal of groundwater amounts to a mining of reserves that took a long time to accumulate because of the minimal amounts of annual precipitation. Large volumes go to irrigation. As use continues to rise and supplies diminish to critical levels, the agricultural industry and related economies will suffer.[15] Thus, while the United States as a whole seems to have ample water to meet future needs, some areas face serious problems in satisfying the demand unless the government or private industry finds effective ways to redistribute the available water from regions of plenty to areas with shortages. Such matters could become major policy issues in the late 1980s and into the next century.

SOURCES OF WATER

Precipitation within the conterminous United States averages about 15,141 GL/day.[14] This amounts to an average annual input nationwide of 760 mm, but it is not uniformly distributed geographically. With the exception of the Pacific Northwest and Hawaii, eastern regions receive the most. The amounts range from 760 to 1524 mm in the Northeast to over 1525 mm along the Gulf Coast. Abundant precipitation occurs in this region during all months, though with some seasonal variation in the South. The nation's heartland through the Great Plains receives an average annual precipitation of about 510–760 mm. It falls mostly from April through August during times of active plant growth. In the northern Rocky Mountains, at some higher elevations in the southern Rockies, and along the California coast the precipitation averages between 510 and 760 mm. Most comes in winter. Along the coast of northern California, in Oregon and Washington, and northward to southern Alaska the yearly total reaches as much as 2540 to over 5080 mm. The smallest amounts fall during summer. Precipitation here averages more than at any other place in the continental United States, except at a few localities along the Gulf Coast. In Hawaii, as much as 11680 mm fall on parts of Kanai. In contrast, within the Great Basin of Utah, Nevada, Arizona, western New Mexico, and southwestern Colorado little rain falls during summer. The yearly amounts total less than 255 mm.[9]

These precipitation patterns make the United States a land of contrasts, in both climate and vegetation. Regional differences have profound effects upon water yields and potentials for development and land use. Generally, tree cover occurs in those regions that receive at least 510 mm a year, though trees will transpire greater amounts if available.[2] Regions having less precipitation will not support extensive forest cover, except in localized places along watercourses or at high elevations where cooler temperatures reduce evapotranspiration losses.

About two thirds of the water falling as precipitation returns to the atmosphere through evaporation and transpiration.[15] The remaining 5300 GL infiltrates the soil or runs into surface lakes, streams, and rivers, and ultimately into the oceans. Much of it collects temporarily in the soil and surface water bodies. This supplies the fresh water that is obtained directly from the surface sources, or by deep wells that tap underground reserves. Since 98 percent of all precipitation in the United States falls onto land, the surface conditions influence greatly the quality and quantity of water running off into ponds and streams, or percolating into the subsoil and deep aquifers.

The total land area of the United States approximates 912.5 million hectares. Forests cover about 33 percent of this and rangelands another 36 percent.[15] Croplands and pasture dominate the remainder, with smaller amounts in urban and industrial uses. Collectively the nonurban areas serve as the watersheds of the country.

About 3197 GL of water flow each day out of the United States, accumulate in storage within the soil or deep aquifers, or are lost to consumptive uses. More than 60 percent of the annual runoff originates from forested lands, and 90 percent of that water is of high quality. Nonforest range yields much less annually, and the water that flows from rangeland contains five to six times more sediment than the water from forested lands.[15] The precipitation falling on roads, highways, parking lots, and other paved surfaces in urban areas usually runs off directly into streams. Or it may flow by indirect routes through storm sewers and canals. This water may also become available for use, though often it is of undesirable quality.

Generally areas receiving less than 510 mm of precipitation annually do not yield appreciable amounts of water. Evaporation and transpiration take much of the amount that falls to earth there. Grassland communities and other types of range usually occupy these regions. Therefore they yield the least water of all our natural watersheds, except for deserts. Farmlands of the humid east also receive abundant precipitation and yield considerable amounts of water. The runoff, however, often has a high silt content due to surface erosion. Further, quantities that run off or percolate into groundwater may have high concentrations of nutrients from agricultural chemicals, fertilizers, and animal wastes. Along with urban sources of runoff, that from agricultural land requires treatment to make it suitable for human consumption and many industrial uses.

Forests make excellent watersheds. They yield water of minimal silt content and which contains few harmful contaminants (Figure 8–2). The forests also afford generous supplies of water, in part, because they receive relatively abundant precipitation or have a surplus of supply relative to losses by evaporation and transpiration. On the average, the forests of the United States receive annually about 1065 mm of water as precipitation, compared with only 610 mm for other types of land. Of this, forests yield an average of about 430 mm as runoff, while nonforest areas give up only about 100 mm annually.[13] Nationwide more than 60 percent of the annual runoff originates in forested watersheds, and in some regions the water from these sources may prove essential for human needs. In fact, more

Figure 8–2.
Forests yield generous supplies of pure water, with about 60 percent of the annual water runoff for the United States originating from forested lands. *(U.S. Forest Service.)*

than 90 percent of all usable water available in the 11 western states originates from high-altitude forests.[15] While other types of land also contribute importantly to supplies at other places, where forests occur in abundance they serve as the major watersheds.

HOW FORESTS AFFECT WATER

If you have ever sought shelter under a tree during a shower, you remember that the leafy canopy caught a substantial amount of the precipitation. If you remained under the tree very long, the water eventually dripped through the wetted foliage. If it was a light shower, you stayed dry for a long time. In a heavy shower, water came through sooner. And if the trees had no leaves, you became wet almost immediately.

The leaf cover of a forest stand can exceed by 20 times the area of land underneath it. This foliage intercepts water or snow as it falls to earth. Some of the precipitation passes through directly or drips from the leaves. More of it flows down the branches and trunks to the ground. But not all of the water reaches the soil. As much as 0.5–2.5 mm may be held in the leafy canopy and on the tree bark before the rain begins to drip through to the ground. From the tree, it becomes absorbed into the plant tissues, or evaporates directly into the atmosphere.[9] Hydrologists call this interception.

Interception

Several factors influence the proportion of precipitation intercepted and held from the ground by forest growth. The amount generally varies inversely with the frequency, intensity, and duration of precipitation. The type of vegetation also has an affect, as does the season. For example, deciduous trees intercept little during seasons when they are leafless, and when precipitation comes as snow rather than rain. Generally coniferous trees intercept more than hardwoods, and with considerable uniformity throughout all seasons. The many needlelike leaves may each capture a droplet. In total, they hold a considerable proportion of the precipitation that falls. By comparison, during a summer storm an aspen-birch type may intercept 10 percent of the rainfall, northern hardwoods 15 percent, red pine 29 percent, and a spruce-fir stand as much as 32 percent.[9]

Beneath the forest's leafy canopy, a layer of twigs and leaf litter blankets the ground. This duff layer may be rather thin, although on some sites it accumulates in thickness to 12–75 mm or more. The litter includes old organic matter in various stages of decay. Such material acts as a sponge. It breaks the impact of droplets as they strike the ground surface. It also intercepts water that drips through the canopy or melts from the snow cover. The litter holds this water until the layer becomes saturated. Then additional water will percolate through the soil.

The litter under a forest may hold as much as 2.5 milliliters (ml) of water per 25 m of litter depth. Further, one study in a region with 1245 mm of rainfall showed that average evaporation loss directly from a 25-mm-deep litter layer averaged 3.0 percent of precipitation.[1] With 914 mm of litter, the loss was 5.3 percent. In total, interception by the litter layer, coupled with that of the overhead crown canopy, may account for as much as 25 percent or more of precipitation, especially in regions where summer rains occur as light scattered showers.[9] Together interception by the crown and by litter may profoundly affect the amounts of moisture available to recharge the soil.

Fog Drip

A forest's leafy canopy will not always subtract from the amounts of moisture reaching the ground. In humid areas in which there is considerable fog, the leafy foliage may become cooler than the moisture-laden air. Then dew condenses on the leaves or needles, forms as droplets, and eventually drips through to the ground or is directly absorbed into the plant tissues. These drops from fog importantly supplement rainfall in some regions. To illustrate, near San Francisco little summertime precipitation falls. One study has indicated that over a 39-day period in midsummer about 1493 mm of water dripped off a single open-grown tan oak tree as moisture-laden fog blew through the leaves. In a closed forest under a 61-m-tall redwood tree, the dripping measured 46 mm.[10] The effect occurs in grasslands, too. One study in Ohio revealed that condensation of fog on grass plants added 100–150 mm annually to rainfall. Overall, in regions where fogs and mists frequently occur, water trapped by vegetation may double or triple the amount that falls as precipitation in the open.[6] Such additions could mean the difference between merely adequate and ample supplies of moisture to sustain plant growth.

Snow Accumulation

Forests also influence the accumulation of snow, and its subsequent melting. In contrast to rain, snow floats to earth as rather lightweight flakes. These flakes blow about in the wind so that localized patterns of air movement influence where the snow accumulates. Snow readily becomes lodged on objects. Conifer trees, especially with their horizontal branches and persistent year-round needles, intercept and hold considerable amounts. For brief periods, the crowns of conifers may trap up to about 5 mm of water equivalent. Ultimately part of this reaches the ground as wind blows the snow off the branches. Also, when crowns becomes sufficiently loaded, the snow slumps off and falls through to the ground.[1]

Deciduous trees intercept relatively small amounts of snowfall. More filters through the branches to accumulate on the ground. Therefore snow usually gets deeper under hardwood stands than under conifers. That is one reason deer congregate under conifers in winter when snow cover becomes sufficiently deep to impede their travel in the open or through hardwood stands. However, openings in conifer forests have more snow than is found underneath the trees. The eddying of winds within small openings of conifer forests may result in deeper snow accumulation there than in the open.

Evaporation of newly fallen snow is usually slight, because the whiteness reflects much of the incoming solar radiation. But after snow becomes covered with dust and debris, the surface absorbs more radiant energy. Then evaporation and melting increase, especially during warmer weather. Evaporation off the forest canopy may occur two to three times faster than from the ground surface.[8] On warm days, some moisture may even condense on the snow within the pack as warmer air blows over the crowns. Total amounts lost from the intercepted snow, therefore, depend upon the warmth of the day, the brightness of sunlight, the condition of the snow, and other factors. In total, however, the loss to evaporation may amount to 6–10 percent of precipitation.[1]

Due to interception losses, snow will not become as deep in conifer forests as in the open nearby. On the other hand, shading by the trees during springtime slows melting. This effect may lengthen the time when snow cover persists under the forest by as much as 40–50 percent. Thus a watershed characterized by a combination of openings and closed conifer forest will have snow melting at different times. Earlier yields will originate from the open spaces, whereas the forest retards melting. In effect, this prolongs the period of spring runoff, reducing peak flows in nearby streams.[2] Thus, by creating a mixture of forest conditions within a watershed, managers can influence the accumulation of snow and the duration of melting.

To determine the amount of water held in the snowpack, forest hydrologists periodically can measure the depth and moisture content. From these measurements, they can calculate the volume of stored water that will eventually run off to recharge soil moisture. The estimates also help them assess the amount of excess water that should flow into reservoirs to supply domestic and other uses (Figure 8–3). These surveys may even indicate when amounts of water held in the snow exceed the levels that can flow safely through the natural drainages. Snow surveys, then, also have value in warning of dangers from springtime flooding.

Infiltration

In an open farm field, water that strikes the soil soaks in by infiltration, accumulates in small depressions, or runs downhill across the surface. Raindrops that hit bare soil compact the surface and splash mud about. This process, called puddling, fills in the pores, and reduces the rate at which water can infiltrate. Small depressions delay the runoff and hold the water until it soaks into the soil. If the depressions overflow, the water runs across the soil surface and can pick up particles of soil. The faster the flow, the greater is its erosive action. In contrast,

Figure 8–3.
By measuring snow depth and weight using a special sampling tube as shown, hydrologists can determine the amount of water stored in the snow pack and accurately predict the volume that will run off during warm weather. *(U.S. Forest Service.)*

water that infiltrates directly into the soil will flow underground to the streams and lakes, and empty into surface water bodies in a purer form.

Forests provide ideal conditions for maximum infiltration and minimum surface erosion.[4] The tree crowns, understory vegetation, and forest litter break the force of falling drops of water. This prevents soil puddling. The litter also acts as a barrier to slow the flow of the water across the surface, thereby prolonging the time for it to soak in.[3] In the soil and decomposed litter, a host of organisms burrow through and mix organic and mineral particles. This aerates the soil, helps improve the structure, and opens channels for water percolation. Dead and decayed roots also provide channels for free water movement to a greater depth through the soil layers. As a result, soils under closed forests usually have an infiltration capacity sufficient to preclude surface runoff, even during flood-prone rainfall.[2] Further, water from forests generally causes little surface erosion and remains clear of sediment.

Forested areas often serve as natural filters to improve the quality of water flowing out of them. Chemicals in water may become bound to soil and organic particles and be taken up into the vegetation. Additionally silt and other material carried in the water from disturbed places will settle on the litter layer, leaving the freshly filtered water to move through the soil to streams and ponds. This capacity was demonstrated by experiments in Pennsylvania, where researchers sprayed municipal sewage effluent in forests over a 12-year period. Approximately 95 percent of the effluent flowed into groundwater reserves in a condition suitable for reuse for municipal purposes.[12] Nitrate-nitrogen concentrations did increase, but not beyond levels deemed safe in drinking water. As a secondary benefit, tree growth accelerated by 80 percent among mixed hardwood stands, and by 186 percent in red pine. The results suggest a potential value of forested lands in cleaning wastewater. Findings also indicate the beneficial role forests play in filtering runoff and precipitation.

Freezing and frost help keep forest soil porous and open to infiltration.[9] Generally frost occurs in two forms: concrete and honeycomb. Concrete frost binds the soil together into an almost impermeable layer that precludes infiltration of water. On the other hand, honeycomb or stalactite frost generally develops under the forest and breaks the soil apart. This opens channels for water to enter at almost the same rate as in unfrozen soil. Thus, in the spring when melting releases water from the snowpack, the honeycombed forest soil remains open to infiltration. Water released from the snow then moves into subsoil layers with minimal runoff.

The litter layer in a forest also acts as a mulch. At northern latitudes, when temperatures drop in early winter, the insulating effect of the litter layer slows soil freezing. In fact, if snow blankets the ground early enough, it may trap the heat in the soil, delay freezing, and keep the frost from getting too deep. However, as the snow does let considerable solar energy penetrate, soil may actually thaw underneath the snowpack after it has frozen. This improves its permeability to melt water in the springtime.

Evapotranspiration

Once water reaches the ground and infiltrates, some flows beneath the surface to streams and ponds. The rest remains in the soil and litter until evaporated off the surface or absorbed by plants. The portion taken into the vegetation will either be stored in the tissues, utilized in photosynthesis, or transpired out of the leaves. In combination, the evapotranspiration reduces the amount of water held within the soil and litter and the quantity that may flow through the soil to streams and ponds.

Of the heat absorbed by trees from incident solar radiation, barely 1 percent goes to produce food through photosynthesis. Another 40 percent evaporates water from the leaves. In the process, tensions develop in the foliage, pulling water from the soil through the roots. This water moves up the tree trunk to the leaves in copious amounts. Estimates suggest that of the water that falls on the 48 contiguous states, about two thirds evaporates or transpires back to the atmosphere. One estimate suggests that for every ton of wood produced, the forest takes up about 2000 metric tons of water.[9] Most evaporates directly to the atmosphere by transpiration,[11] which may account for 80 percent of water vaporization from a hardwood tract, or 60 percent in conifers.[2]

Movement of water from the soil through the plants serves several functions. First, nutrients required in photosynthesis, growth, and reproduction processes come diluted in soil water. Water transpiring from the leaves creates the negative pressure to draw this water into the tree roots and through the tissues. The loss of water through open leaf pores, called stomata, also cools the tissues and prevents overheating in intense sunlight. In fact, temperatures of rapidly transpiring leaves can be as much as 27 °C lower than the surrounding air.[9] At night and during cool periods, the stomata close. This reduces transpiration. Thus, as the plants take on water, this offsets deficits that developed during daytime.[7] At times of inadequate moisture supply, the stomata also may close and transpirational use slows. Such characteristics differ among species, making some better adapted to survive on sites with moisture deficiencies.

MEASURING WATER IN THE HYDROLOGIC CYCLE

Precipitation, infiltration, subsurface water movement, overland flow, and evapotranspiration compose a system of water inputs and outflows called the hydrologic cycle. It involves the movement of water into, within, and out of an ecosystem in liquid, solid, and gaseous forms. Further, the nature of an ecosystem depends upon the net balance between inputs and losses. Deserts result where evaporation consumes most of the input from precipitation, and forests develop at places with a net balance sufficient to support transpirational use of the trees. Marshes and ponds accumulate an excess of input compared with drainage out of them. However, worldwide the amount of water received at the earth's surface equals the losses to the atmosphere through evapotranspiration. In balance, this results in a continuous cycle between the earth and the atmosphere surrounding it.

Forest hydrologists, watershed managers, and other professionals concerned with water and its management for human benefit have developed ways to monitor the cycling of water. Through such an accounting, they have come to understand the role of water in forest ecosystems. Consequently they have devised techniques to manage forests to improve water yields, and have developed a capacity to predict amounts available for human consumption.

Plant scientists can conduct experiments to separate losses to evaporation and to transpiration. To do this, they can plant trees in a container with a cover over the soil surface. Only the plant stems and foliage protrude. Losses of water from the soil result almost entirely from transpiration through the foliage. Thus they can measure the use by periodically weighing the containers. Conversely moist soil in open containers with no plants growing in them would lose water solely to evaporation. Again they weigh the containers to measure the loss. In both cases, losses usually occur most readily from moist soil. Apparently the water loss does not take place at a uniform rate, but depends upon the amount of moisture held within the soil.

To obtain a more natural measure of the hydrologic cycle, forest hydrologists use lysimeters. In essence, these are large tanks filled with soil from the forest or grassland, and with as natural a structure as possible. The hydrologists can grow various grasses, trees, and shrubs in the lysimeters and thereby simulate a natural ecosystem. They measure rainfall reaching the surface of the lysimeter, the amounts that run off the soil surface, and that which flows from the tank beneath the ground level. By comparing these measurements, they can determine what proportions of rainfall would reach underground water supplies and how much would flow out into streams and ponds. The difference between input and outflow represents evapotranspiration use.

Data from lysimeter tests concerning water use by plants usually prove more representative of an ecosystem that do pot tests. Still results do not exactly represent what happens over large watersheds. Conditions impossible to represent by the lysimeter may alter the hydrologic cycle. Also, effects over large areas may differ from those of the rather small test plots. Consequently, for many needs, hydrologists monitor entire natural watersheds. They determine the outflow by measuring the water that flows through a weir or flume built into the stream channel at the low end of the watershed. The evaporation is the difference between precipitation falling on the watershed, and the volume running across the weir (Figure 8–4). By installing automatic recording devices, hydrologists can even detect changes in streamflow from hour to hour. Then by applying different management treatments to similar watersheds, they can determine how different conditions of vegetative cover affect water use and yields.

Results from various studies indicate that plants transpire water abundantly as long as the soil contains generous amounts.[4] This transpirational use reduces the wettness of the soil. As a result, air may enter, increasing the supply of oxygen to roots and for use by decomposing organisms. Removing tree cover from poorly drained areas may reduce transpirational loss sufficiently to raise the water table. In some cases, the soil may become so saturated that regeneration of a new stand may not occur.

APPLICATIONS IN WATERSHED MANAGEMENT

Information gleened from studies of the hydrologic cycle have taught watershed managers how to influence water yields. These practices have special importance in regions with limited amounts of precipitation and limited water supplies. For example, in southwestern United States, watershed management often involves a treatment of what hydrologists call riparian vegetation—plants that grow along streams and ponds and use large quantities of moisture in transpiration. They will not survive farther away from the streams where the soil contains less moisture during most of the year. They include shrubs and trees, as well as smaller plants, that transpire water in generous quantities, especially during hot summer days. To reduce these losses, the watershed manager may cut the riparian vegetation and

Figure 8–4.
Recording devices indicate the water level that backs up behind the V-notch, providing the watershed manager with a measure of stream flow through the weir. *(U.S. Forest Service.)*

encourage its replacement with species that use less water. This reduces the transpirational loss and increases the streamflow. However, the lush plant communities that naturally develop along streams and ponds provide essential habitat for many species of wildlife. Thus the treatment of the riparian vegetation to increase water yields can have a negative impact that conflicts with other needs and values. Managers therefore must assess the appropriateness of a treatment by balancing the different needs and uses against each other and determining how to realize an acceptable solution that provides the best overall benefit.

In moister regions that support forest cover, watershed managers can also temporarily increase stream flow by cutting the forest. The cutting reduces the amount of water removed from the soil for transpiration, thus increasing the surplus of water available to drain through the soil into streams and ponds. Greatest increases result from cuttings that remove all the trees, rather than only a part of the stand. In partially cut areas, the residual trees draw freely from soil moisture reserves and continue to transpire abundant water into the atmosphere. Clearcutting substantially reduces transpirational loss, however, leaving more water to drain through the soil. The effect does not persist. Deciduous forests of the eastern United States regenerate new trees, shrubs, and herbaceous vegetation rapidly. Within five years, transpirational losses again parallel those measured before cutting. In drier regions, regrowth is slower and the increased water yield may persist for a longer time. By being aware of the patterns of regrowth, transpirational use, and precipitation watershed managers can develop strategies for managing vegetation. In many cases, through silvicultural practices, they can effectively influence water yields and improve supplies for various uses.

The quick rise of streams and lakes during heavy storms results primarily from runoff across the surface of saturated or impervious soils. With the porous condition of forest soils, much of the precipitation infiltrates. Little runoff results. Instead the water flows through the soil. Once it infiltrates, the water moves much more slowly than it would across the surface. Thus the flow may not reach such high levels in the streams, and peak flows take longer to develop. Because of these effects, programs of watershed management often include tree planting projects designed to reforest open lands or speed regeneration following timber harvesting.

Watersheds with full vegetative cover have more porous soils than nearby cultivated fields. Forests especially have favorable watershed conditions. Their advantage was noted in an experiment in New Jersey. There a forested area absorbed 3810 mm of cannery wastewater without surface runoff. In contrast, surface flow commenced from a cultivated field after research workers sprayed on only 51 mm.[9] Such experiments have helped to demonstrate that forests reduce chances of flooding. Their removal or absence often result in heavy runoff following heavy storms.

Watershed managers can manipulate vegetation to improve snow accumulation. Indirectly this influences water yields. For example, if they thin stands, the more open crown canopy lets more snow fall through to the ground. The managers can also cut patches or strips throughout a forested watershed. As the air currents eddy in them, the snow falls to the surface. Such openings can increase snow water storage by as much as 40 percent. In addition, planting or leaving trees to shelter windswept areas will reduce air movement and cause the snow to fall to the ground. These vegetative windbreaks may also save as much as 50–75 mm of water per year from evaporation loss.[1] These benefits may prove substantial in regions that depend upon winter snow accumulation to recharge soil moisture and provide water for streams and ponds.

CAREER OPPORTUNITIES WITH WATER RESOURCES

Forestry provides many opportunities for working directly or indirectly with water, its management, and its use. These include positions in forest and range management, research, policy development, program administration, and related efforts that deal with water resources. Professional opportunities for careers directly related to water include hydrology, watershed management, public administration, and pollution abatement. Those who generally engage in managing forests, grasslands, and agricultural crops cannot avoid involvement, nor can those responsible for timber harvesting and wood products manufacture. In fact, any natural resource-related activity that can potentially influence water yields or quality must give attention to the potential impacts. Requirements of the Federal Water Pollution Control Act of 1972 make water resources the business of all these people.

The Weeks Law of 1911 included provisions to protect the headwaters of navigable streams. This helped to focus the attention of forestry upon water resources. In the years that followed, concerns for water became well developed in schools of forestry, as well as in those offering programs in engineering. This interest persists and has manifested itself in courses in hydrology, watershed management, water resources policy and administration, soil and water conservation, and watershed protection. Elements of these often appear in required soils, timber harvesting, policy, management, and administration courses to give all students at least an introduction to water resource concerns and opportunities. Elective courses include hydrology, geology, soil science, forest influences, vegetation management, meteorology, and public policy. Through these, interested individuals can strengthen their understanding and orient their academic preparation toward career opportunities related to water resources.

Many schools offer graduate-level instruction in forest hydrology and watershed management. These may feature an engineering orientation or be directed toward management. Programs in managerial science and policy also may afford opportunities for graduate studies related to water resources. Both approaches would provide studies leading to the master of science and the Ph.D. degrees. Course work would build upon the basic undergraduate preparation and allow specialization of many types. These programs usually require independent research leading to a thesis. However, for some M.S. programs the student may substitute an internship or special project, and increase the amount of course work. Such alternatives normally become available in the policy and administration areas of water resources specialization. Graduates with advanced degrees will find a wide array of employment opportunities, including teaching and research.

Federal and state agencies employ vast numbers of persons who work in soil and water programs. Resource management activities are concentrated at the national level in the Departments of Agriculture and the Interior, with notable programs under administration of the Forest Service, Soil Conservation Service, Bureau of Land Management, and others. These agencies, in turn, cooperate with companion programs in the states that are directed toward soil, forests, and water. Efforts in flood control and navigation come largely under the auspices of the Corps of Engineers, while major irrigation programs in the western United States are administered by the Bureau of Reclamation. Such agencies as the Tennessee Valley Authority and the river basin planning commissions often work across these federal and state administrative jurisdictions. Efforts at water quality management remain largely with state governments, although the federal Environmental Protection Agency has national responsibility for implementing nationwide programs. Other opportunities could be listed, but reference to these few illustrates a major point. Attention to water and soil resources has become a part of active programs at all levels of government and in private industry. As a result, many employers seek persons interested in pursuing careers related to these essential national resources. Demand for wise use and management of our soil and water should increase in future years, opening even more opportunities for professional involvement unparalleled in U.S. history.

REFERENCES

1. H. W. Anderson, "Storage and Delivery of Rainfall and Snowmelt Water as Related to Forest Environments," in J. M. Powell and C. F. Nolasco (eds.), *Proceedings of the Third Forest Microclimate Symposium,* Seebe, Alberta, Canada, 1969, Canadian Forest Service, Calgary, 1970.
2. H. W. Anderson, M. D. Hoover, and K. G. Reinhart. *Forests and Water: Effects of Forest Management on Floods, Sedimentation, and Water Supply,* U.S. Forest Service General Technical Report PSW-18, 1969.
3. N. C. Brady, *The Nature and Properties of Soils,* 8th ed., Macmillan, New York, 1974.
4. E. J. Cleary, *The ORSANCO Story: Water Quality Management in the Ohio Valley Under an Interstate Company*. The Johns Hopkins Press, Baltimore, 1967.
5. E. A. Coleman, *Vegetation and Watershed Management,* Ronald Press, New York, 1953.
6. J. Kittredge, *Forest Influences,* McGraw-Hill, New York, 1948.
7. T. T. Kozlowski (ed.), *Water Deficits and Plant Growth,* vol. 1, *Development,*

Control, and Measurement, vol. 2, *Plant Water Consumption and Response,* Academic Press, New York, 1968.

8. R. E. Leonard and A. R. Eschner, "Albedo of Intercepted Snow," *Water Resource Research,* 4:931–953 (1968).

9. H. W. Lull, "Ecological and Silvicultural Aspects, Section 6," V. T. Chow (ed.), *Handbook of Applied Hydrology,* McGraw-Hill, New York, 1964.

10. G. Y. Oberlander, "Summer for Precipitation on the San Francisco Penninsula," *Ecology* 37 (4):851–852 (1956).

11. Soil Conservation Society of America, *Proceedings: Water and America's Future,* Ankeny, Iowa. 1967.

12. W. E. Sopper, "Renovation of Municipal Wastewater for Groundwater Recharge by the Living Filter Method," in J. Tourbier and R. W. Pierson (eds.), *Biological Control of Water Pollution,* University of Pennsylvania Press, Philadelphia, 1976.

13. W. E. Sopper, *Watershed Management,* Natural Water Commission, National Technical Information Service No. PB 206670, Springfield, Va., 1971.

14. U.S. Council on Environmental Quality, *Environmental Quality.* Sixth Annual Report on Environmental Quality. Executive Office of the President, Council on Environmental Quality, Washington, D.C. 1975.

15. U.S. Forest Service, *An Assessment of the Forest and Range Land Situation in the United States,* U.S. Department of Agriculture, Forest Service, FS-345, Washington, 1980.

CHAPTER **9**

RANGELAND

Numerous archeologic relics reveal that early organized societies kept domesticated animals for food, by-products, and religious ceremonies. These show evidence of goats in Jordan as early as 7000 B.C., in Persia by 6000 B.C., and throughout much of the Middle East between 3000 and 2500 B.C. Archeologic finds also provide evidence of sheepherding and breeding in northern Persia by 6000 B.C. As early as 4970 B.C., domesticated animals served as a principal source of meat in the area now known as northern Iraq, and by 5000–4000 B.C. people had herds of both sheep and cattle in Turkistan, and the area now called Iraq and Iran. In Africa, the Tasian culture of 5000–4500 B.C. also kept sheep and goats, and by 4500–4000 B.C. Fayum people had domesticated cattle and pigs as well. Even as early as 2500 B.C., well-characterized breeds of cattle existed, and archeologists have found evidence of early domesticated cattle in India and Mesopotamia dating back to the fourth millenium B.C.[24] As skills at domesticating and breeding improved and spread, people came to depend less and less upon hunting and trapping of wild animals. Today, the domesticated herds provide the bulk of animals utilized as food and hides, for work, and for many other purposes.

Altogether almost one half of the earth's surface presently supports some type of controlled grazing. Worldwide these lands feed about 1118 million cattle, 1072 million sheep, and 348 million goats.[17] Much of the area provides for local grazing on a noncommercial scale. The bulk of commercial use remains limited principally to tropical savannas of Africa, southern Brazil, and northern Australia. A commercial industry also flourishes in temperate zone grasslands and open forests of the Andes and southern South America, Australia, New Zealand, southern

Africa, and in the southeastern and western United States and adjacent regions of Canada and Mexico. Most of these areas, however, do not lie near major population centers that will use all of the meat produced. This is true only in North America. Elsewhere commercial grazing provides a surplus for export to the more populous countries where domestic grazing cannot satisfy the demand.

GRAZING IN THE UNITED STATES

Commercial grazing actually has a relatively short history in the United States compared with other parts of the world. Records indicate that Columbus brought horses and sheep to the Caribbean area in 1492. Ponce de Leon introduced livestock to Florida in 1519, and Spanish conquistadores brought horses, sheep, and cattle to southwestern United States and Mexico in 1540. These multiplied into sizable herds that grazed the western rangelands well into the 1800s. The colonists imported additional species when they settled the eastern seaboard, and cleared forests for pasture to support grazing for food and work animals. With the westward expansion, cattle herding began to flourish in the South Central States, so that by 1850 New Orleans was the biggest beef-exporting center in the United States. However, it was not until the 1870s that the nation awoke to potentials for commercial grazing on the open rangelands west of the Missouri River.[1,23]

Prior to the boom of commercial grazing late in the 19th century, the western ranges of the continental United States supported sizable herds of wild animals. Apparently seasonal migration, predation, disease, accidents, and other natural occurrences kept their numbers in balance with the capacity of the native vegetation to support them. Yet animals were abundant. Seton estimated that about 50 million bison, 10 million elk, 40 million antelope, and 50 million deer grazed the western range.[14] One study compared these herds with current domesticated ones by using a measure called an animal unit. This equals 450 kg of live weight per animal unit, and the weight of an animal divided by 450 gives the number of animal unit equivalents in the beast. In those terms, the wild herds included about 67.1 million animal units of freely roaming animals. By contrast, in 1959 the horses, cattle, and sheep in the same general region totaled about 50.2 million animal units.[17] So, clearly the virgin range had a great meat-producing potential just among the animals that lived there as a part of the natural ecosystem.

The great expansion of western grazing changed these rangelands. This was brought to the attention of the Congress in the 1930s by a report from the Secretary of Agriculture. He advised that overgrazing had destroyed palatable and nutritious plants and reduced the vegetation density throughout the western ranges,[19] especially in Kansas, Nebraska, eastern Colorado, the Red Desert of Wyoming and Utah, and across Montana and the Dakotas. Further, conversion of prairie to farming had reduced the extent of range significantly. This change followed the building of the western railroads that encouraged both irrigated and dry land farming, and with them the plowing of vast areas of prairie and range. Farming spread especially in the Dakotas, New Mexico, northwestern Texas, and Colorado.[1] Gradually the extent of native prairie and grassland declined. And by overuse the capacities of the remaining area to support grazing diminished. The exploitation of the last century left a legacy of abuse and change that remains apparent even today.

The extensive loss of range and grasslands continued into the 1900s, and some still takes place. Statistics indicate that since 1880 the area decreased by more than 123.5 million hectares. Still, within the 50 states, remaining rangelands total about 332.0 million hectares, mostly west of the Mississippi (Table 9–1). Another 34.0 million hectares of cropland pasture also provide forage for livestock throughout the nation.[23] However, because of unsuitable climate, inaccessibility, or its use for other purposes, only about two thirds of our rangeland currently supports commercial grazing. Most of the use for livestock remains limited to the 48 contiguous states.[23]

Throughout the eastern seaboard, lands used for grazing are mostly improved pastures rather than rangelands. The rangeland has a vegetative cover comprised principally of native plants, while the pastures have undergone cultivation and seeding to introduced species. These cultivated areas, plus substantial amounts of natural grasslands in the southern and north central regions, provide considerable grazing resources east of the Mississippi. They support a major grazing industry that thrives in the South and Central States. Nevertheless, most of the nation's rangeland lies to the west. In Alaska, Arizona, Montana, Nevada, New Mexico, Texas, Utah, and Wyoming rangelands cover more than 50 percent of the land area and dominate much of the landscape. These states alone account for 45

TABLE 9–1
RANGELAND RESOURCES OF THE UNITED STATES, 1980

Region of the United States	Hectares of rangeland	Percent of total	Percent privately owned
North	59,433	*	94
North Central	669,069	0.2	83
Southeast	905,506	0.3	91
South Central	41,251,253	12.4	99
Rocky Mountains and Great Plains	165,335,498	49.8	58
Pacific Coast, including Alaska	105,947,780	31.9	8
Pacific Southwest, including Hawaii	17,816,882	5.4	42
All regions	331,985,421	100.0	46

Source: Adapted from U.S. Forest Service [23]. Used by permission.
*Less than 0.1 percent

percent of the nation's grazing resources. In total, rangelands from the Great Plains westward represent about 88 percent.[23]

As of 1976, non-federal landowners held 46 percent of the country's rangeland. Federal ownerships are mostly concentrated in the Great Plains, western rangelands, Alaska, and Hawaii. This includes about 64 percent of the rangelands in the 49 contiguous states, 97 percent in Alaska, and 31 percent in Hawaii.[23] Among these, the federal government has jurisdiction over 1 percent of the prairie grasslands, 6 percent of plains grasslands, and 16 percent of mountain grasslands. In contrast, it has 82 percent of the sagebrush lands and 70 percent of the desert shrub areas.[21] Private individuals and corporations thus own the most productive range resources, principally in the Plains States and in the South.

Nationwide the forest and open rangelands provided 213 million animal unit months of grazing in 1976. Only about 14 percent of this use occurred on federal lands (Figure 9–1). Estimates developed for the 1975 Renewable Resources Assessment suggest that future demand will be even greater. Beef and lamb consumption should rise from about 60 kg of carcass weight per person in 1976 to as much as 67 kg by 2020. On the other hand, use of lamb and mutton should

Figure 9–1.
Range includes natural open lands with plants suitable for grazing and that are not more than 10 percent stocked with trees, but characteristically these lands have soils or climate unsuitable for cultivated crops or forests. *(U.S. Bureau of Land Management.)*

decline from just under 1 kg in 1976 to less than one half that amount.[23] Imports from other countries will help satisfy some of the projected increase.

Still, demand for domestically produced meat will place great pressures upon the grazing resources of the country. In fact, Government statistics suggest that by 2020, range grazing should increase by as much as 38 percent over the 1976 level. Fortunately the nation's range has a biologic potential to support up to 566 million animal unit months of grazing, or more than 2½ times the grazing that took place in 1976. To realize the potential, however, will require more intensive range and livestock management.[23]

CHARACTER OF RANGELAND

Rangelands characteristically have rugged topography with stony, infertile, shallow, or poorly drained soil. The climate may prove unsuitable for cultivated crops or forest. Generally the natural ranges receive no more than 380–500 mm of precipitation annually, with minimal growing season rainfall. This often results in distinct wet and dry seasons in some regions, and loss of moisture to evapotranspiration leaves insufficient amounts to support more than a scattered growth of trees. The native plant communities developed over the millennia under these conditions, plus the influence of surface grazing by wild herbivores. They can provide good forage. However, the poor environmental conditions make much of the range sensitive to overgrazing, and it has deteriorated from overuse by domesticated animals.

Early settlers who moved west following the railroads attempted to farm the range across the plains. During the early 1900s, they plowed millions of hectares and sowed wheat and other crops. But often they failed because of the lack of growing-season precipitation. Those with access to water for irrigation fared better. They could grow good crops on the better soils, especially along the rivers. Farther away from the water, they could manage only wheat and other dry farming crops, plus grazing. Success required much toil, wise use of the land and its meager resources, and good management of cattle and capital. Many gave up and the land reverted to wild vegetation. Today grazing, combined with some farming for wheat and other grains, persists. Along major rivers, good crops flourish due to advances in irrigation techniques and equipment. But over most of the area, the ranchers still face frequent crop failures, uncertainties of grain and beef markets, severe weather, and other adversities. Wise and skillful management often makes the difference between success and failure.

By definition, rangeland encompasses more than the plains. Pastures and cultivated fields that landowners frequently fertilize, irrigate, or subject to other types of regular cultural treatment do not qualify as rangelands, even if grazed. Range includes the native grasslands, savannas, shrublands, most deserts, tundra, alpine meadows, coastal marshes, wet meadows, and forestland that is less than 10 percent stocked with trees. These communities have natural vegetation predominantly of grasses, grasslike plants, forbs, and shrubs suitable for grazing and browsing. They may have undergone treatments to revegetate the land naturally or artificially, but the landowner manages them similar to natural range[21] (Figure 9–1).

Rangeland in the United States includes the natural grasslands, shrub communities, and open forests of the western states and Great Plains, plus some grazing areas created from the presettlement forests of the East. The semidesert areas of the Great Basin from southwestern Arizona northward to southeastern Oregon and southern Idaho also serve as rangelands. These areas receive less than 250 mm of precipitation annually. In addition, higher elevations in the Rocky and Sierra Nevada mountains have grasslands that are used for grazing. These mountainous areas get up to 500 mm of precipitation and have lower rates of evapotranspiration due to the cooler temperatures. But throughout the western ranges, wet and dry years recur periodically. Less-than-average precipitation falls in three to four out of 10 years.[17,19]

Tall-grass Prairie

The tall-grass prairie lies west of the eastern deciduous forests from Illinois northwest to Canada, and southwestward to Texas. It extends westward to meet the short grasses in central portions of the Dakotas, Nebraska, Kansas, and Oklahoma (Figure 9–2). Rolling topography prevails throughout the region and the species of grasses present vary with the topography and moisture.

Annual precipitation typically measures less than 1010 mm in the tall-grass prairies, with dependable rainfall to support plant growth through the four- to nine-month growing season. With the deep fertile soils and favorable climate, the region supports good growth of grasses.[17] In fact, the virgin prairie had scarcely a plant species that grazing animals did not eat. These included big and small bluestem, Indian grass, wild-rye, switchgrass, sloughgrass, needle-grass, and side-oats grama.

Productivity of these lands made them the best of the western rangelands, and they covered as much as 102.0 million hectares.[19] Great herds of bison originally grazed throughout this region, but most of the land now supports farming. It comprises some of the richest cropland in America and among the finest in the world. Grazing occurs primarily on lands unsuitable for field crops.[17,21] Overall the prairie ecosystem today covers only about 15.4 million hectares, with bluestem grasses making up about 70 percent of the vegetation. Woody plants are sparse in this region.[21]

Short-grass Plains

West of approximately the 100th meridian, the tall-grass prairies disappear and the vegetation changes to sod-forming grasses that attain only shorter heights. This area reaches from the Texas Panhandle to the Canadian border, and westward to the foothills of the Rocky Mountains.[17,19] Annually 250–650 mm of precipitation

Figure 9–2.
In the central Dakotas, Nebraska, Kansas, and Oklahoma the short-grass plains and tall-grass prairie converge, providing excellent grazing in both vegetation types for an important livestock industry.

falls on the short-grass prairies, mostly between April and September. Rains generally come as showers that wet the soil only to shallow depths.[17] Moisture limits plant growth in much of the region.

Originally the short-grass prairies covered about 113.4 million hectares. Though interrupted by areas of sagebrush, wheat-grass, June-grass, and other herbaceous plants, the short-grass plains originally supported a dominant cover of grama and buffalo-grass that could withstand the drier conditions.[19] Most of the plants in these short-grass ranges are palatable and their cured leaves provide nutritious forage throughout the year.[17,19] While blue grama and buffalo-grass remain dominant, wheat-grass and needle-grass grow more abundantly in the northeastern part of the region. Shrubs such as juniper, sagebrush, rabbit bush, and mesquite occur in the northern or southern reaches. Dry farming utilizes extensive areas of the high-quality land in this region. Still, about 70.0 million hectares of rangeland remain,[21] making the short-grass plains the most important livestock range in the nation.[17]

Mountain Grasslands and Open Forest

Within the Rocky Mountains and west to the Cascades and Sierra Nevadas, and from western Montana to the Pacific, settlers found another 247.0 million hectares

Figure 9–3.
Within the Rocky Mountains and to the west open forests and interspersed grasslands provide excellent grazing throughout the year.

of bunchgrass lands that provided excellent grazing for early herds of livestock. These were interspersed with areas of open forest that covered another 53.0 million hectares (Figure 9–3). There grasses grew underneath scattered tree cover.[19] Today open range in the region comprises about 32.3 million hectares of mountain grassland, 10.6 million hectares of desert grassland, and 7.7 million hectares of mountain meadows and alpine vegetation.[21]

Only 200–500 mm of precipitation falls annually in the region. Most comes as snow in the winter. Grasses become dormant and dry out in early July and do not resprout until the following April. Yet they provide good forage even when dormant. Overgrazing has fostered a spread of sagebrush with a subsequent decline of the nutritional grasses that once covered the area.[17] Such changes reduced its capacity to sustain the amount of grazing it once supported.

Within this broad region, cattle can graze throughout the year in the low-elevation grasslands and under stands of open-grown ponderosa pine. These forests support low grasses and shrubs, interspersed with meadow areas. Higher in elevation along the central Rockies, areas with more favorable water balance support closed stands of Douglas-fir and aspen, or spruce-fir. Grassy mountain meadows and alpine zone vegetation interrupt the stands of trees, and cattle roam from one ecosystem to another foraging on a variety of plants. In combination, the open and forested areas provide good forage during summer months.[17,19]

However, timber values become high in some mountain areas and logging may dominate activity where the forest cover becomes more closed and more dominant.[17]

Desert grasslands occur in the arid, rough country paralleling the Mexican border from Texas to Arizona, and northward into Utah and Nevada. Originally desert grasses covered as much as 37.7 million hectares up the broad, flat valleys, low hills, mesa tops, and along lower mountain slopes. Rothrock grama, black grama, and curly mesquite provided the best forage, but thorny shrubs, dwarfed trees, cacti, yucca, and other zeric plants were interspersed throughout.[19] However, overgrazing over the years has increased the abundance of the shrub species at the expense of grasses. Productivity of the range has diminished and grazing occurs mostly above 1300-m elevation, where animals find nutritious shrubs and grasses. Sheep thrive in this range, but cattle remain most abundant.[17]

A different type of grazing land covers about 3.0 million hectares in the central valleys of California.[21] This region has cool moist winters and hot dry summers. Only 10 percent of the annual precipitation comes during the summer season. Here past land use has caused replacement of the original bunchgrass by annual herbs, and less than 5 percent of the herbaceous cover remains in perennials. The annual plants die and reseed annually, and green forage for grazing persists only from February to May. Nongrazing uses dominate the area.[17]

Southeastern Grasslands and Pine Forests

The southern United States, including the eastern parts of Texas and Oklahoma, has about 21.7 million hectares of grassland and prairie. These include 1.4 million hectares supporting wet grasslands, mostly in Florida's palmetto prairie and the Everglades. There sawgrass, saw palmetto, wiregrass, and sweet and red bog are the principal plant species. Precipitation averages as much as 1250 mm annually, and prevailing warm temperatures provide 200–365 frost-free days per year.[17] Throughout the combined southern region, the lands yield abundant browse and herbage. A thriving and growing livestock industry has developed, and projections indicate that demand for grazing in the region could increase by more than 1.8 times by the year 2020.

Sandy soils of the Costal Plain and Piedmont from Virginia to Texas also have pine stands of open grown character that support grasses beneath the trees. Frequent fires have helped to perpetuate the grasses and prevent invasion by shrubs and vines. Altogether about 19.9 million hectares serve for grazing and timber production. Landowners mostly graze cattle, but pigs also find adequate food in the forests. Together the beef and timber crops contribute importantly to the local economy.[17] Potentials for beef production have encouraged landowners in some areas to clear considerable forest and convert the land to grasses.

Shrub Ecosystems and Noncommercial Forest

About 95.6 million hectares of land in the Great Basin and along western Texas support five distinct shrub ecosystems. Some, such as sagebrush (about 38.2 million hectares), southwestern steppe shrub (15.6 million hectares), and Texas savanna (6.2 million hectares) have several grasses growing underneath the shrubs and in open places between them (Figure 9–4). Small areas in Texas, Oklahoma, and New Mexico also support a midgrass prairie (0.8 million hectares) with open to dense stands of deciduous shrubs. The desert shrub associations cover another 24.7 million hectares and generally have exposed soil between the bushes. Throughout the Southwest, woodland types of pinyon-juniper and chaparral occupy another 30.3 million hectares.[21]

Climatic conditions vary greatly across the regions supporting shrub associations. But most have limited precipitation that falls principally during the dormant

Figure 9–4.
Many hectares of shrub range, such as this sagebrush, lack understory plants as a result of overgrazing, thus limiting their value for cattle and wildlife until rehabilitation measures can restore the low vegetation. *(U.S. Bureau of Land Management.)*

season. In desert areas, as little as 20–50 mm may fall annually, with less in some years. Associated noncommercial woodlands generally receive somewhat more precipitation, but it is poorly distributed in the growing season.[17]

All of these associations support seasonal grazing and suffer harm from overuse. Recent comparisons with an inaccessible virgin pinyon-juniper range showed a loss of soil fertility, a shift of plant composition, a loss of some species, and soil compaction following prolonged grazing. Bare ground covered 8 percent of the surface in virgin areas, but 65 percent where continuously grazed.[2] Similar effects probably occurred in other shrub ecosystems as well.

The shrub types have increased in extent compared with virgin range. In 1936, the sagebrush-grass associations occupied about 39.1 million hectares, and even then covered about 2.6 million more than under presettlement conditions. Desert shrub associations originally occupied 20.6 million hectares and the salt-desert shrubs another 16.6 million. Collectively these shrub ecosystems covered about 76.3 million hectares. Another 38.5 million hectares supported pinyon-juniper and chaparral noncommercial forests.[19] Hence, compared with the 1930s, these shrub associations have increased by about 19.2 million hectares, while the noncommercial woodland types have declined by about 8.1 million.

Forage Production

Throughout the contiguous United States, various types of range and forest produce about 440 million metric tons of herbage and browse annually. This amounts to an average of about 907 kg/ha each year. Overall non-federal lands yield about 227 kg/ha more annually than do the government-controlled ownerships.[21] Actual production differs considerably between the various range ecosystems, but none produces at a level equal to the biologic potential. To illustrate, the differences between present production and the amount possible from areas in good condition among four principal range ecosystems include[23]:

Ecosystem	Millions of metric tons of herbage and browse produced annually	
	Present production	Potential production
Plains grasslands	60.2	79.5
Prairie	45.9	58.9
Sagebrush	41.9	54.4
Desert shrub	5.2	6.4

These four types represent the most extensive grassland and shrub ecosystems in the country and show the importance of rangeland improvement in satisfying projected increases in demand for meat production.

Of all rangeland ecosystems, the wet grasslands produce the highest levels of herbage and browse, averaging about 5783 kg/ha. These cover about 1.7 million hectares, with 83 percent of them in the South and South Central States. The desert shrub and desert grasslands produce only about 282 and 348 kg/ha respectively, and rank among the least productive of the rangeland ecosystems.[23] Generally the shrub ecosystems have lower productivity than grasslands, and the open western rangelands outproduce the forested ecosystems by about 40 percent. The productivity of Alaska's range remains largely unknown, though the short growing season would seem to limit it. Currently Alaska supports few domesticated cattle.[21] Forested ecosystems of Hawaii have a high potential and now produce over 4500 kg/ha.[23] If cleared, many forested areas of the United States would yield an average of 1.5–2.5 more herbage and browse than the natural open range. However, extensive clearing of forest seems unlikely in most regions of the nation.[23]

Though the national output of forage seems high, in 1976 only 14 percent of the rangeland was in good condition. These areas presently produce at least 61 percent of the potential for the site. Another 31 percent yields about 41–60 percent of the potential for herbage and browse. The other 142.1 million hectares remain in poor or very poor condition. Also, 11.3 million hectares still have no fire protection. However, if intensively managed, the total range could sustain more than twice the current level of grazing.[23]

RANGE MANAGEMENT

Range management involves the manipulation of range ecosystems to optimize returns and sustain production of livestock, milk, cut forage, and other benefits to society. It applies scientific principles of ecology and animal husbandry to increase the long-term returns from livestock, water, and other range resources. Generally effective programs strive both to maintain the health and vigor of the vegetation and to keep livestock use in balance with the productivity of the land. Range managers give attention to wildlife, water, and recreation resources as well. Hence balanced management programs assure production and use of high-quality forage, consistent with these other land uses and values.[6,13,17]

As early as 1913, assessments of American ranges revealed important deterioration due to overgrazing and exploitative use. Early management proposals included recommendations for controlling fire, keeping herds in balance with the productive capacity of the vegetation, prohibiting grazing, and reseeding to improve natural vegetation or speed recovery of the range.[1] Later the Secretary of Agriculture's report to Congress in 1936 suggested that most of the 291.9 million hectares of range west of the Mississippi River would regain full productivity if landowners simply controlled animal use. Nearly one seventh had deteriorated so badly, however, that closure and special rehabilitation with artificial reseeding seemed needed.[19]

As of 1976, fully 86 percent of the nation's rangelands produced less than 60 percent of their potential.[23] Long-proven range improvement practices have gained little use. For example, programs that combine vegetation and livestock management have been applied to only about 10 percent, and exploitative grazing still occurs on about 20 percent. Most range management involves informal approaches to maintaining or improving productivity of the vegetation communities.[23] Raising productivity to levels nearer the biologic potential would require more intensive practices and new technologies. These would increase the costs of grazing, probably by more than twice the current levels.[23]

Rangeland Inventory

An old adage in range management says that the condition of the livestock reflects the condition of the rangeland. Animals that feed on healthy, productive ranges gain more weight than livestock that graze on depleted areas, or on lands that lack adequate forage due to seasonal variation in the vegetation. But such informal approaches reveal little about needs or opportunities for managing the vegetation, or for making decisions about integrating grazing with other alternative uses. Consequently planners need a full range of data pertaining to vegetation, soils, water, grazing, wild animals, and other features that might influence management decisions.[9,12]

Several types of inventory aid in this process. Traditional ones include on-the-ground measurement of attributes such as plant species, density of vegetation, signs of animal feeding, and volume of plant biomass.[17] These reflect the condition of the range and indicate the nature of forage available for animal use. By adding information about soil series and water resources, managers can judge the type and density of vegetative cover the area will potentially support. By using these data, they can stratify the land into productivity classes and judge the capacity for plant growth and grazing. Further, by inventorying the numbers and migration patterns of wild animals, they can determine the amount of competition for available forage. Collectively these measurements reflect the capacity of the land to support domesticated livestock,[12] and indicate the management methods that would preserve or enhance the productivity of the range.

In addition to ground plots, rangeland inventories may also include study of small- and large-scale aerial photographs and satellite images. These can provide a fairly detailed picture of range attributes at specific places. They also give a broad view of the patterns of vegetation, water, soils, and topography over large areas. Geographers and range managers have used them successfully to separate different vegetation communities, characterize features of topography, delineate vegetation density, separate plants by species, pick out concentrations of rodent disturbance, locate animal carcasses, and identify areas improved by range rehabilitation practices.[6,8,17] When employed in conjunction with field plots that provide on-the-ground verification, the photography and imagery give range planners and managers powerful tools for amassing information needed to manage and use range resources wisely.

Deliberate inventory principally provides a means of keeping track of the nature and character of the ecosystems. It yields estimates of range condition by area, and of the quantity, quality, and juxtaposition of various necessary resources.[9] These inventories can detect watering places that exist or appear suitable for development, frequently used trails, areas where animals tend to congregate and overuse the vegetation, and places with potentials for important erosion. The data can also portray plant density and diversity, the relative dominance of different species, and the proportions of plants eaten. From this information, managers can determine the degree to which animals utilize available forage, and keep track of long-term changes in range condition.[17] Range managers and photogrammetrists must often pool their skills to design the inventory and plan for the collection of critical data. Then they can translate the data into useful information to help guide management planning.

Grazing Management

Results from more than 70 years of research by the U.S. Forest Service and various universities have given range managers the capacity to organize grazing management programs that balance demand for livestock use, water resources, wildlife habitat, and other rangeland opportunities while still maintaining a productive vegetative community.[8,13] This usually entails a twofold program that combines livestock and vegetation management. The grazing system controls animal use to promote and maintain a healthy range, while still allowing a high level of meat production.[8] It generally includes a plan for regulating the intensity, timing, distribution, and duration of grazing, and for rotating the use among various sites to prevent overbrowsing.

In general, some approach other than continuous year-round grazing will best safeguard and improve the condition of vegetation on most western ranges.[8] Ranchers should move the livestock between range areas from season to season. Further, they must determine the best time to graze a place by judging the type of vegetation present. Communities of shrubs and other plants on arid ranges withstand grazing best when dormant. Thus managers should restrict their use during the growing season. Range grasses readily regrow after cropping. They serve as good forage both when green and when dormant, and the cattle can graze upon them safely in any season. On the other hand, forbs disappear during winter, and so must be grazed during the growing season. For all these vegetation types, however, ranchers should restrict grazing during early stages of the growing season to allow the plants to resprout and maintain vigor. If this is impractical, they should rotate spring grazing from year to year so that no single place serves repeatedly as spring range.[17]

Regardless of the system employed, managers must take steps to ensure uniform grazing. Animals tend to congregate rather than to spread out across the range.[6] Such factors as the presence of water, the topography, the type of vegetation, prevailing winds, and even the kind of livestock influence the ways the animals naturally disperse.[17] Poor distribution of water frequently causes the greatest problem in effectively spreading the animals out to achieve uniform grazing (Figure 9–5). Range managers may put out salt licks, develop dispersed water resources, subdivide the range with fencing, herd the animals from place to place, and use other measures to control the animals. These must be fitted to the kind of livestock grazing the area.[8,17]

As a rule of thumb, managers do not let livestock graze off more than half of the average annual forage production. That preserves the vigor of the vegetation so it can produce seed or resprout during the next growing season. They control the use by rotating the grazing, limiting the number of grazing days, regulating the

Figure 9–5.
Development of water resources, along with fencing and other methods of herd management, enable the range manager to obtain more uniform use of the available forage. *(U.S. Bureau of Land Management.)*

number and kind of animals, and restricting the grazing season.[8,13,17,18] Through such measures, managers hope to develop and maintain healthy stands of perennial grasses where conditions of climate and physical environment permit.[13] Generally it takes an integrated, well-balanced program of vegetation management coupled with several of these livestock grazing controls to assure the most effective use of range without jeopardizing the productivity and the value of the associated resources.[8]

Range Improvement

Areas protected from grazing often will gradually revegetate to a good growth of natural plants suitable to be grazed. But in other cases, the range has deteriorated too much to make natural rehabilitation possible in a reasonable period of time. Then range managers must supplement grazing controls with special measures to enhance the process. These may include brush control, irrigation, rodent and rabbit control, elimination of poisonous plants, fertilization, and even reseeding. All facilitate reestablishment of vegetation better suited to long-term grazing.[8,13,17,18] Burning may prove useful if the range has enough fuel to sustain a fire of sufficient intensity to kill or set back the undesirable plants.[13] From among these treatments, the manager can choose those fitted to the situation and which would prove most attractive economically.

The types of range improvement practices employed depend upon the condition of the range, the physical environment, and the vegetation community involved. In general, successful range rehabilitation will require one or more practices, such as brush control, site preparation and seeding, fertilization, irrigation, and rodent and rabbit control.[13] Although some of these work in many types of range ecosystems, skilled managers will need to choose carefully among the alternatives available in order to gain full benefit and ensure success under different circumstances. They will also need to restrict grazing until the new vegetation community becomes well established and able to withstand browsing again.

Some ranges have deteriorated to a point that natural restoration would seem unlikely in the absence of supplemental treatments. Prolonged overgrazing or natural plant succession may have eliminated the desired species, creating a community with little value to livestock. In these cases, the range manager will need to take rather drastic action to establish a new collection of plants of value. This may entail efforts to eliminate the existing vegetation by burning, mechanical means, chemically, through biologic agents, or a combination of these.[17] Afterward the prepared sites may be reseeded to a useful species. The particular combination of measures used will depend upon whether the manager needs simply to modify the existing plant cover, or to replace it altogether. In some circumstances, even costly approaches such as site preparation and seeding may seem justifiable if natural rehabilitation would not likely succeed or would take too long.[13]

The way to rehabilitate overgrazed ranges will differ with the ecosystem as well as with the degree of deterioration. This depends upon the nature of the physical environment and the type of vegetation that dominates the plant community. For example, experiments in a sagebrush-grass ecosystem in Utah showed that plowing depleted area and seeding with crested wheat-grass increased the productive capacity tenfold (Figure 9–6). Alternatively the manager could prohibit spring grazing for about 15 years and artificially control the sagebrush. In combination, these would restore the vegetation to a productive status comparable to that of the artificially seeded areas.[10] In other ecosystems, converting pinyon-juniper woodland to grassland required clearing, site preparation, and seeding. These resulted in sufficiently better forage to make otherwise submarginal ranching more profitable, but importantly affected wildlife habitats. The decision to convert such sites thus requires careful coordination with wildlife needs.[13]

In ponderosa pine-bunchgrass, rehabilitation may involve site preparation, seeding, and some control of unwanted species and competing vegetation.[11] Also, managers must generally fence or otherwise control animal use to protect seeded areas for as long as two to three years. During this time, the new plants develop good root systems and become well established.[13,17] In alpine rangelands, such techniques do not work well.[3,13] Instead managers often must scarify or reshape the surface, sow the seed, and mulch to inhibit frost action or summer drying. In

Figure 9–6.
In the Great Basin clearing the shurbs plus seeding to crested wheat-grass can increase the productive capacity tenfold, or in some areas natural revegetation will follow shrub eradication if the range manager prohibits grazing for a period of years.

addition, they must sow species that colonize under harsh conditions, and sometimes even transplant seedlings from a nursery instead of relying on seeding. Other examples would portray even a wider variety of range improvement practices. But these few illustrate several typical ways in which range managers can restore depleted vegetation to improve its value for grazing.

For forested areas, silvicultural practices such as thinning and timber harvesting may help to improve forage production. By applying deliberate silvicultural systems, foresters can provide for both timber and grazing uses. For example, clearings made in southwestern ponderosa pine-bunchgrass areas have increased forage production to 680 kg/ha annually, compared with just 57 kg under stands of dense tree cover. Reducing stand densities to a basal area of less than 14 m^2/ha would also increase forage and improve sawtimber yields as well.[15] At the same time, light to moderate grazing during the growing season causes negligible damage to ponderosa pine regeneration, making it unnecessary to exclude animals during periods of forest regeneration.[5] Similar practices should have benefit in other forest types. However, heavy grazing in the deciduous forest of the eastern United States may damage timber, soil, and water resources.[22] There good management must often exclude exploitive grazing to prevent damage to the forest ecosystem.

EFFECTS UPON OTHER RANGE RESOURCES

Good range management usually provides protection to soil and water resources and also maintains productive vegetation of economic value. Practices that contribute to good range use will also reduce the potential for soil erosion and improve watershed values. For example, the lush vegetation in riparian zones along watercourses is attractive to livestock. Unless restricted, the animals will tend to overuse these areas, reduce the density of plants, disturb the ground surface, and compact the soil. These may trigger erosion directly into streams and adversely affect water quality.[13] Also, heavy animal trampling there and farther upslope, plus compaction of the soil by rain splash between the sparse vegetative cover, can reduce the infiltration capacity. Under such conditions, surface runoff will increase and speed up, as will the potential for soil movement during periods of rainfall and snow melting.[13,17] The effects become manifest as higher, shorter-duration peak flows in the streams—which magnifies the chances of flooding. Further, reduced infiltration results in less soil moisture storage, exaggerating the dryness of these sites.[17] All of these reduce the quality of the watershed and adversely affect the water resources. With proper grazing management, however, the deterioration becomes less likely, and the more thrifty vegetation lends greater protection to soil conditions.

While livestock grazing has great economic value, rangelands also have importance as wildlife habitats. As people seek more opportunities to view and hunt wild animals for recreation, these give the wildlife economic value, too. The issue has intensified where range improvement practices alter the vegetation to the detriment of wild animal populations. One important debate centers around the clearing of shrub and woodland communities in the Southwest to transform these ecosystems into grasslands. Alternatives such as converting the pinyon-juniper in alternate strips or patches, leaving northeastern slopes and steeper topography in existing cover, and limiting clearings to no more than one-quarter mile wide have helped to resolve the conflict.[13] Still, more comprehensive planning will be needed to assure multiple values from rangelands and to balance demands and opportunities to serve a wide spectrum of interests.

Interests in recreation, expanded mining, and grazing often conflict over the use of land. The mobility provided by off-road vehicles, and the damage these do to the delicate plant communities, introduce real controversy between recreationists and persons involved with grazing and preserving the range. Additionally, increased awareness of the nation's dependence upon foreign oil has given impetus to exploitation of the coal resources underlying the rangelands of the short-grass prairies and other areas. Stripping the land surface to expose those reserves temporarily would destroy large areas of natural grasslands. The mining companies would need to undertake major reclamation projects to reestablish an appropriate vegetative cover and restore the original contour of the land. Solutions to such questions concerning the utilization of these rangelands necessitate wise and comprehensive planning to identify effective ways to integrate the uses of the land and sustain its productivity over the long run.

CAREER OPPORTUNITIES RELATED TO RANGE MANAGEMENT

In 1905, the Congress transferred the Forest Reserves of the United States to the Department of Agriculture. In that reorganization, the Forest Service became responsible for management of the grazing resources on National Forests. In many places, the forage had proved more valuable than the timber, but federal agencies exercised no control over grazing. Consequently the Forest Service had to establish a permit system and to charge fees for grazing privileges. It assigned grazing areas to each permittee and determined how many animals the range would support. Such programs continue, and governmental foresters become involved with them.

Forest resource managers working in governmental agencies at the state and national level in many western and southern states inevitably find themselves involved with these and similar range resource programs in some way, and with efforts to integrate forest and range uses. Water, wildlife, and the recreational use of rangelands call upon the skills of hydrologists, watershed and wildlife managers, recreation specialists, biometricians, persons skilled in photogrammetry, economists, ecologists, plant scientists, and many other disciplines related to forestry.

Land grant colleges and universities have special involvement with range resources and their management. These institutions provide both the research and the education critical to the future mangement of rangelands. Faculty and staff have contributed significantly to the understanding of vegetation and livestock management, as well as of the relationships with wildlife, water resources, and recreation. They provide educational opportunities to students who choose courses of study focused upon range resources, and they conduct public education programs for ranchers and others directly involved with the use of rangelands.

Centers of range management research and education are found mostly in the western states where rangelands are an important part of the landscape. Persons wishing to study in range science or to emphasize range management as a part of their forestry education should seek out educational opportunities in those parts of the country. Many schools of forestry offer at least introductory courses in range management. Land grant colleges with schools of agriculture may even offer opportunities for students to strengthen skills in the plant and animal sciences that form the foundations for range management.

For the most part, forestry students will look for courses in range management to supplement their general education. This will help them to fulfil responsibilities if they obtain employment in regions where livestock grazing has high economic value. At the same time, those interested in wildlife, water, soils, meteorology, biometrics, and other disciplines may also find opportunities to apply their specialized skills to rangeland resources. All of these individuals will need to develop and maintain communication with range managers and ranchers and to work jointly with them in programs of comprehensive planning for range resources. The basic courses taken during undergraduate studies will often make this communication more effective.

REFERENCES

1. W. C. Barns, "Western Grazing Grounds and Forest Ranges." *The Breeder's Gazette,* Sanders, Chicago, 1913.
2. C. Baxter, "A Comparison Between Grazed and Ungrazed Juniper Woodland," in *Ecology, Uses, and Management of Pinyon-Juniper Woodlands,* U.S. Forest Service General Technical Report RM–39, 1979.
3. R. W. Brown and R. S. Johnson, "Rehabilitation of Disturbed Alpine Rangelands," in *Proceedings of the First International Rangeland Congress,* 1978, pp. 704–706.
4. M. Clawson, *The Western Range Livestock Industry,* McGraw-Hill, New York, 1950.
5. P. O. Currie, C. B. Edminster, and F. W. Knott, *Effects of Cattle Grazing on Ponderosa Pine Regeneration in Central Colorado,* U.S. Forest Service Research Paper RM–201, 1978.
6. R. S. Driscoll, *Managing Public Rangelands,* U.S. Forest Service AIB–315. 1967.
7. Ibid., *Color Aerial Photography—A New View for Range Management,* U.S. Forest Service Research Paper RM–67, 1971.

8. S. L. Ellis, "Shrubland Classification in the Central Rocky Mountains and Colorado Plateau," in *Integrated Inventories for Renewable Natural Resources: Proceedings of the Workshop,* U.S. Forest Service General Technical Report RM–65, 1978.
9. R. E. Francis, "Current Grassland Inventory Methods—Compatibility Toward an Ecological Base?" in H. G. Lund, V. J. LaBau, P. F. Ffilliot, and D. W. Robinson (eds.), *Integrated Inventories for Renewable Natural Resources: Proceedings of the Workshop,* U.S. Forest Service General Technical Report RM–55, 1978.
10. N. C. Freschknecht, "Effects of Grazing, Fire, and Other Disturbances on Long-Term Productivity of Sagebrush-Grass Range," in *Proceedings of the First International Rangeland Congress,* 1978, pp. 633–635.
11. H. F. Heady, *Range Management,* McGraw-Hill, New York, 1975.
12. F. E. Kinsinger, "Data Requirements for Management of Rangelands," in H. G. Lund, V. J. LaBau, P. F. Ffilliot, and D. W. Robinson (eds.), *Integrated Inventories for Renewable Natural Resources: Proceedings of the Workshop,* U.S. Forest Service General Technical Report RM–55, 1978.
13. H. A. Paulsen, Jr., "Range Management in the Central and Southern Rocky Mountains," U.S. Forest Service Research Paper RM–154, 1975.
14. S. T. Seton, *Lives of Game Animals,* vol. 3, *Hoofed Animals,* Literary Guilde of America, New York, 1909.
15. G. H. Schubert, "Silviculture of Southwestern Ponderosa Pine: The Status of Our Knowledge," U.S. Forest Service Research Paper RM-123, 1974.
16. Society for Range Management, *A Glossary of Terms Used in Range Management,* Society Range Management, Portland, Oregon, 1964.
17. L. A. Stoddard, A. D. Smith, and T. W. Box, *Range Management,* McGraw-Hill, New York, 1975.
18. G. H. Turner and H. A. Paulsen, Jr., "Management of Mountain Grasslands: Status of our Knowledge," U.S. Forest Service Research Paper RM–161, 1976.
19. U.S. Forest Service, *The Western Range,* letter from Secretary of Agriculture, U.S. Department of Agriculture, 74th Congress, 2d Session, Senate Doc. no. 199, 1936.
20. Ibid., *Annual Grazing Statistical Report 1968,* U.S. Department of Agriculture, Washington, 1969.
21. Ibid., *The Nation's Renewable Resources—An Assessment,* U.S. Department of Agriculture, Forest Service, Washington, 1976.
22. Ibid., *Final Environmental Statement and Renewable Resources Program, 1977–2020,* U.S. Department of Agriculture, Forest Service, Washington, 1977.
23. Ibid., *An Assessment of the Forest and Rangel and Situation in the United States,* U.S. Department of Agriculture, Forest Service, FS–345, Washington, 1980.
24. F. E. Zeuner, *A History of Domesticated Animals,* Harper & Row, New York and Evanston, 1963.

CHAPTER **10**

WILDLIFE AND FISH

Wild birds, mammals, and other animals have a real but largely indirect influence on human welfare in the United States. Most people do not depend upon them for food, or as beasts of burden. Domesticated livestock have long supplanted dependence upon wildlife for these purposes for most people. Yet society finds value in wild animals through recreational pursuits, and as indicators of the ecologic health of the land. Wild animals also are a critical element of forest, range, and water ecosystems. They contribute to the function, integrity, and stability of these natural systems.

Current statistics indicate that the resident and migratory vertebrate species in the forest, range, and inland waters of the United States include 175 amphibians, 1000 birds, 1000 fishes, 430 mammals, and 300 reptiles.[29] Altogether, within the 48 contiguous states, the large mammals such as deer, elk, moose, bighorn sheep, mountain goats, and bear totaled about 5 million animals in 1975. Songbirds, lesser mammals, upland game birds, reptiles, amphibians, and fishes occurred in untold numbers.[28] Additionally, invertebrates by the billions coexist with these larger creatures and fulfil many inconspicuous functions critical to the health and integrity of the ecosystems in which they live.

All of these animals depend upon land and water resources for their habitat. About one half of the vertebrates essentially live in terrestrial ecosystems, and approximately one third in aquatic ones. Some tolerate a variety of conditions and can live in more than one ecosystem. Of the terrestrial species, about 1831 live on rangelands and 2536 in forested habitats. About 1451 species can use both effectively.[29] In regions where the different kinds of habitat adjoin each other, these species may roam back and forth to satisfy their needs for food, cover, and other essentials.

VALUE OF WILDLIFE

In the United States, we do not depend upon wild animals for meat production, though hunting does provide supplemental food for some families in rural and remote areas of the country. On the other hand, free-swimming fishes of the seas and inland waters still serve as a vital source of food, as they do throughout the world. In fact, since 1950 the nation's commercial fishing has taken just under 2.3 million metric tons annually, mostly from waters off the continent.[29] Commercial ventures to raise fish for food in ponds and other controlled habitats produced about 22,700 metric tons.[28] Other direct commercial income from wildlife accrued primarily from trapping 24 species of furbearing mammals. In 1975–1976, trappers took 13 million pelts from these animals, worth about $123 million. This represented a twofold increase in the number taken just five years earlier.[29]

While saltwater fishing does have considerable economic value, commercial exploitation of wild animals and inland fishes has generated only limited personal or business income. Instead many people spend and earn money related to wildlife directly or indirectly through recreational pursuits. To illustrate, federal estimates suggest that in 1970 about 4,500,000 people devoted time to photographing wildlife, and another 7,000,000 engaged in bird watching.[28] In 1974, direct expenditures for the enjoyment of upland nongame birds totaled about $500 million, probably accounting for one sixth that spent by all hunters.[25] In 1970, the 14.3 million hunters spent nearly $2 billion on their sport. By themselves, the 2.9 million who hunted waterfowl spent about $10 per day, which amounted to $245 million. In addition, nearly 21 percent of the nation's population over the age of 12, or about 33 million people, fished for sport, spending another $5 million.[28] Thus collectively, leisure activities related to wildlife brought $5 billion in business to the recreation industry of the United States in that year.[29] Further, in 1975, the market value of meat from deer alone was worth $134 million to hunters, and that of fish caught equaled $1.3 billion. These represent the amounts that would have been spent to purchase equivalent quantities of meat at prevailing local prices.[29]

The benefits of wildlife include other noncommercial ones difficult to describe in tangible terms. These social values accrue from the way wild animals help to make the land more interesting, esthetically enhance our surroundings, and influence our feelings and attitudes—values to which we cannot assign a specific dollar value. We can only speculate about the worth by considering what we would miss if wild creatures disappeared from our surroundings, or if their number and diversity diminished to an important degree. In this same regard, some attribute great worth to just knowing that wild animals roam the nearby forests, ranges, and inland waters. They feel that wild creatures help us to recall our cultural heritage and the struggles of our ancestors to develop the land. Some believe that wildlife helps us to appreciate the fundamental organization of plants and animals in the different ecosystems that surround us. Further, by exercising self-restraint in hunting and fishing, others learn the importance of using resources wisely.[20] For most people, this intangible worth of wildlife is attributable to their culture. Their feelings reflect a general attitude of society, and develop and grow through their upbringing and experiences.[14,15]

Wild creatures also have biologic and scientific worth. Several examples help to illustrate this. Biologic benefits accrue from the feeding by wildlife on other vertebrates and insects. This helps to control the population of the prey species, and to maintain a balance within the ecosystem. The predation also may regulate pest species that bother people or their commerce. Wildlife helps to sanitize the landscape by eating the remains of dead animals. It aids in soil building, and carries out other functions vital to maintaining the viability and stability of natural and altered ecosystems. In addition, some wild creatures represent a genetic resource that breeders potentially could use for hybridization and domestication. Wildlife also has value in the study of animal behavior and of population dynamics. Further, it may serve as an early-warning signal of changes in environmental quality.[14,15]

The potential of wildlife as an indicator of environmental quality has gained considerable attention among scientists and other observers. Many feel that by monitoring wildlife, they can detect a deterioration of the physical environment that might prove important to us all. For example, sewage, silt from erosion, pesticides, and thermal pollution can kill wildlife directly, interfere with its food sources, alter other critical elements of the habitat, and affect animal reproduction systems.[2] As any of these threats seriously impact an ecosystem, the effect becomes manifest in an eventual change in the abundance and species composition of the animal life.

Wild animals have special value in this regard, since various species react differently to environmental change. Some tolerate only limited alteration of the habitat and become endangered or threatened fairly readily (Figure 10–1). Both a species' behavior and its physical sensitivity influence its adaptability, and therefore its survival.[21] Thus the loss of a particularly sensitive species might give early warning of an environmental change due to abuse, overuse, or careless exploitation [7,12,28] The disappearance of several species might signal an imminent hazard to people, or to domesticated animals.

Future shifts in our wild animal populations will certainly occur, further altering the abundance of many species and threatening some. Accelerated danger to wild animals has come primarily from land development.[12] But public sentiment, as reflected in the Endangered Species Act of 1973, has established a national policy of preserving each species of wild animal and plant, and maintaining existing species diversity across the nation. Although the program recognizes that some localized shifts and losses will take place, it gives primary attention to

Figure 10–1.
The distribution and abundance of a species such as these elk in Yellowstone National Park indicate the extent and suitability of desirable habitat, whereas loss of a species from an area may signal an important change in the quality of the natural ecosystem that sustains it.

monitoring the habitat for species considered endangered or threatened, or that have shown great sensitivity to changes in habitat condition. Constant monitoring will keep track of these animals, and provide an early warning of the condition of our natural ecosystems.

As of 1978, federal lists included 195 species considered threatened or endangered, with 54 found among the Caribbean and Pacific islands. The list has six amphibians, 70 birds, 40 fishes, 32 mammals, 15 reptiles, and 31 invertebrates.[29] Efforts are being made to manage or preserve critical habitat and to protect these species to prevent their eventual extinction, and to increase their abundance to a safer level.

Clearly wildlife has much worth, even though its value cannot be easily quantified in dollars. Overall, society places great importance upon wild creatures and seems willing to spend public and private money to preserve them and to improve their habitats. Yet many individual landowners and businesses suffer considerable economic loss due to the activities of wild animals. For example, some animals destroy fences and structures. Others feed upon crops or eat vegetation the landowner wants to preserve for livestock. Some species damage forests and can inhibit successful regeneration of new tree crops. A few large mammals prey upon domesticated animals. Some interfere with waterways, ponds, and irrigation systems.

These and other nuisance activities can result in lost production, destroyed crops, or damaged improvements. Current statistics indicate that loss of agricultural crops to wildlife exceeds $100 million annually, and that predation on livestock by wild animals approaches $170 million.[29] Such costs often provoke conflict between the individuals who suffer the loss and those who want to preserve the wildlife for its social value. By handling the creatures and their habitat wisely, it is hoped the nation can learn better to sustain the economic use of its forests and ranges while still maintaining a quality environment that assures the survival of all living things that make up the biotic component of our natural ecosystems.

THE CHARACTER OF WILDLIFE

As considered here, wildlife includes all free-roaming species that basically derive their food and other life essentials without deliberate human assistance. Among these are many mammals, birds, fishes, reptiles, and amphibians that scientists more broadly classify as vertebrates. Although many biologists disagree, some include the wild horses, pigs, donkeys, cats, dogs, and individuals of other normally domesticated species that now live and reproduce in wild environments. In either case, the point is that wildlife has a basic characteristic of wildness and of independence from human aid for survival and reproduction. And by including a wide variety of species, the definition reflects modern concepts of wildlife management. Such programs seek to maintain or enhance the broad range of values that society derives from the total resource of animals that inhabit the diverse ecosystems of the earth.

Wildness does not mean shyness. It includes a capacity to recognize others of similar kind, to communicate by some mechanism, to reproduce, to find food and shelter within natural ecosystems, to survive the rigors of climate throughout the year, and to successfully avoid or fend off predation. The pheromones mentioned earlier represent a key group of chemicals that furnish these links between individuals of the same species and provide for communication.[31] Some act as signals emitted in response to danger. These warn others nearby. Some serve as territorial markers placed by a dominant male, and others may mark trails. Pheromones also act as tags for individuals, such as odors that enable a fawn to recognize its mother. In these and other diverse ways, pheromones allure, repel, or otherwise stimulate specific responses among the animals. They have profound influences upon the social behavior of most species, and on their capacity to survive and reproduce in a wild environment.

Sight and sound also affect animal behavior and enable many species to find food and detect enemies.[24] Some species have unusually keen senses of seeing and hearing. To illustrate, falcons and other birds of prey have highly developed vision and can recognize even small rodents moving along the ground at great distances. The hearing of many animals detects wavelengths and sounds of lower volumes than humans can hear. Bats have a type of sonar they use to detect objects and locate prey as they fly about at night. All of these characteristics contribute to the capacity of the different wild animals to find food and survive.

The response of wildlife to changes in seasons and in the length of the day is another way in which they adapt to a natural environment. Responses differ among species. Bears, for instance, crawl into a protected place during winter and hibernate there until the weather moderates toward spring. They emerge again as higher temperatures begin to melt the snow and stimulate activity among plants and insects. Seasons bring other responses as well. Each fall, and again in springtime, geese and ducks migrate between winter and summer ranges. In this way, they avoid the harsh climatic extremes of either range. Anadromous fishes have a similar habit. They swim widely in the ocean. Then at the proper season upon reaching maturity, they return to spawn in the river, and even in the same brook, of their birth. In Alaska, the caribou migrate over long distances in response to the changing seasons, and with them come the predators that depend upon the herds for food. Through these and other behavioral characteristics, wild animals have adapted to the wild environments that support them, and thereby maintain their independence of human assistance for survival.

DYNAMICS OF POPULATIONS

Wild animals that live effectively in a suitable environment usually develop into healthy individuals. They mature and reproduce normally. Each female produces several offspring in her lifetime, so a population may increase as long as food and other life-supporting resources remain adequate. However, in many natural ecosystems, a population of animals may grow sufficiently large to eventually overtax the habitat. Then some animals do not find enough food or lack other essential elements to keep them healthy. As they weaken, they become more prone to disease and illness, less able to escape predators, and incapable of surviving accidents or extremes of climate. These die or are killed, and the population declines.[14,17,26] Gradually the food supply or other requisite resources will improve sufficiently to support more individuals. Then the population may once again increase, only to reach another peak, and eventually decline once more. Such dynamics typify many wild animal populations.[14]

In some cases, the changes in an animal population may result from a greater complex of factors than noted here. Also, the numbers of some species may fluctuate more widely than others, and to some degree behavioral attributes of the animals temper the population dynamics.[10] Nevertheless, by some balance between births and the losses to natural catastrophe and accidents, wild animal populations generally remain within the capacities of the ecosystem to support them, at least over the long run. Further, populations of predators and disease organisms also fluctuate, usually in response to the change in numbers of host organisms. This results in a complicated pattern of interdependent cycles within a natural ecosystem.[17]

A predator generally lives on the surplus of its prey. As a result, it normally does not seriously deplete the breeding population of its target species. Periodically, as the prey population becomes more abundant, the numbers of predators may also increase. After a while, predation and other factors will again reduce the target population. Then the predator must find some other species to eat, or its numbers will also decline. This interaction between predators and prey varies with the species and circumstances. Usually the predator will not decimate the host population it depends upon, but instead will help to keep its prey in check and thereby play a vital role in helping to keep wild animal populations in balance with resources of the habitat.[18]

The ratio of males to females and the ages of individuals may also influence wild animal population fluctuations.[26] Populations with large proportions of reproducing females usually have high growth rates. On the other hand, ones comprised mostly of nonreproducing females or of males would tend to decline. Wildlife professionals use such understanding to judge how populations will replenish annually to compensate for natural losses and hunting.

Many other types of population controls operate in wild environments, some of which are more passive in nature. For example, male birds of many species establish a territory by singing. This performance drives off less aggressive males and attracts a mate. Then nesting begins. Throughout this time, the nesting pair will keep intruders of the same species from invading their territory. Such mechanisms help to maintain the population density for a species within limits of the habitat.[22]

Such dynamics and behavior typify wild animal populations. Attempts to deal with wildlife, then, must address the characteristics of each resident species, the individuals that compose the populations at a given place, and the interrelationships among the different animals that coexist in the same habitat. To succeed, a plan for management should take cognizance of the variety of species present, how they will react to possible changes induced by management, and whether the enhancement of conditions for one species might trigger undesirable effects upon the other animals. Wildlife managers must deal with the characteristics, survival requirements, and interdependence between species to effectively maintain a suitable environment for the populations of wildlife they hope to encourage.

WILDLIFE HABITAT

Each species of animal needs a certain area of land on which to live, and the amount differs among them. Within this territory, an animal must find food, cover, and water in sufficient quantities to sustain its life functions.[3,5,16] Further, the critical elements must be dispersed over an area sufficient to satisfy the space needs of the creature.[3,14] And if a species has special needs for survival, such as a particular nesting or hibernating site, the habitat must meet these as well.[3]

Wildlife biologists call this combination of requisite physical and biologic features and the environment in which they occur the species' habitat. To provide for the highest state of productivity, all these habitat elements must be interspersed in proper quantities and balances within the animal's cruising range.[14,15] The habitat must also have a good intermixing of conditions to serve the functional needs of animals of different ages and sex. And accessible food, cover, and water must remain available in at least minimal amounts to ensure survival and good health throughout the year.[14,15] Hence the juxtaposition of essential elements, as well as their abundance, determines the suitability of a place to support an animal.[16] The diversity of those conditions determines the numbers of species that will live there (Figure 10–2).

The vegetation often represents the most critical element of a habitat. Conditions of soil, the landforms, and the climate also play important roles in determining the adequacy of a place to support a species.[5] And the animals, of course, also influence the vegetation. Wildlife, then, represents a product of the land and

Figure 10–2.
Diversification of habitat conditions with the requisite food and cover, such as this mixture of stands in Alaska, tends to favor most wildlife. *(U.S. Forest Service.)*

reflects the condition and productivity of the habitat, its misuse, or both.[7] Any major environmental change affecting even one plant or animal species may impact others as well.[3]

Often we forget that wild creatures do not roam aimlessly across the landscape. Rather each animal frequents a particular geographic area, and this home range must provide the diversity of food, cover, water, and other essentials needed throughout its lifetime. Mobility will differ between species, but each has a distinctive limit wherein it will roam. The limits of suitable habitat and the territoriality of other animals influence these ranges. And while several individuals may have overlapping home ranges, they are not likely to leave them unless the competition for food and other resources becomes intense enough to force emigration. Biologists call this the concept of home range, and use it to assess whether an area can satisfy all the seasonal needs of an animal.[14,16,22]

The carrying capacity of a habitat also is important in judging wild animal populations and the potential of an area to support them. It represents the maximum number of animals a specific area can sustain in equilibrium with essential habitat elements. Numbers may temporarily increase above this level, but eventually decline to what an area can sustain in perpetuity. Theoretically, as conditions of the habitat improve or decline, so will the capacity to support a given number of animals.

Neither the capacity nor the population will necessarily remain static. These may fluctuate over time, and from season to season. Over the long run, a place may attain a balance in the numbers it will support, given no major change in the ecosystem. However, the introduction of a new species, a change in the vegetation, alteration of the habitat by people or other animals, and other stimuli that temporarily or permanently affect the abundance and availability of essential habitat elements will also affect the carrying capacity for a species.[8,9,14,16,26]

Understanding these relationships helps wildlifers to anticipate population changes that may follow different types of natural disaster, or the development of vegetation and other habitat conditions after a change in land use. Historic experience illustrates the point. For example, extensive post-Colonial clearing of the eastern deciduous forest followed settlement over an extensive area. Later westward expansion drastically reduced the abundance of forest even more. Then in the 20th century, as owners abandoned their attempts to farm areas of low productivity, the old fields began to revert to brush and young forests. This altered the admixture of habitat conditions and brought a gradual increase in forest-dependent species, with concurrent losses of those that require open fields. Likewise, a wildfire in an old-growth Rocky Mountain forest may kill the trees and cause a conversion to young growth. Then species that use more open land or stands of young trees may temporarily migrate into the burned area until the forest grows tall enough to establish closed conditions again. Or logging may cut off the tall trees or reduce their density. This temporarily alters the habitat in a manner similar to that of a wildfire, blowdown, or insect infestation. These occurrences all affect the combination of wildlife species and their numbers at a given place. In combination the many natural and human forces continually act to temper the conditions of food, cover, and water. As these change, so will the populations of animals.

WILDLIFE MANAGEMENT

In a broad sense, wildlife management encompasses all the activities used to increase the values people derive from wildlife. These include protecting crops, livestock, and property from intolerable damage by wild creatures, as well as steps to enhance the health, abundance, diversity, and dispersion of such animals. Wildlife management involves both science and art. It draws upon scientific disciplines and principles to devise methods for managing wild animal populations and their habitats, and capitalizes upon the experience of professionals to skillfully apply these to achieve the objectives. In this context, wildlife management also means working with the people who look to the wild creatures for pleasure, convenience, and use.[1,3,14]

Wildlife management, from a professional viewpoint, involves making decisions about what to manage for, and how to do it. These can include both active and passive measures designed to control wild animals and the benefits people derive from them. Active measures mean ones to increase, decrease, or stabilize populations by working with the habitat or the animals themselves. Passive action will entail measures to let natural developments take their course, as when a program strives to preserve habitats in their natural state and allows predation, disease, and other natural forces control the number and kinds of animals present. With either system, the wildlife manager must identify the objectives, delineate the activity needed to bring them to fruition, and implement any treatments called for in the management plan.[14]

Population and Habitat Analysis

As with any other of the natural resources, professionals cannot make wise decisions about wildlife management and use without knowing the condition, extent, and health of the animal populations and their habitat. From this base of information, they can decide what combination of active and passive measures will achieve their purposes. The process begins with the gathering, classifying, and analyzing of inventory facts. Both the relative abundance of a population and the carrying capacity of the site have importance in determining the long-term potential for sustained yield management.[4] These data should have a sound ecologic basis and may need to reflect both the habitat conditions for specific target animals, and more broadly based ecologic assessments that delineate potentials for a broader array of species. The end use will determine the exact types of information needed and the accuracy required.

Wildlife managers have available several techniques for taking a direct census of animal populations or estimating their levels (Figure 10–3). Direct counts usually entail sampling the geographic area of interest and enumerating the animals seen in it. These may utilize drives of larger mammals, wherein observers attempt to count the numbers that run or fly out of a specifically bounded site ahead of the drivers. For some species, the workers often walk along predetermined courses and count the numbers flushed, seen, or heard. Other less direct approaches that have proved useful with small animals such as songbirds involve sitting and listening for different species during a given time interval, and tabulating the frequency at which each species sings or emits a clear sound. For small mammals the census may require capturing the animals in traps, marking them for future identification, releasing them, and observing the numbers of new and repeat captures over a specified time interval. Most recently, the use of helicopters, fixed-winged aircraft, and radar and similar instruments has aided census taking.[13,14,24] With the enumerations acquired through such sampling techniques as a basis, biometricians can use sophisticated statistical methods to estimate the total population within the area sampled.

Figure 10–3.
Wildlife managers employ several techniques for studying the populations and habits of animals, such as this banding of an Aleutian goose to mark it for future identification. *(U.S. Bureau of Sport Fisheries and Wildlife.)*

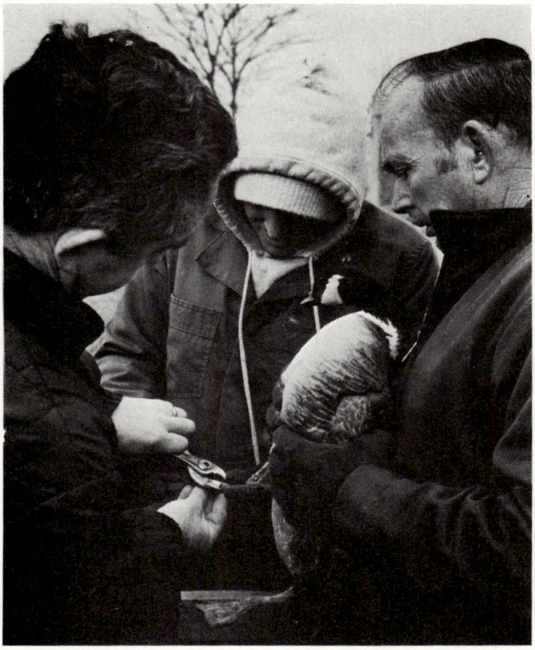

For some species, wildlifers relate direct counts and other information about animal biology to the frequency with which they find signs of the animals. Such signs include tracks, houses or nests, and pellets and other fecal material. Use of these data requires clear understanding of the biologic habits of a target species. From this information, wildlifers can determine how much the animals use and forage in an area.[7,13,33] However, such data may prove less reliable than direct counts for some census purposes.

Other estimates of the size and characteristics of a population come from harvest counts. For these, many states annually mail questionnaires or use mandatory registration to tabulate numbers of animals killed by persons who hunt or fish. Voluntary approaches pose problems with interpretation, since some people may not bother to respond and so bias the sample. Sometimes biologists also establish check stations and by autopsy and other observations determine the sex, age, and health of the animals they examine. These observations aid considerably in judging the sex ratio, vigor, and age structure of target populations, and provide valuable information wildlife managers can use in planning effective management programs.[13]

Previous chapters described vegetation and habitat inventory methods used in evaluating rangelands and their capabilities to support grazing. Similar methods have usefulness relative to wildlife. These usually involve a reconnaissance or more formally designed inventory, and the taking of measurements and observations to identify the condition of various habitat elements encountered. Accessibility to food and cover, use by wildlife, abundance and condition of requisite plant species, and availability of other habitat essentials all provide an indication of the potential of an area to support wildlife, and ultimately will be evident in the numbers and health of the animals themselves. In these analyses, the assessment of interspersion of habitat elements and their extent also is important in helping the wildlife manager to determine the needs for management programs.[7,13]

Habitat Manipulation

The inventory of habitat conditions will reveal factors that might limit the population of a target species. On the basis of these assessments, wildlife managers can formulate plans to improve the status of the habitat and overcome any deficiencies. Their philosophy is that by improving the conditions of the habitat, they can also increase its carrying capacity. As the animals reproduce and survive more effectively, their population will gradually increase.[9,19]

Managers can use a variety of techniques for manipulating habitat, depending upon the nature of the ecosystem and the animal species involved. These will help to maintain a given set of habitat conditions, restore ones that have deteriorated with overuse or abuse, or add others not present.[32] For example, measures to improve the habitat in farmlands might include planting trees and shrubs, controlling grazing in woodlands, designing artificial hiding places, propagating food patches, controlling erosion, and developing watering holes. In woodlands, the foresters can use thinnings and harvest cuttings to create openings and to stimulate understory vegetation (Figure 10–4). They might also improve the water resources, protect the forest from grazing and fire, and leave dead trees for cavity-nesting species.[27] Generally, in most ecosystems the management practices will combine techniques to improve food production, water availability, and cover.[32,33]

Edges between adjoining plant communities often have the highest productivity of some species. Basically the edge represents a transition zone, or ecotone, between one kind of land use and another. Along these edges, the wildlife have easy access to a variety of habitat conditions within close proximity, and so both their abundance and diversity may be greatest here. Knowing this, managers frequently will endeavor to increase the amount of edge in an area by creating clearings in the forest, interspersing plantings of trees and shrubs in open areas, and establishing mixes of plant species that maintain multispecies plant communities that abut each other.[14] If forest and wildlife managers work together, they usually can satisfy a variety of goals and provide multiple benefits from the land.

Figure 10–4.
In forested areas timber harvesting such as this patch clearcutting will create openings and stimulate understory vegetation to improve cover and browsing for animals.

Population Control

Wildlife management does not always try to increase the number of animals. Under some circumstances, a program may attempt to limit them consistent with other uses of the land.[14] This might mean controlling predators in rangelands, reducing large herbivore populations that damage trees in timber production areas, or limiting populations of various animals in depleted habitats until the vegetation regrows to a healthy state. Damage control measures may reduce the number of animals, restrict their access by fencing and other structural measures, frighten them away from selected places, encourage adequate harvest by hunting, adjust the condition of vegetative cover and food availability, or divert the animals to special feeding and resting areas.[6] The exact means employed will depend upon the local needs and circumstances.

Hunting offers one direct approach to manipulating wildlife populations, and some populations lend themselves to such control. Hunting acts as a surrogate form of predation, and can prove useful where human action has removed the natural enemies. In these cases, managers can act in concert with state regulations to encourage or restrict hunting by regulating the number and kinds of permits issued, by opening or closing specified areas for hunting, by changing the length and time of the hunting season, and through other measures that influence the intensity and duration of hunting. They can both limit the harvest and encourage it.[14]

More extreme control measures such as trapping, poisoning, herding, and introducing predators often prove costly and require careful assessment. Biologists must weigh the costs against the benefits to assess the wisdom of these measures. The choices may involve moral as well as economic assessment; thus decisions whether or not to move ahead may require prior consultation with local people and special-interest groups. Managers must learn to understand the different views held and look for ways to compromise in seeking consensus. In fact, many aspects of wildlife management involve this interaction between wildlife professionals and lay persons. These contacts come in the seeking of public opinion, in carrying out education programs, and through law enforcement. They require serious attention to developing and maintaining good interpersonal relationships with various clientele groups in the daily routine of business, in the setting of policy and administrative directives, and in controlling the use of wildlife for hunting and other recreational pursuits.[14,23]

Refuges and Preserves

The concept of wildlife refuges where management protects the animals from hunting, trapping, and fishing began in Europe centuries ago. It became manifest in the United States as early as 1870. Theodore Roosevelt created the first federal refuge by executive order in 1903, setting aside Pelican Island off the Florida Coast.[19] Such sanctuaries may give total protection to the wild animals, and fill an important role in safeguarding threatened and endangered species. Additionally many levels of government maintain game management areas, parks, and preserves intensively managed to provide public recreation and hunting opportunities. Private shooting preserves often receive the most intensive management, supported by fees imposed for hunting privileges. National Forests, state-owned forests, and the vast acreage of private nonindustrial and corporate lands also are important as wildlife habitat for recreational purposes.[14] People will often pay to hunt on those lands where the owner limits access. Consequently, in some regions, private landowners can take advantage of this market and lease hunting rights, even if they carry out no special wildlife management on their property.

FISH MANAGEMENT

An ecosystem approach also has long served in the managing of fish and aquatic life in inland waters. Basically fishery mangement usually strives to optimize the yield of fish to people. It takes cognizance of the complexity of aquatic environments. Success may depend upon identifying and overcoming factors limiting the spawning, hatching, and survival of fish populations to keep them in balance with capacities of the ecosystem to support them. Management may include stocking of hatchery-reared fish, habitat manipulation, and population control.[11]

People concerned with the management and use of woodlands become involved with fishery management because of the water that forms in forest streams, rivers, ponds, and lakes. These serve as headwaters of major rivers and their condition influences the waters that flow downstream. Management programs for aquatic ecosystems normally include protection, restoration, enhancement, and maintenance of essential habitat. Most efforts focus on stream banks, beds, currents, and streamside vegetation, but erosion control and the abatement of pollution should be considered as well. All depend upon a clear understanding of stream and pond ecosystems. Among the objectives of such programs are to provide more living space and concealment, to improve reproductive habitat, and also water temperature and flow rates.[30] Forest managers will have special concern for protecting the integrity of stream banks and retaining shading over the water by controlling cutting of streamside trees.

Clearly the complexity of fishery management deserves more attention than given here. However, many of the ecologic principles already reviewed relative to terrestrial animals also apply to aquatic ones. This brief account serves only to point out that multiple-use programs for forestlands include management of fishery resources, and that the fish species have an important role in forest environments. Forestry professionals must recognize this when planning the management of forest resources.

CAREER OPPORTUNITIES WITH FISH AND WILDLIFE MANAGEMENT

Wildlife and fishery managers work in the forests, on the range, and on inland waters to promote the welfare of animal species, and to enhance their contributions to the people. Their work includes the many functions of habitat assessment and management, the census and enumeration of animal populations, and the manipulation of habitats and animals. But wildlife managers also work directly with the people who use, appreciate, and benefit from wild animals, as well as with the landowners who control the habitats upon which wild populations depend. Consequently wildlife management involves concern both for the animals and for their habitat, and also for helping people to increase their awareness of wildlife. It requires the practitioner to take to the field when and where the animals gather their food, build nests, rear their young, and teach their offspring independence, and at other critical times in the life cycles of the target species. While trees and other forest resources remain stationary, animals do not. Thus, to study them, the wildlife professional must follow them where they go.

Because of the linkage between forests and wildlife, most schools of forestry offer at least basic courses in zoology and wildlife management, and may require these of all students. Forestry students can then elect more advanced courses to strengthen their understanding of wildlife and ways to manage the habitat for it. Some colleges also offer special wildlife management programs for persons interested in pursuing careers as wildlife professionals. These may differ from the forestry curriculum and require greater preparation in animal biology, ecology, animal physiology, pathology, botany, animal behavior, and other offerings in basic biology. In general, individuals who wish to pursue careers in wildlife management will need to develop their skills beyond the undergraduate degree and work toward the master of science degree in wildlife management. Some special employment opportunities in research and teaching will demand studies in preparation for the Ph.D. degree.

Federal and state agencies and universities offer the greatest number of opportunities for careers in wildlife management and science. The Fish and Wildlife Service of the Department of the Interior is the federal agency assigned responsibility for managing, protecting, and promoting use of wildlife. This agency also administers the programs related to rare and endangered species, and monitors the health of our wildlife populations. Individual states carry out active wildlife management programs, and have the responsibility for controlling public hunting and fishing. State governments also provide technical assistance to private landowners. Along with universities and the federal government, some states maintain active research programs that offer career opportunities to qualified persons. In addition, many large corporations hire wildlifers to help in planning the management of their holdings. Other opportunities develop with private foundations, public interest groups, and conservation societies.

REFERENCES

1. H. E. Alexander, "Changing Concepts and Needs in Wildlife Management," in *Proceedings Annual Conference, Southeastern Association of Game and Fish Commissions,* 16:161–167 (1962).
2. D. E. Arnold, "Environmental Pollution—Effects on Fish and Wildlife Management," in R. D. Teague and E. Decker (eds.), *Wildlife Conservation Principles and Practices,* The Wildlife Society, Washington, 1975.
3. G. V. Burger, "Principles of Wildlife Management," in R. D. Teague and E. Decker (eds.), *Wildlife Conservation Principles and Practices,* The Wildlife Society, Washington, 1979.
4. H. P. Caulfield, Jr., "Political Considerations," in R. D. Teague and E. Decker (eds.), *Wildlife Conservation Principles and Practices,* The Wildlife Society, Washington, 1979.
5. H. N. Coulombe, "Toward an Integrated Ecological Assessment of Wildlife Habitat." in H. G. Lund, V. J. LaBau, P. F. Ffilliott, and D. W. Robinson (eds.), *Integrated Inventories of Renewable Natural Resources: Proceedings of the Workshop,* U.S. Forest Service General Technical Report. RM-55, 1978.
6. M. W. Cummings, "Wildlife Damage Problems," in R. D. Teague and E. Decker, (eds.), *Wildlife Conservation Principles and Practices,* The Wildlife Society, Washington, 1979.
7. A. DeVos and H. S. Mosby, "Habitat Analysis and Evaluation," in R. H. Giles, Jr. (ed.), *Wildlife Management Techniques,* The Wildlife Society, Washington, 1971.
8. L. L. Eberhardt, "Population Analyses," in R. H. Giles, Jr. (ed.), *Wildlife Management Techniques,* The Wildlife Society, Washington, 1971.
9. R. Y. Edwards and C. D. Fowler, "The Concept of Carrying Capacity," *Transactions of the North American Wildlife Conference,* 20:589–602 (1955).
10. P. L. Errington, "Factors Limiting Higher Vertebrate Populations," *Science,* 124:304–307 (1956).
11. H. W. Everhardt, "Fishery Management Principles," in R. D. Teague and E. Decker (eds.), *Wildlife Conservation Principles and Practices,* The Wildlife Society, Washington, 1979.
12. C. W. Faucett, "Vanishing Wildlife and Federal Protection Efforts," *Ecology Law Quarterly* **1**(3):520–560 (1971).
13. A. D. Geis, "Population and Harvest Studies," in R. D. Teague and E. Decker (eds.), *The Wildlife Conservation Principles and Practices,* The Wildlife Society, Washington, 1979.
14. R. H. Giles, Jr., *Wildlife Management,* W. H. Freeman, San Francisco, 1978.
15. R. T. King, "Wildlife and Man," *N.Y. Conservationist,* 20:8–11 (1966).
16. Ibid., "The Essentials of a Wildlife Range," *Journal of Forestry* **36**(5):456–464 (1958).
17. E. J. Kormondy, *Concepts of Ecology,* 2d ed., Prentice-Hall, Englewood Cliffs, N.J., 1969.
18. R. M. Latham, "Predation—Problems and Principles," in R. D. Teague and E. Decker (eds.), *Wildlife Conservation Principles and Practices,* The Wildlife Society, Washington, 1979.
19. A. Leopold, *Wildlife Management,* Scribner, New York, 1933.

20. Ibid., *A Sand County Almanac,* Oxford University Press, 1949.
21. Ibid., "Adaptability of Animals to Habitat Change," in J. A. Bailey, W. E. Elder, and T. D. McKinney (eds.), *Readings in Wildlife Conservation,* The Wildlife Society, Washington, 1974.
22. A. N. Moen, "Animal Behavior," in R. D. Teague and E. Decker (ed.), *Wildlife Conservation Principles and Practices,* The Wildlife Society, Washington, 1979.
23. W. B. Morse, "Law Enforcement—A Tool of Management," in R. D. Teague and E. Decker (eds.), *Wildlife Conservation Principles and Practices,* The Wildlife Society, Washington, 1979.
24. S. W. Overton, "Estimating the Numbers of Animals in Wildlife Populations," in R. H. Giles, Jr. (ed.), *Wildlife Management Techniques,* The Wildlife Society, Washington, 1971.
25. B. R. Payne and R. M.DeGraaf, "Economic Value Associated with Human Enjoyment of Non-game Birds," in *Proceedings of the Symposium on Management of Forest and Range Habitats for Non-game Birds,* U.S. Forest Service General Technical Report. WO-1, 1975.
26. R. D. Taber and K. J. Raedeke, "Population Dynamics," in R. D. Teague and E. Decker (eds.), *Wildlife Conservation Principles and Practices,* The Wildlife Society, Washington, 1979.
27. R. E. Trippensee, *Wildlife Management,* McGraw-Hill, New York, 1948.
28. U.S. Forest Service, *The Nation's Renewable Resources—An Assessment,* U.S. Department of Agriculture, Forest Service, Washington, 1975.
29. Ibid., *An Assessment of the Forest and Range Land Situation of the United States,* Review Draft, U.S. Department of Agriculture, Forest Service, Washington, 1980.
30. R. J. White, "Stream Habitat Management," in R. D. Teague and E. Decker (eds.), *Wildlife Conservation Principles and Practices,* The Wildlife Society, Washington, 1979.
31. E. O. Wilson, "Chemical Communication Within the Animal World," in E. Sondheimer and J. Simeone, (eds.), *Chemical Ecology,* Academic Press, New York, 1970.
32. J. D. Yoakum, "Habitat Improvement," in R. D. Teague and E. Decker (eds.), *Wildlife Conservation Principles and Practices,* The Wildlife Society, Washington, 1979.
33. J. D. Yoakum, and W. P. Dasmann, "Habitat Manipulation Practices," in R. H. Giles, Jr. (ed.), *Wildlife Management Techniques,* The Wildlife Society, Washington, 1971.

CHAPTER **11**

PROTECTING FORESTS

Throughout the geologic ages, forests have withstood many kinds of destruction. These included the uplifting and subsidence of land, effects of glaciation, significant changes in climate, the influence of volcanic eruption, and many other natural forces. If the event created conditions unsuitable for plants and animals, the forest disappeared. After surface conditions, moisture, temperature, and other features of the site once more became favorable, plants colonized the area again. Through natural succession, trees eventually took root, replacing the pioneering plants, and giving rise to a new forest. Likewise, as the vegetation developed, animal life migrated back. Eventually a forest ecosystem with all the plant and animal components once more filled the site. Effects of the catastrophe that had originally killed the trees and transformed the landscape no longer seemed apparent.

IMPACTS UPON THE FOREST

The degree to which the ecosystem changes and the time it takes for the forest to return depend upon the nature and severity of the disturbance. After a major devastation, the regrowth may take centuries. In other cases, only a relatively short time elapses before a new stand of trees covers the landscape. Between the major disturbances, other less catastrophic environmental agents work on the forest. These include fire, insects, diseases, animals, drought, flooding, and human activity. Though less destructive than an earthquake or volcanic eruption, these do profoundly affect the trees. Some reduce the abundance of a species. Others alter their economic value. But all are a natural part of the forest ecosystem.

They recur periodically and in that way help to shape forest development over eons of time.

Resistance to Destruction

Despite these ever-present dangers, the forest survives. Many mechanisms help. For one, the forest as a natural system has genetic diversity. It contains a variety of species that differ in their resistance to destructive agents. If one harmful organism becomes epidemic, it may seriously affect one species but not the whole forest. This intermixing of species also tends to slow the spread of many insects and diseases, compared with stands where only one kind of tree grows. Also, within a single species, the trees have differing degrees of susceptibility to various natural agents. Some will have great resistance and survive effects that might destroy other trees nearby. Further, this mix of species and genotypes occurs in a random pattern across both space and time. That adds to a community the resilience to withstand many endemic agents, and buffers the forest against a rapid multiplication of these into epidemic proportions. However, if people or major environmental events upset this balance, the system may suffer. The problem may rapidly multiply and losses mount quickly. Thus foresters must remain alert to the possibility that their management might weaken the forest's natural resilience and must devise contingency measures for corrective action should the need arise.

While natural forests and the species of plants and animals that compose them generally withstand the impact of both endemic and epidemic levels of insects and diseases, outbreaks periodically may cause important loss of products and other values. And forests established by planting may suffer even more. These stands generally lack the diversity and buffering capacity of natural stands due to the homogeneity of species, spacing, ages, and genetic composition. Generally they have greater susceptibility to insects and diseases. However, some destructive agents, such as fire and windstorms, may prove equally destructive to all types of forests. They may quickly wipe out the stand and trigger a shift in species composition as a result of the changes they bring to the site.

Types of Losses

Insects, diseases, and fire do many types of damage. Among the more obvious are the loss of wood, forage, wildlife, recreational opportunities, stabilizing effects on the soil, beauty landscape, and yields of high-quality water. Not all kill the trees; some only cause injury, deformities, or rot. Tree health may suffer, but not to the extent that the tree dies. Growth may slow, delaying the healing and recovery. Secondary insects and diseases may weaken the tree further. In some cases, these damage a tree so extensively that it no longer is useful for timber or other products. In other cases, the attack only deforms the stem enough to limit the types of products it affords. While they do not pose a major threat to the overall survival of a forest stand, such endemic and persistent action may reduce the economic value of individual trees. If widespread, the harmful agents may even limit the potential of entire stands.

The forest provides a home for a variety of harmful organisms. But although many do damage the trees, virtually all contribute in some way to the overall condition and perpetuation of the forest itself. For example, fire may reduce thickly accumulated organic matter that would prove unsuitable as a seedbed for regeneration of some tree species. Actions of wind, disease, and insects in unmanaged forests contribute to successional changes in the stands. They also may create environmental conditions needed to perpetuate a specific type when it becomes overmature and would otherwise succeed to another mixture of species. Any devastation of a forest in a well-advanced stage of development would give rise to new stands of young trees that might provide better habitats for some wildlife. Also, decadent and dying trees furnish dens and feeding sites for birds and rodents. In fact, the endemic presence of these many agents may prove critical in maintaining less obvious natural balances within the forested ecosystem, therefore assuring its presence into the future.

Practicality and " ecologic reality," therefore, have taught foresters to seek control to a point that minimizes losses and prevents their untimely occurrence. The aim of control is to prevent minor pest problems from developing into major ones by creating and maintaining a healthy, resilient forest. In some instances, managers use the natural forces as a silvicultural tool. For example, they might set fires deliberately to reduce the possibility of an untimely wild fire, or to prepare sites for regeneration. Entomologists have even introduced one species of insect in the hope of controlling the possible outbreak of another. And foresters design cuttings to capitalize upon wind to help disseminate seed. In all their management, however, foresters hope to prevent the occurrence of a major disaster primarily by maintaining a forest with sufficient natural resistance to prevent a serious pest buildup.

Effects of Age

Many trees and forest stands become susceptible to insects and disease by virtue of age or poor vigor. Forest trees, like other long-lived plants and animals, have no fixed life span. Some species, such as aspen, rarely live more than 70–90 years in the East. But others, such as sugar maple, may live for over 300 years and Douglas-fir for over 500. Individual giant Sequoia have survived for 4000 years, and gnarled and weather-beaten bristlecone pines in Arizona took root more than 4600 years ago. Few trees of any species die because of old age alone. Instead they become broken by storms or struck by lightning. Decay may also weaken the stems of even the most long-lived species, and they eventually fall. Others decline in vigor due to suppression by larger neighbors, or because they get insufficient light, nutrients, or moisture. As these damaged and low-vigor trees suffer

suffer attacks by insects and diseases, they decline in health even further and ultimately die. Such events represent a continuing natural process that makes more resources available to survivors, and opens space where new trees become established and eventually replenish the forest.

Groups of trees growing together in forest stands also have no fixed life span. But even-aged stands eventually reach a physiologic maturity. Then their gradual decline initiates a sequence of events that leads to the replacement of the old stand by a new one. The process may follow a course something like this. First, as the stand matures, individual trees slow in growth and decline in vigor. Those with smaller crowns and less exposure to direct sunlight often suffer sooner than the more dominant members of the community. Eventually the suppressed trees die, while neighboring trees continue to grow and benefit from the extra space and site resources that become available with their death. Over time, as the stand matures, more and more trees die because of declining health and vigor. Others blow over or succumb to various insects and diseases, and the stand gradually begins to deteriorate. This process may take a long time, but eventually defects in individual trees, the fallen stems, and the broken tree tops will give evidence of the decadence.

In mature even-aged stands, many trees may weaken at about the same time. This creates a habitat where populations of insects and disease organisms can build up to outbreak proportions (Figure 11–1). These may cause a sudden loss of all the

Figure 11–1.
In mature even-aged stands such as this lodgepole pine in Wyoming, many trees may weaken at about the same time, thus fostering a buildup of insects, diseases, or loss to fire followed by formation of a new even-aged stand over the area.

old trees at once, followed by the formation of a new even-aged stand of young trees across the entire area. Such open conditions may prove favorable to the establishment of species that do not grow well under partial shade. At the same time, the sunlight will dry the surface litter and dead branches, creating conditions conducive to the rapid spread of wildfire that might follow a lightning strike. If a fire occurs, it may change the seedbed sufficiently to set the stage for regeneration of a species that otherwise would not succeed in the undisturbed litter. Thus the events that take place as the stand matures will help determine what species will take root and survive on the site.

In some stands, individual trees gradually die while others live for a long time. Then small clusters of regeneration form underneath the dead trees, resulting in a more gradual and less spectacular replacement of the old stand. Under these circumstances, the old even-aged stand may gradually give way to an uneven-aged one comprised primarily of species that have a high tolerance to shading. These usually are representative of a more climax stage of forest succession. In either case, the selection pressure of insects and diseases in unmanaged stands serves as a natural mechanism for the periodic replacement of the old forest with a new, more vigorous one.

The Influence of People

Not all of these changes result from insects and disease. Wildfires have also historically played an important role in provoking forest change. They periodically burned among North American forests and rangelands long before European settlers arrived. These fires started not just in the declining old-growth timber, but also among younger stands following times of prolonged dryness that made surface conditions right for rapid burning. Indians set some fires to clear the forest for settlements or crops, or to drive out animals for hunting. Other fires escaped from Indian encampments as a result of carelessness. As the White settlers pushed westward, the larger population and more diverse human activity triggered a dramatic increase in the threat from fire caused by people, whether deliberately or by accident. Many fires burned unchecked over wide expanses of forest and range. With the vastness of resources yet untapped, the losses seemed unimportant. Today, however, we would consider the destruction inestimable.

History has amply demonstrated that human activities pose a threat second to no other agent in the capacity to consume and destroy. These include dangers from careless use of fire, chemical pollution, discharge of waste substances, and dumping of other by-products by our highly industrialized society. If released in large amounts that can cause harm, the chemicals may change the natural ecosystems, transform them, and reduce their productivity or stability. At the same time, judicious management can help to safeguard the land against such threats and even find ways more fully to capture its productive potential for human benefit. To accomplish this goal, we must learn to appreciate the consequences of various actions and to control the harmful ones effectively.

LOSSES TO HARMFUL AGENTS

Probably the most dramatic losses of timber and habitat result from wildfires. The most destructive ones often start in remote areas following periods of prolonged dryness, and may explode into conflagrations of major size. Yet many smaller wildfires burn in forested ecosystems and rangelands each year. They also cause loss and destruction. During 1969, for example, forest fires burned almost 2.7 million hectares across the United States,[1] and from 1960 to 1970 fires burned an average of 1.6 million hectares annually within the 48 contiguous states.[35] Between 1957 and 1971, the number of fires increased from about 82,000 to 112,000. Few started naturally; people caused about 91 percent, and most fires burned private and state-owned lands.[36] Currently wildfires burn about 1.5 million hectares annually, with losses of property and timber, plus costs of suppression, ranging upward of $600 million.[36]

While most wildfires never develop into spectacular conflagrations, some large and especially destructive ones trigger side effects that add to the economic and other losses. A couple of cases illustrate the point. In 1970, for example, nearly 217,611 ha burned in one fire in California. This fire destroyed 900 homes and killed 14 people.[1] Later, in 1977, fires in the central and northern parts of the state each burned across 28,000–30,400 ha, destroying timber valued at more than $140 million in one case, and $175 million in the other.[19] Less direct losses also result, such as in 1945 at Salt Lake City following a 253-ha fire that destroyed vegetation and litter over a nearby watershed. Effects due to subsequent flooding from that small watershed caused about $347,000 in damage the following year. Another small 91-ha fire near San Antonio, Texas, led to flooding damage worth $47,000.[9] In current dollars, similar floods would result in much higher losses today. And when all types of losses are combined, the value becomes great.

Collectively insects and diseases do more harm by killing trees and inflicting injury than do all other destructive agents, including fire. They do not generally cause property loss as will some forest fires. Still, estimates place timber losses to insects and diseases in the United States at upward of 75 million cubic meters annually. This equals about 40 percent of the total harvested each year, and represents a loss of more than $500 million in stumpage value alone.[36] Additional impacts upon forests and urban trees from various abiotic stresses, such as nutrient deficiencies and toxic materials in the environment, greatly compound the economic loss.

Again, a few cases illustrate the magnitude and importance of insects and diseases as destructive agents. These represent direct financial loss related to timber values of killed and injured trees. For example, in 1970 three types of insects brought extensive damage to forests across the United States.[27] Douglas-fir tussock moth reached epidemic proportions on 328.9 million hectares in the Pacific Northwest. In the South, the southern pine beetle damaged or threatened timber on another 19.2 million hectares from Virginia to Texas. And in the northeastern states, the gypsy moth defoliated more than 683,200 ha of hardwood forests. Oaks form a principal component of these forests, including many urban trees. During the early 1970s, these three insects alone had importantly impacted a total of more than 405 million hectares, causing great economic loss plus incalculable effects on less tangible values for recreation, habitat, and other uses.

Losses attributable to diseases usually do not result from major outbreaks, though annual losses due to death or damage to individual trees almost defies estimation. At times, major disease epidemics have eliminated or drastically depleted representation of some species in the forest. For example, through much of the Northeast, a fungus and scale insect association has greatly reduced the abundance of American beech as a commercially available species. Eventual consequences appear uncertain at the present time. Such epidemics bring incalculable losses as a result of the disappearance of a species from American forests.

Domesticated and wild animals also adversely affect forests and rangelands, primarily in the ways they injure individual trees or harm the plant community through overgrazing. About 40 percent of the 160.9 million hectares of forest ecosystems in the eastern United States supports grazing by livestock, and on about one half that area the grazing has damaged the forest resource. This results in impaired individual tree growth and reduction of soil and watershed values.[36] In contrast, of the 40.5 million hectares of forests currently grazed in the West, only about 1.5 percent suffers from exploitative grazing. In other areas, grazing by wild animals may affect the prospects for successful regeneration of forest stands, and thereby foster problems that often are overlooked.

Generally programs of forest protection will concentrate on preventive measures in the hope of reducing the need for control of an outbreak or catastrophe. Where possible, these include measures to keep the forest healthy, productive, and free of conditions that would enable various harmful agents to develop into major problems. If the prevention fails or if conditions deteriorate due to uncontrollable natural phenomena, the foresters, entomologists, pathologists, fire control specialists, and others who work with forests and ranges will undertake control by direct attack. Then they will attempt to stop the spread and to minimize the losses.

FOREST FIRES AND THEIR CONTROL

For most people, mention of a forest fire conjures up thoughts of a cataclysmic event that leaves vast destruction as an aftermath. Blowup fires do present an awesome sight. They travel across the landscape at speeds up to 6 km/hr. Flames may leap 50 m into the air, releasing energy that sends a convection column of smoke and embers hundreds of meters upward. Burning embers carried a kilometer or more downwind may ignite new fires that exaggerate the conflagration.[3] Such fires defy our capacities to extinguish them, and eventually die down only when a change in weather or a lack of fuel supply reduces their energy sufficiently to permit direct control. Fires of this type cause major burns and great damage, and take large expenditures for control. However, most fires never attain such ferocity.

Many more fires burn with far less intensity. Some creep slowly across the surface at a rate of only tens of meters per hour, consuming all or part of the litter and generating comparatively less heat. Trees may suffer damage that leads to decay problems, but still survive. In other cases, flames and the heat they generate will kill seedlings, young trees, and some timber. Hot fires may also destroy elements of wildlife habitat, kill soil organisms that help in the decay of organic litter, and expose mineral soil to increased danger of erosion. The extent of damage will depend upon the dryness and type of fuel present, and the atmospheric conditions at the time of burning.

Kinds of Forest Fires

Three types of fires occur in forested areas. The most common burn only the dry litter and vegetation on the surface of the ground. These surface fires usually do not consume all the organic layer (Figure 11–2), and may not expose mineral soil or kill soil-inhabiting organisms. They have little effect on the physical characteristics of the soil itself. Also, surface fires frequently do little damage to large trees. They are common in hardwood forests in late fall or in springtime, where they burn

Figure 11–2.
Surface fires usually consume the dry fuel in the upper litter layers and on the surface, and may kill understory shrubs and small trees, but usually do little harm to the larger timber.

only the dried and newly fallen leaves that have not become matted upon the surface of the older litter. Partly decomposed organic materials below the surface litter will usually contain sufficient moisture to prevent ignition of deeper layers.

Ground fires, a second type, have a greater effect upon forests. These burn the entire organic litter, and in forests where much organic material accumulates, they may burn to a depth of a meter or more. But ground fires burn slowly, creeping across the stand at less than walking speed. In deep litter, they may even smolder for weeks, occasionally flaring up into a more rapidly moving surface fire, and then subsiding again. Many times, ground fires only become extinguished after periods of rain, when they run out of suitable fuel, or if dug out. They often cause tree mortality because the heat kills roots in the upper soil and organic layers. The heat may also alter soil properties near the surface, and kill soil-inhabiting organisms. Ground fires normally occur following prolonged dryspells, during which the surface litter is desiccated to a considerable depth.

Crown fires represent the most spectacular and most dangerous of fires. These burn in coniferous forests, become exceedingly hot, travel rapidly, and may flare up into conflagrations of enormous size. Crown fires are most likely during periods of extreme dryness. They often start after heat from a surface or ground fire dries the needles in the tree crowns and heats them to the ignition point. Once burning within the tree crowns, these fires generate great heat that rapidly dries the foliage ahead of the flames, causing the fire to spread. Soon the fire in the crowns will outrun the surface fire that accompanies it. In their wake, crown fires create a swath of dead trees, burned surface litter, and general devastation that may leave the forest devoid of most living organisms. Fortunately such catastrophic fires occur with far less frequency than the other types, though losses from crown fires usually outweigh those from the others.

Essentials for Fire

All these fires depend upon the essential combination of dry and abundant fuel, oxygen, and a source of heat sufficient to raise fuel temperature to the ignition point. Oxygen, of course, surrounds us. Forests and rangelands also usually have abundant fuel, though some forest types have more combustible materials than others. However, these fuels will burn only when dry. Then, if sparked by lightning, a carelessly thrown match, or perhaps a spark from machinery, the dry fuel may ignite and the flames spread.

In many parts of the United States the climate serves as the most effective means of fire prevention. Areas of frequent precipitation benefit from periodic wetting of surface fuels, and conditions for burning may occur only during brief periods of the year. Additionally high atmospheric humidity also acts to thwart combustion and further reduce the threat of wildfire. Such conditions predominate in much of the northeastern states throughout most years, and in the South during most months. In the western states, precipitation occurs primarily during winter, with only light, infrequent rain falling in the warm summer months. Lower elevation forests, open woodlands, and shrublands subjected to little rainfall and a rapid summer drying from evaporation pose the greatest fire danger. Wildfires take place frequently in these regions, although they do not necessarily develop into cataclysmic events in all years. Nevertheless concerns about wildfires and their effects play a major role in the planning for and managing of forest resources in many western regions, and in the southern states during the driest months.

Forest Fire Management

American forestry has traditionally included strong programs of forest fire protection, with remarkable accomplishments. Through early detection and effective suppression, local, state, and federal fire control agencies have reduced the area of forest and range burned from a total of about 12–16 million hectares at the turn of the century to less than 1.5 million at present. Even greater reductions seem possible through the application of advanced technology for preventing, detecting, and extinguishing wildfires across the United States. However, the greatest gain would come from reducing the number of fires that start annually.

Humans start over 90 percent of all forest fires. In part, this results from carelessness and maliciousness. But in some regions, rural people have burned the forest for decades as a means of controlling pests, reducing understory vegetation, and promoting forage for browsing. Burning has become a part of the rural culture with no intent to use it for destruction or harm. On the other hand, malicious incendiary fires account for about 30 percent of those started by people. The proportion has risen in the past two decades, presenting a major problem for fire prevention personnel and pushing the losses and costs of forest fires upward. These and other unwanted people-caused fires require imaginative efforts to prevent them. Education, aggressive law enforcement, and other preventive measures have all been part of traditional programs, and have helped to reduce the incidence of forest fire in the United States in past years. Yet agencies must develop new approaches that address the root causes for incendiary fires and those that start from careless burning of debris. With these fires on the rise, the prevention programs will need a new direction if the responsible agencies expect to realize even more progress in their reduction.

Increased appreciation of fire as an essential ecologic process in the management of our forests and rangelands has broadened the scope of this field during recent years. Fire does have beneficial effects in many cases. It influences the physical-chemical environment, aids in reducing dry matter accumulation, helps in nutrient recycling, prepares seedbeds, and has other uses in silviculture. Fire also controls some insects and diseases, and reduces their impact. For example, it provides the best approach to reducing dwarf mistletoe, a parasite that develops slowly in stands protected from fire.[21] It tempers plant communities, setting the stage for some species that otherwise might not regenerate or persist in its absence. Fire may even beneficially affect many characteristics of wildlife habitat and thereby influence the populations of creatures that utilize different areas.[37] Such an evolution does not deemphasize the need for efficient and effective prevention and control of unwanted destructive fire. Rather it recognizes that fire may also serve as a valuable management tool, if used properly. In cases of selected forest ownerships, including some wilderness and parklands, the management plan may designate areas where the forester will use only limited kinds or levels of fire suppression. Instead naturally occurring fires may be allowed to burn as they did before effective fire control programs were instituted. This approach allows selected fires to influence the character and make-up of the natural ecosystem in the same fashion as they did prior to settlement. It represents an important change in the attitude about forest fires, and is an important aspect of a fairly new technical field called fire management. As such, fire management includes not just the traditional activities related to fire suppression, but also its use as a tool in managing the forest ecosystem.

Presuppression Activity

Fire control still represents an important aspect of fire management. It involves two primary activities. First, control programs strive to prevent fires from occurring to the degree possible. Second, they work toward early detection and the rapid suppression of those that do start. In this, education plays an important role, primarily by increasing public awareness of dangers from forest fires and encouraging care with the use of fire and its causes. As a more direct prevention effort, agencies may close the forest to recreational use and logging during periods of high danger, require spark arresters on machinery, control debris burning, and in other ways limit activities that might cause fires.

Effective fire prevention and control also require rapid deployment of suppression crews, especially in areas of high danger. By analyzing records of former fires, specialists can identify places where they frequently occur and maintain surveillance for early detection. This might include the use of observation towers, overflights by aircraft, road patrols, or even satellite monitoring. The agency may construct fire roads and lines to facilitate access to danger areas, or to impede the advance of a fire. The type and abundance of fuels can be mapped out and measures taken to reduce them. Also, control agencies can stockpile equipment and supplies at strategic locations, ready for use when needed.

Careful monitoring of fuel and weather conditions also plays an important role in presuppression activities. By following changes in temperature, humidity, fuel moisture content, time since last precipitation, time of day, and similar factors, specialists can effectively predict the likelihood that fires will occur. They can also forecast the probable rate of spread in different types of fuel. Given such information, fire control specialists can judge the danger, and take measures to intensify detection and patrol when necessary. Also, if fires do start, these same indexes provide clues to the difficulty of suppression and help the analysts to determine the size of crew and kind of equipment needed to bring a fire under control. Coupled with experience and seasoned judgment, these aids prove valuable in fire control programs.[10]

Fire Suppression

Fire suppression includes all the activities to control a fire, from the time of its detection until it is extinguished. These may involve relatively simple procedures and rapid suppression, or major logistic efforts that last for weeks. Early detection often proves critical. If the fire is discovered before it spreads, one person or a small crew may bring it under control in a short time. For these small fires, attack with shovels, rakes, water from backpack pumps, or other simple tools usually proves sufficient. The crew will first work on hot spots, then cut off fuel at the head of the fire, and eventually work around the flanks by digging out or raking away the fuel ahead of the flames. When the fire reaches the fire line, it burns out from lack of fuel.

Prompt action will allow containment of slowly moving ground fires before they spread. Nevertheless they often prove difficult to extinguish because they burn so deeply in the organic litter. With surface fires that do not spread rapidly, the crews can employ the method described above. However, with fast-moving surface fires and ones that reach heavy fuels, they need a more indirect attack. The crew will begin at the flanks and move around the edges to the point of the fire. This often proves the only safe approach. In some cases, the firefighters go out well ahead of the flames, prepare a fire line, and let the flames burn to meet it. They may even set a fire inside the line and let the flames burn back to the main fire, consuming the fuels along the way. To save time, crews may employ bulldozers and other machinery to push away the fuel ahead of the flames, or to plow up the mineral soil and bury the fuels. Actual techniques depend upon how quickly the crews reach the fire, how much it has spread before they arrive, and the intensity of the combustion due to fuel load and dryness.

Crown fires pose the greatest threat and may prove most difficult to control. These usually become less intense during the coolness of the night when higher relative humidity naturally dampens the flames. By extinguishing the surface fire along the ground, the crew may cause the crown fire to die out. Otherwise firefighters must create fire breaks in the vegetation or find natural ones such as streams and highways. Then they can set back fires to burn out the surface ahead of the crown fire. This reduces the buildup of heat and the chances of crowning. When such techniques fail, the crews move farther ahead of the flames to create new fire breaks, and try again.

Large fires take heavy equipment. Bulldozers, large plows, tank trucks with hoses, and other machines all have a role in suppression activities. In many areas, fire control organizations have large aircraft equipped with water-holding tanks.

These fly over fires and bomb the flames with water and fire retardants such as sodium calcium borate and benonite clay in slurries.[25] Other chemicals, such as silver iodide or dry ice, seeded into clouds may even help to dissipate electric charges that cause lightning.[27] Helicopters serve many functions for observation, transport of crews, evacuation of people, and delivery of special smoke jumper crews by parachute into otherwise inaccessible or remote areas.

Prescribed Fire

Not withstanding the disastrous effects of uncontrolled fire, foresters in America have gradually learned to use fire as an important management tool. Such fires differ from wildfires in that the forest managers carefully prescribe conditions for setting them and the reasons for their use. The prescriptions will limit burning to times when conditions favor combustion, but will not encourage flare-up into major conflagrations. The time of day and the weather at the time of ignition influence the rate and intensity of the burn. Relative humidity and wind speed are especially critical. Carefully constructed fire lines will contain the flames in a desired area, and different methods for igniting the fire, against the breeze or with it, influence the burning rate. By using such carefully planned methods, American forest managers annually burn about one million hectares, mostly in the southern states.

Prescribed burning has several uses. It disposes of logging slash to prevent later unwanted fire or to reduce obstruction to regeneration operations. In southern pine, foresters also use fire to kill understory hardwoods and favor regeneration of fire-resistant pine species. It also lessens damage by brown-spot needle disease.[20] Prescribed burning has been used among coastal Douglas-fir stands to prepare them for regeneration following logging.[5] Fire has also proved useful in thinning overdense ponderosa pine stands, in reducing the fire hazard in many western areas, and in perpetuating certain types of vegetation in our National Parks.[17] Other examples would reflect even broader benefits from the use of fire in many different forest types. However, these few illustrate the modern role of controlled fire as a tool to manipulate vegetation and alter conditions of the physical environment.

PROTECTION FROM INSECTS

More trees die from insect attack than from any other cause. In addition, insects adversely affect tree growth by feeding on the foliage, eating the fruits, or chewing on the roots and bark. By selective attack, one insect may virtually eliminate a tree species from a stand, or prevent its regeneration. Others that bore into the wood often open entrances for wood-rotting fungi, weaken the stem, or render the wood almost useless. Some insects even help to spread diseases that in turn kill the tree. In these ways, insects may upset planned cutting cycles and programs, interfere with recreational use, unfavorably modify wildlife habitat, affect water regimes in streams, and generally alter the overall environment.

Insects and Their Effects

Five types of insects cause damage to forest trees. These include sucking insects that penetrate the bark and leaves and draw out sap. They do not kill directly, but weaken the trees and often facilitate an attack by other insects and by disease. Defoliators strip a tree of foliage, thereby killing it or greatly reducing its vigor. The serious defoliators include certain moths, sawflies, and some beetles. These insects can strip the trees over large areas if their populations reach epidemic proportions. Another group of insects bore into the wood and destroy its product value. These attack many forest types, and even sawn wood products. The bark beetles chew into the bark and cambial region, and thereby girdle the tree (Figure 11–3). The bark beetles have a great effect on timber quality. Also, stain and decay fungi gain entry through their galaries, further compounding the problem. The final group of insects feeds on buds and the terminal shoots of trees. This impairs shoot elongation and deforms the tree. Many of these remain fairly inconspicuous and escape detection, except by skilled observers.

Figure 11–3.
The adult southern pine beetle bores into a tree and chews out a tunnel-like gallery between the wood and bark, where the female lays eggs. After hatching, the larvae widen the gallery and finally move into the inner bark. If large numbers of beetles attack the same tree, their collective feeding can effectively girdle the tree and kill it. *(U.S. Forest Service.)*

Forest insect populations increase rapidly periodically, and then cause heavy damage. Later their numbers decline to an endemic level. Outbreaks of this kind usually trigger a corresponding increase in predators and disease organisms. These eventually check the growth of the pest population and cause its collapse. The defoliators and bark beetles tend to fluctuate most widely in population levels. They may live in the forest for a long period and cause no appreciable widespread damage. Then, when conditions foster a rapid buildup, the numbers suddenly explode into a devastating outbreak. Of all these insects, the ones introduced to an area pose the greatest danger. They usually lack natural enemies, so once a population develops, it may remain at high levels for extended periods. Introduced insects such as the gypsy moth have persisted at epidemic levels for such a long time that they have impacted a large area rather significantly. In fact, entomologists lack an effective means for reducing the population or preventing its spread into areas not previously threatened.

Aside from forest-dwelling insects, others attack wood already in service. Termites, carpenter ants, and powderpost beetles especially may cause major damage in homes and other wooden structures. Powderpost beetles also work in tool handles and furniture, leaving behind their telltale small round holes that detract from the serviceability and beauty of the wood. However, no insect can survive in wood at less than 6 percent moisture content, and even 15–20 percent often provides good protection against them. Proper ventilation usually inhibits the spread. Other insects bore into green logs and lumber, degrading the material or carrying in stain and disease fungi that discolor the wood. These species may cause significant losses to unprocessed logs and lumber. Prompt use of logs and rapid drying of the wood usually will prevent such attacks.

Preventing Insect Problems and Damage

To prevent all insects from feeding on trees would prove impractical and might upset natural ecologic processes. But a forest manager who understands the life cycles and habitat requirements of important local insects can use various silvicultural measures to help minimize their impact. These might include periodic thinning to help keep stands thrifty and free of diseased and weakened trees. Properly planned cuttings and other practices can also reduce the food and other habitat of some insects, and encourage predators and parasites that attack the target insect. Direct control might be necessary if the other measures fail to work effectively. Also, applied genetics programs make available resistant trees for planting in areas where pest problems prove serious impediments to the development of naturally available stock.

With all of these measures, forest managers try to keep the stands vigorous through proper management practices. They use short rotations and harvest the trees before senescence begins. They also remove weakened trees and reduce the amounts of debris that might serve as breeding grounds for pests. Such measures often help to preclude the growth of a pest population, thereby keeping losses within bounds. Usually these indirect means help prevent insect problems at a lower cost than with the application of pesticides or use of unusual biologic control techniques after an outbreak occurs.

Forest managers also can minimize their insect problems by the way in which they choose to regenerate stands.[8] In particular, the choice of a species and the methods employed to reforest an area artificially offer great potentials for minimizing losses. In selecting a species to plant or seed into an area, managers can choose ones that do not suffer from attacks by locally important insect pests. Or they can use genotypes with a proven resistance to damage. In addition, they can limit the planting to species that grow well in that physical environment. Moreover, they can use appropriate methods to prepare the site in advance of seeding or planting. These techniques destroy the habitat for many insects.

In some instances, mixing tree species within the same stand or in adjacent stands provides another practical long-term means for minimizing the dangers of insect problems. Large-scale plantings of a single species over large areas may provide an abundant source of food or other habitat needs for a particular insect. At peaks of its population cycle, the insect may spread rapidly and remain at epidemic levels for long periods. By intermixing species to break the continuity of the host tree, forest managers can often prevent this rapid spread and thereby minimize the losses that might otherwise occur.

Direct Control of Insects

After World War II, DDT (dichloro-diphenyl-trichloro-ethene), other chlorinated hydrocarbons, and various highly toxic broad-spectrum insecticides became available for use in controlling forest insect pests. Field tests with ground and aerial spraying stopped outbreaks of defoliators within minutes of application. But follow-up studies revealed possible side effects, including evidence that widespread use was harmful to some birds, fishes, and reptiles.

These discoveries led to more recent efforts to accomplish biologic and biochemical controls that affect only specific target insect pests. Many approaches have been tried, with varying degrees of success. Of some 520 predatory insects introduced over a 10-year period, only 20 have proved effective. Use of viruses and bacterial diseases with specific prey and host requirements has succeeded to some degree, such as the bacterial insecticide *Bacillus thuringiensis* for controlling certain defoliators.[15] Rearing and release of large numbers of sterilized males also has worked against a few species that concentrate in a relatively small area and where females mate only once. But despite the encouraging leads, dependence upon biologic controls remains more a promise than a reality for many needs. Their use requires further research and testing.

Biochemical means also offer some hope as alternatives to chemical insecticides. These include the pheromones that insects emit and use to communicate with others of the same species. Entomologists have tried sex attractants in the hope of upsetting behavior and interfering with population growth. Other chemicals prevent metamorphosis, and methods of using them to inhibit development of a species have been devised. These and other achievements in chemical ecology afford exciting new possibilities for insect control. When applied in conjunction with other direct and indirect control measures, they offer promise for stemming outbreaks of both native and exotic insects.

Monitoring and Prevention

Surveillance and early detection serve as critical elements in pest management programs.[23] By monitoring the status of known insect pests and keeping abreast of habitat conditions conducive to their spread, entomologists can often forecast potential problems in time to take preventive measures. If detected early, a nucleus population can be controlled before it spreads. And entomologists can warn forest managers of needs for salvage and sanitation operations in dead or weakened timber, thereby reducing habitat conditions that might foster population growth into epidemic levels.

State and federal agencies have primary responsibility for the monitoring of insect populations and problems. They employ entomologists and pest management specialists to undertake the work. Increasingly these people use modern surveillance techniques such as remote sensing and aerial photography as well as mapping and field surveys.[8] Often close working relationships between pest management specialists and forest managers are an important link in communications to alert entomologists of the need for direct measures to control developing insect problems. If given early warning, pest management specialists usually have greater flexibility in developing appropriate ways to protect the forest or prevent serious attack by potentially important insect pests.

PROTECTION FROM DISEASES

Tree diseases seldom cause the sudden spectacular losses that wildfire or insect outbreaks do. Much of the damage by disease remains hidden inside the infected trees as rot, or it results in reduced growth. These follow infestation by foliage diseases and stem cankers. Exceptions to the rule include introduced fungal diseases, such as the chestnut blight brought to the United States in 1904. It essentially eliminated the American chestnut as a commercial tree throughout its natural range.[6] Dutch elm disease, brought in about 1930, also has killed significant proportions of American elm across the continent, and threatens the survival of that species as a major commercial tree.[32] Blister rust brought in from Germany on nursery stock has also spread and has reduced potentials for white pine. Yet most disease organisms have less dramatic effects. They largely impair the soundness, quality, and growth of individual trees in forest stands or along urban streets and in open spaces.

Parasitic Diseases

Tree diseases result from parasitic organisms and abiotic agents. The parasites live within or on another organism and derive sustenance or protection without benefitting the host. They include bacteria, mycoplasts, fungi, viruses, nematodes, and mistletoes.[2,11,21] Among them, fungi cause greater losses in trees than does any other parasite. Nematodes mostly destroy seedlings in nurseries, although pinewood nematode can kill larger trees.[21] Mycoplasma cause such diseases as phloem necrosis of elm and the brooming of black locust. Dwarf mistletoe presents one of the most serious problems of conifers in the west.[21] All can kill trees.

The rust fungi illustrate one group of plant organisms of special importance. These are specialized fungi, such as the white pine blister rust that attacks all five-needled pines. It infects the tree through the needles, spreads through the branch to the main stem, and causes a canker that girdles the tree.[12] Once in the tree, it cannot spread directly to another pine. Instead spores must first infect currant or gooseberry plants, and spread from these bushes back to a pine. Thus control of the alternate host would seem to provide a feasible means of preventing the disease from spreading. But efforts to eliminate currants have generally not succeeded. As a result, the U.S. Forest Service no longer plants western white pine on its lands, or favors the species in natural regeneration.[16] Eastern white pine seems less susceptible, but landowners who have the species may still suffer considerable economic loss due to the blister rust.

Other parasitic diseases move from tree to tree on the bodies of insects. As insects move about feeding, they pick up the organism in one place and serve as the mechanism for inoculating other trees. The Dutch elm disease exemplifies this type. The fungus acts as a parasite on living trees, and persists in the dead ones to rot the wood. Its yeast-like spores disrupt water movement through the tissues and cause the death of the tree. Adult beetles lay eggs in recently killed trees and in logs. The larvae eat galaries in the area between the bark and wood. Then, as the larvae emerge, they pick up the fungus on their bodies and carry it with them to another tree. Later, when feeding on twigs of a healthy tree, the insect may inoculate it, thereby spreading the disease. In this case, elimination of dead and dying elm wood helps to control a spread of the disease. Spraying a biologically degradable insecticide such as methoxychlor has some effectiveness. Also, researchers have used sex pheromones to allure and trap male beetles in experiments to curb the growth of insect populations and slow their spread.[18,26]

Other Diseases

Still another important group of fungi attack tree roots and cause mortality. Root fungi such as *Armelloria mella* or *Fomes annoses* also may decay the aboveground parts of trees. This reduces the vigor, weakens the stem, and the tree may break. Once weakened, the trees also become more susceptible to attack from harmful insects and other disease organisms. These root rots prove especially important in plantations, and can have great impact during early stages of stand development. The close spacing and uniformity make conditions ideal for their spread through inter-root contact between the trees. Often planting at wider spacings will help deter the spread, as will the intermixing of species in the same stand. Some of these organisms can also infect stumps and then spread to living trees. Consequently, in highly dangerous areas, the forester often prescribes chemical treatment of cut stumps as an adjunct to thinning operations.

Dwarf mistletoe is another important disease in western conifers. These parasitic seed plants grow on the branches. Their rootlike outgrowths called haustoria penetrate conductive tissues of the tree, and the plant draws off moisture and nutrients from the host. Tree vigor will decline and mortality may follow. Spreading from tree to tree can take place due to an unusual mechanism of seed dispersal. When the seeds mature, the fruit opens in a way that forcibly ejects the seed as far as 7–15 m horizontally. Other trees within that distance become especially susceptible, but the seed may be spread by animals as well. As a control, foresters use clearcutting and fire to help sanitize infected stands.[23] Pruning off branches and cutting trees with extensive infection will also help reduce the spread.

Wood Decay

Another group of fungi called saprophytes do not kill trees. Instead these organisms invade dead woody material and rot it. Saprophytic decay fungi occur abundantly in fallen trees and downed tree branches, in the organic litter that covers the forest floor, and even on dead portions of standing trees. Some wood-rotting fungi will also cause considerable damage to wood in service. In fact, as much as 10 percent of the country's annual timber production will replace wood that decayed after being put into use in products and structures.[2] In these cases, just as in natural environments, invasion by the fungus depends upon moisture in the wood. Good drying before use and proper ventilation afterward help to minimize losses to these wood-rotting fungi. Where the wood material will have contact with the ground, as with railroad ties and poles, the products can be impregnated with such preservative chemicals as creosote, zinc chloride, or pentachlorophenol to inhibit or restrict fungal development and prevent damage.

Wood decay fungi of living trees cause more damage to standing timber than do any other destructive agent of the forest. Wind-blown spores of these fungi infect standing trees through such injuries as broken branches, fire scars, insect entry holes, logging wounds, or any opening through the bark that exposes the wood. As these fungi grow in the tree, the fungal hyphae break down the cellulose and lignin of the cell walls. Though they generally do not kill the tree, the fungi render the wood useless for lumber. The rot may also so weaken the main stem that it will break from the force of the wind, heavy snow loading, or other pressures.

In well-advanced stages, decay fungi produce fruiting bodies that appear as easily recognized conks on the standing trees. Spores dropping out of these may carry long distances in the wind to infect wounds on other trees. Forest managers and pathologists have limited means for controlling such organisms, except through indirect silvicultural methods to keep trees healthy and minimize wounding that might open a tree to infection from fungi attack.

The process of wood decay in some cases appears to progress in stages, and may involve more than one organism. Only after one group has acted on the wood do others take over to continue the decay process.[29] Acquaintance with the ways fungi work in trees and the external signs of their presence inside a tree enable the experienced forester to judge the extent of decay (Figure 11–4). By using these signs, the forest manager can select infected trees for felling during thinnings and other improvement cuttings. This helps foster growth of trees free from disease. Also, by knowing the extent of decay in trees, foresters can judge their merchantability and determine the amounts of usable volume they might recover during harvesting operations. Detection with devices such as electric resistance meters may indicate more precisely the extent of decay in infected trees, and aid management decisions about them.[30]

Figure 11–4.
Wood decay progresses in stages and may involve more than a single organism, as with this rot that developed in a sugar maple following a basal wound to the tree. *(U.S. Forest Service.)*

Abiotic Diseases and Injury

Abiotic diseases also cause injury and death to forest and urban trees. These include effects of insufficient or excessive water, extremes of temperature, salt spray from highways and the ocean, absorption of toxic chemicals, and other nonbiologic factors that cause stress in trees. Their effects often prove difficult to verify. Damage sometimes develops slowly, but in other cases mortality may show up suddenly. In some instances, the damage may cover a large area, as with the extensive mortality of ponderosa pine in 1969 along the San Bernadino Mountains of California. Studies indicated that automobile exhaust gases contributed to the losses.[4] Likewise, tests in the copper basin of Tennessee have linked sulfur dioxide fumes from the processing plants with mortality of trees nearly 15 km from the smelter.[14] Other investigations of declines and diebacks in the Northeast have attributed these, at least in part, to environmental stresses resulting from air pollution.[31] Such evidence suggests that environmental pollutants may contribute to widespread losses not previously associated with any particular causal organism.

Conditions of the physical environment not linked to pollution also may stress trees and cause declines in growth, and even mortality. These may become manifest during years of extreme dryness, in times of excess moisture, or as an aftermath of temperature extremes. Red pine trees planted in shallow or poorly drained soils of the East, for example, often have small and poorly developed root systems that pull in insufficient amounts of moisture to satisfy transpirational and physiologic uses during warm and dry periods of summer. On such sites, the trees grow slowly, have poor vigor, and may eventually die.[21] On deeper, well-drained soils, this same species grows well and has proved highly productive.

A deficiency of nutrients may affect the health and vigor of virtually any species. This shows up as a discoloration of the foliage, as a reduction in growth, and in abnormalities in fruit production. Again these symptoms often appear in trees planted on sites where they would not normally grow. In some cases, foresters have devised preventive measures, such as fertilization of sites, to overcome certain nutrient deficiencies. In other cases, they may inoculate the seedlings with mycorrhizal organisms prior to planting. These infect the roots. They do not harm the tree, but tend to extend the absorptive surface and facilitate nutrient uptake.[22] Generally a careful matching of the species to the site will prevent these and other types of root-related maladies, and pay off in the long run in better growth and higher yields from a stand.

Many types of abiotic problems defy human control. Pathologists neither understand the cause in many cases, nor have a remedy to correct the ill. Foresters try to anticipate what problems might develop and use management practices that maintain healthy, vigorous conditions in the forest. Among these, careful control of the species composition often proves most critical. Natural forests and those of locally adapted species usually withstand both the abiotic and biotic diseases best.

On the other hand, plantations of poorly suited trees tend to have particular susceptibility to such things as extremes of climate, soil problems, and many of the biologic diseases. Even in areas threatened by chemical pollution, use of carefully selected species that tolerate the situation may circumvent a problem. Generally the forester will find it less costly to control the species composition at the beginning of a rotation, than to correct any difficulty that might arise later on.

WEATHER AND ITS EFFECTS

Several other common types of forest damage problems warrant mention. These do not qualify as diseases, but do cause important losses in many areas. Again they frequently result from weather-related events and often the forester can mitigate their effects by wise choices during the stand regeneration period. On wetter sites, for example, frost action may heave seedlings of any species from the soil during winter. This usually happens during the first year after planting, and becomes less likely as the root systems grow and anchor the trees better. In most instances, the forester can minimize the problem simply by planting in spring, rather than during fall. Root elongation during the summer will usually prove sufficient to hold the tree in place the following winter. Later in the life of a stand, high winds may uproot the trees or break them off (Figure 11–5). But management schemes that

Figure 11–5.
Strong winds may break or uproot large trees, but management schemes that take into account the dangers can substantially reduce potential losses on wind-prone sites.

maintain a proper density and preserve the integrity of the crown surfaces along the edges of the stands may effectively reduce the possibility of damage. Also, wet snows and freezing rains may overload the tree crowns and cause breakage. For sites prone to these weather conditions, use of narrow-crown species and maintenance of mutual support between adjacent trees will help them withstand loads of snow and ice without harm. Hence, by knowing the potential environmental conditions and adjusting species selection and subsequent management to meet any adversities, the forest manager can often circumvent possible problems that otherwise might prove substantial.

Some types of environmental forces will remain beyond human control. These include the widespread effects of severe weather—extreme cold, high temperatures, hurricanes, and flooding. These happen at unpredictable intervals and often cover great expanses of the countryside. Their force or impact can become so powerful that the resultant destruction or damage takes years to repair. Their extent causes some unusual economic problems as well. To illustrate, a hurricane that forms in the Atlantic off the Carribbean islands may build up tremendous force and move northward up the coast or farther inland through the Gulf States. Some pass offshore and do not hit land until New England. The winds can reach speeds of 50–80 km/hr. They can rip roofs from houses, overturn trailers, and tear down powerlines. In forests, they blow down the timber and snap off tree tops. Foresters can only enter the woods after the destruction, try to salvage the timber, and take steps to regenerate new stands of trees to replace the losses. However, the damage often covers so large an area that the mills cannot take all the timber. Landowners who move quickly will find markets for the trees, but the others will suffer losses that can be financially devastating. We cannot control or protect against such catastrophic effects, but can only react by organizing efforts to rebuild for the future.

PROTECTION FROM ANIMALS

Animals of all types are an integral part of forest and range ecosystems, and affect the plants in them in many ways. They also perform a variety of beneficial functions, such as helping to disperse seeds and controlling the number of other animal species. But wild animals also injure trees and other vegetation. They browse on the twigs and foliage, uproot entire plants, and dig into them for feeding and nesting. In most cases, the damage is not extensive overall, except where unusual conditions or human intervention upsets the ecologic balance. An animal population then may increase, overtax the capacity of the ecosystem to support it, and become destructive. Theoretically at least, predation should eventually equalize the situation and a balance between plants and animals develop again. However, if people reduce the predation or destroy the natural balance in some other way, the population of a normally harmless animal may multiply and seriously impact the forest or range.

Since wild forest animals must derive their food from that environment, it often comes in part from woody vegetation. They eat leaves, twigs, buds, bark, fruits, and fungi that grow on or in living and dead trees. In the process, the feeding inflicts some injury on the vegetation, though normally not enough to threaten the plant's survival or subsequent growth. As long as the number of animals remains in balance with the capacity of the forest to provide this food, few problems develop. But at peaks of their population cycles, or when for some other reason the number of wild animals using a forested area increases dramatically, feeding and other uses may prove harmful to the trees.

Different creatures feed upon trees in various ways. Mention of a few will illustrate the diversity of this use. For example, squirrels eat the fruits, and during times of high population levels may consume most of the acorn crop in an oak forest. Mice, chipmunks, and birds also eat tree seeds in large quantities, and some pull up and eat young seedlings. Deer and rabbits feed upon twigs. White-tailed deer tend to browse selectively upon some species, and in the East they eat maples, ashes, and birches in preference to beech and spruce.[33] Rabbits, mice, squirrels, porcupines, deer, and elk eat tree bark as well. Ruffed grouse, squirrels, porcupines, deer, rabbits, and many other animals also feed upon buds. Beaver will cut off entire trees to get at the bark and twigs, and will dam up swampy areas and streams to inundate adjacent forest stands and kill the trees. In these and many other ways, wild animals continually affect forest ecosystems. Mostly the animal damage proves of little consequence overall. Integrated application of basic principles of wildlife and timber management will provide an adequate solution to most problems that develop.

For most timber management purposes, the greatest threat from animals normally comes when the forester tries to regenerate a stand. Then heavy feeding on the seeds can prevent regeneration, and overbrowsing may destroy existing seedlings. As a remedy, the forest manager may promote hunting and trapping, or employ other methods to control the population. Often, if the forester and wildlife manager work together, they can find a solution that will meet the timber production needs without seriously affecting the long-term welfare of the animal population. The controls may stir controversy among special-interest groups, necessitating communication with them to explain the needs and solicit their support. The program must also respect state and federal laws that regulate hunting and trapping or the use of other control methods. However, where foresters and wildlife managers have implemented reasonable programs, effective control has resulted during the regeneration period and has led to the successful establishment of new seedlings of suitable species.

Domesticated livestock present a different situation. They can cause unusual pressures on a forested ecosystem unless managed properly. Livestock has few natural enemies, and economics dictates keeping herds of cattle, sheep, or goats in fairly large numbers concentrated in a relatively small area. Unless regulated, they will consume available forest browse in a relatively short period, strip the land of understory vegetation, and prevent its regrowth. Milling animals also trample small plants and compact the soil. This may reduce the infiltration capacity, soil aeration, and porosity. Root growth may suffer and soil moisture-holding capacity decline. Surface erosion may increase. In turn, these changes will adversely effect tree growth and overall stand productivity.

Not all domestic grazing is harmful. Areas of open conifer cover can serve both for timber production and grazing, and tend not to suffer in the ways described. In southern conifer forests, for example, the low-density timber stands support considerable grass and herbage between the trees. Landowners can graze cattle to take advantage of the potential for forage. The feeding also reduces the ground-level fuels that might otherwise prove a fire hazard, and does not harm the trees in the process. Grazing of sheep and goats may eliminate unwanted understory hardwoods in areas scheduled for conversion or replanting to pines. Extensive areas of open forests also are found throughout the Rocky Mountains and are important to the livestock industry as grazing lands. In regions where the forests form dense stands that cast dense shade, however, the animals will find little forage in the understory. These stands can support relatively few cattle per hectare, and overbrowsing often leads to tree damage as well as destruction of many features of habitat essential to indigenous wild animals. In these cases, forest managers will need to exclude domesticated animals by fencing, or regulate their number and duration of use to prevent the harm that might follow uncontrolled grazing.

PROTECTION FROM PEOPLE

The release of chemical wastes through industry and domestic activities also cause the mortality or damage of trees. These problems frequently develop in urban and industrial areas. Another class of people-related damages result from mechanical rather than chemical causes. They show up even in remote forests, and in any place that people frequent. Some come from the exploitation and use of resources in commercial ventures, and some from carelessness in recreational use of forests and range. All have protection implications for foresters.

Any earth moving or construction that alters the natural drainage through soils may affect tree growth. Usually this is caused by the blocking of drainage, but an activity that improves water flow out of a swampy or boggy area also could upset the vegetation and animals there. Foresters may avoid these problems by the proper location and design of roads and similar improvements. Other losses result from the careless clearing of land in exploring for oil and minerals, in strip mining, and in extracting many types of products. The exploration and mining may also release chemicals and wastewaters and could affect the quality of water in streams and ponds nearby unless properly handled. People can cause still other problems, as when impounding waters for lakes and ponds, clearing land for recreational sites and structures, and converting the natural vegetation to an artificial type.

While these often do not impact large areas, they all alter the ecosystem. If carelessly undertaken and not controlled in their extent, many could precipitate needless losses of an important, if intangible, nature. These, too, require solutions that permit efficient operation while maintaining safeguards against unnecessary environmental harm.

The rapid growth in outdoor recreation since the 1950s has increased protection problems in many forests. Their more frequent use by people raises the threat of fire through carelessness with campfires, cigarettes, and the discarding of combustible materials, and from other causes. Widespread use of off-road vehicles in summer and snowmobiles in winter has brought extensive damage to trees, roads, and trails in many areas. The effects on roads has triggered erosion in some hilly terrain, and overtaxed maintenance budgets to repair the damage. This harm to soils has special importance in light of the federal government's efforts to address water-quality needs through the Water Pollution Control Act. Increased utilization of forests for recreation has also left behind large quantities of trash. While not necessarily harmful to the environment, the refuse may interfere with other visitors' enjoyment. Like most people-caused problems, these reflect carelessness. Solutions often prove costly, and in some cases foresters have had to close the forest to recreational use.

Probably the greatest incidence of damage by people occurs in urban areas. Aside from the industrial causes already alluded to, many come simply from the intense activity in the cities. People trample the ground, compact the soil, and break the vegetation in parks and plazas. They run into trees and shrubs with their cars and other machines. They plant trees on sites and under conditions poorly suited to the species, and in soils compacted and otherwise interfered with during construction. The exhausts from chimneys and vehicles cause numerous abnormalities in tree growth, and kill some trees. People even poison their trees by applying too much fertilizer, carelessly using weed killers, and inadvertently harming the root systems and foilage. These all require unique solutions. It takes proper planning and subsequent caution in development and landscaping. Self-restraint will often serve as an excellent means for protection thereafter.

CAREER OPPORTUNITIES IN FOREST PROTECTION

As a result of the diverse ways in which forests suffer harm, many different scientific disciplines contribute to forest protection activities. These include fire protection, fire behavior, entomology, pathology, physiology, and plant nutrition. Additionally most forest managers invest at least part of their time in addressing forest protection problems. They become involved with both prevention by direct and indirect means and control when outbreaks occur. Consequently forestry schools include courses to familiarize all graduates with basic approaches to forest protection, and how to avert problems through proper management. These can include work in entomology, pathology, fire prevention and control, or integrated pest management that combines elements of all three. Students interested in directing their careers toward work in some specific area of forest protection can later strengthen their capabilities through elective courses and graduate study.

In many parts of the country, foresters will find opportunities to work within fire management. These usually are with state or federal agencies, though some large companies train fire crews for emergency service on their own lands. Professionals would have responsibility for planning and supervising prevention programs, and for overseeing suppression when fires occur. In many cases, individuals can develop their skills through in-service training. In addition, they can pursue college courses that involve meterology, fire science, personnel organization and management, equipment use and maintenance, forest engineering, and other fields that strengthen their management and supervisory skills. Some universities even offer opportunities for graduate study through the Ph.D. degree related to fire science or ecology. Persons with these advanced degrees might work in research and teaching rather than with fire management agencies.

Most state departments of conservation and natural resources hire forest entomologists and forest pathologists. The federal government offers opportunities in the Bureau of Plant Quarantine, the U.S. Forest Service, and other agencies of the Departments of Agriculture and the Interior. These include positions for people schooled at the undergraduate and graduate levels. Persons holding advanced degrees would receive the more scientifically specialized positions, and work in research and teaching as well. In these cases, the students would need to strengthen their capabilities in the basic biologic sciences. Courses might include plant and animal taxonomy, physiology, ecology, and management. Since most career opportunities would require graduate study, many interested individuals can first complete a forestry degree, and then concentrate on the entomology or pathology.

Whether they pursue it as a specialized career field or through more general responsibilities as a forest resource manager, most foresters become involved with forest protection. Through good management, they can institute practices that will prevent many problems from developing on a catastrophic scale. When outbreaks occur, they will join in efforts to combat the problem. And afterward they will work to salvage the damaged trees and promote the successful regeneration of new crops to replace the one damaged. Foresters must become equipped to handle forest protection and involve themselves with it in the practice of their careers.

REFERENCES

1. J. S. Barrows, "Forest Fire Research for Environmental Protection," *Journal of Forestry* **69**(1):17–20 (1971).
2. J. S. Boyce, *Forest Pathology,* McGraw-Hill, New York, 1961.
3. A. A. Brown and K. P. Davis, *Forest Fire: Control and Use,* McGraw-Hill, New York, 1973.

4. F. W. Cobb, Jr. and R. W. Stark, "Decline and Mortality of Smog-injured Ponderosa Pine," *Journal of Forestry* **68**(3):147–149 (1970).
5. J. D. Dell and L. R. Green, "Slash Treatment in the Douglas-fir Region—Trends in the Pacific Northwest," *Journal of Forestry* **66**(8):610–614 (1968).
6. J. D. Diller and R. B. Clapper, "A Progress Report on Attempts to Bring Back the Chestnut Tree in the Eastern United States 1954–1964," *Journal of Forestry* **63**(3):186–188 (1965).
7. I. S. Goldstein, "Potential for Converting Wood Into Plastics," *Science* 189:847–852 (1975).
8. S. A. Graham and F. B. Knight, *Principles of Forest Entomology,* 4th ed., McGraw-Hill, New York, 1965.
9. R. F. Hammatt, "Bad Business: Your Business," in *Trees, The Yearbook of Agriculture,* U.S. Department of Agriculture, Washington, 1949.
10. C. Hardy and A. P. Brackenbusch, "The Intermountain Fire-Danger Rating System," in *Proceedings of the Society of American Foresters, 1959 National Convention,* San Francisco, 1960, pp. 133–137.
11. G. H. Hepting, *Diseases of Forest and Shade Trees in the United States,* U.S. Department of Agriculture, Forest Service, Agriculture Handbook no. 386, 1971.
12. R. R. Hirt, *Blister Rust, A Serious Disease of White Pine,* New York State College of Forestry, Syracuse, N.Y., 1933.
13. D. R. Houston, E. J. Parker, R. Dorrin, and K. J. Lang, "Beech Bark Disease: A Comparison of the Disease in North America, Great Britain, France, and Germany," *European Journal of Forest Pathology* 11:199–211 (1979).
14. C. R. Hursh, "The Forest Legion Carries On," *American Forests* 38:16–19 (1932).
15. G. W. Irving, "Agricultural Pest Control and the Environment, *Science* 168:1419–1424 (1979).
16. D. E. Ketcham, C. A. Wellner, and S. S. Evan, Jr., "Western White Pine Management Programs Realized on Northern Rocky Mountain National Forests," *Journal of Forestry* **66**:329–332 (1968).
17. B. M. Kilgore, "Integrated Fire Management on National Parks," in *Proceedings of the Society of American Foresters, 1975 National Convention,* Washington, 1975, pp. 178–188.
18. G. N. Lanier, R. M. Silverstein, and J. W. Peacock, "Attractant Pheromones of the European Elm Bark Beetle *(Scalytus milistriatus)*. Isolation, Identification and Utilization Studies," in J. E. Anderson and H. K. Kaya (eds.), *Perspectives of Forest Entomology,* Academic Press, New York, 1976.
19. L. Ledbetter, "Two Million Acres in West Buin. Lightning Storms, Increase Danger," *New York Times,* August 8, 1977.
20. P. E. Lightle, *Brown Spot Needle Blight of Longleaf Pine,* U.S. Forest Service Forestry Pest Leaflet. 44, 1960.
21. P. D. Manion, *Tree Disease Concepts,* Prentice Hall, Englewood Cliffs, N.J., 1981.
22. D. H. Marks, W. C. Bryen, and C. E. Cordell, "Survival and Growth of Pine Seedlings with *Disalithus Ectomycorrhizae* After Two Years on Retereitation Sites in North Carolina and Florida," *Forests Science* **23**:363–373 (1977).
23. C. A. Meyers, F. G. Hawksworth, and J. L. Stewart, *Stimulating Yields of Managed Dwarf Mistletoe-infested Lodgepole Pine Stands,* U.S. Forest Service Research Paper RM-72, 1971.

24. National Research Council, *Renewable Resources for Industrial Materials,* National Research Council, Board of Agriculture and Renewable Resources, National Academy of Science, Washington, 1976.

25. C. B. Phillips, "Fighting Forest Fires from the Air," in *Proceedings of the Society of American Foresters, 1959 National Convention,* San Francisco, 1960, pp. 137–140.

26. L. R. Schrieber and J. W. Peacock, *Dutch Elm Disease and Its Control,* U.S. Department of Agriculture Information Bulletin 193, 1974.

27. K. R. Shea, "Progress on Integrated Pest Management: Douglas-fir Tussock Moth, Gypsy Moth, and Southern Pine Beetle," in *Proceedings of the Society of American Foresters, 1975 National Convention,* Washington, 1975, pp. 218–223.

28. A. L. Shigo, "The Beech Bark Disease Today in the Northeastern U.S.," *Journal of Forestry* **70**:286–289 (1972).

29. A. L. Shigo and E. vh. Larson, *A Photo Guide to the Patterns of Discoloration and Decay in Living Northern Hardwood Trees,* U.S. Forest Service, Research Paper NE-127, 1969.

30. A. L. Shigo and A. Shigo, *Detection of Discoloration and Decay in Living Trees and Utility Poles,* U.S. Forestry Service, Research Paper NE-294, 1974.

31. A. L. Shigo and R. L. Campara, *Forest Diseases Research Priorities in the Northeast,* Forestry Diseases Subcommittee, RP 203, Northeastern Forestry Commission, Publication by U.S. Forest Service and Universities of Maine, Massachussetts, New York, New Hampshire, Pennsylvania, Vermont, and West Virginia, 1976.

32. W. A. Sinclair and R. J. Campana (eds.), *Dutch Elm Disease Perspectives After 60 Years,* Northeastern Regional Research Publication Search, vol. 8, no. 5, 1978.

33. W. C. Tierson, E. Patrick, and D. B. Behrend, "Influence of White-tailed Deer on the Logged Northern Hardwood Forest," *Journal of Forestry* **62**(12):801–805 (1966).

34. M. Tribus, "Physical view of cloud seeding," *Science* 168:201–211 (1970).

35. U.S. Forest Service, *The Outlook For Timber in the United States,* U.S. Department of Agriculture, Forest Service, Forests Resource Report no 20, 1973.

36. U.S. Forest Service, *RPA Recommended Renewable Resource Program Final Environmental Statement to Renewable Resource Program, 1977 to 2020,* U.S. Department of Agriculture, Forest Service, Washington, D.C., 1976.

37. H. E. Wright and M. L. Heinselman, "The Ecological Role of Fire in Natural Conifer Forests of Western and Northern North America—Introduction," *Quarterly Research* **3** (3):319–328 (1973).

CHAPTER **12**

RECREATION AND AMENITIES

People throughout the world depend upon forested lands and rangelands to provide a wide variety of commodities essential to human welfare. They look for many indirect benefits as well. These include landscape and esthetic values, use as a habitat for nongame wildlife, use for recreation, influences on local air quality, and many other intangibles we perceive as important. Many of these are difficult to evaluate, such as the satisfaction derived from leisure time activity. These feelings may develop when tending flowers and landscaping a lot in a highly urbanized environment, or upon penetrating a remote forest that lacks apparent traces of human activity. Forests and rangelands, or even a small collection of trees in a city neighborhood, provide for recreation and amenity needs. They offer a setting where individuals can remove themselves from routines of life, and where the trees and other natural features help to enhance the experience.

SCOPE AND POPULARITY OF OUTDOOR RECREATION

Outdoor recreation provides relief from daily stress, greater self-confidence, improved physical fitness, and better mental health.[21] These have great social worth for personal well-being and a higher level of individual productivity in many ventures. By definition, recreation should provide such renewing experiences, and involve a voluntary investment of time in return for enjoyment rather than material gain. Though great in variety, outdoor recreation commonly includes physical activity that takes place in an outdoor environment, or depends upon that setting. Examples include canoeing and boating, swimming, fishing, hiking, camping, hunting, gathering food and flowers, collecting rock and mineral samples, photographing plants and animals, and bird watching. Many go together, such as picnicking and hiking, or fishing and camping. Actually many activities in which people participate could take place almost anywhere, but are enhanced by the outdoors.[9,21] The setting attracts people who wish to enjoy the landscape as well as the recreational activity itself (Figure 12–1).

The popularity of outdoor recreation shows up clearly in statistics presented by the U.S. Forest Service as a result of the nation's first assessment of forests and rangelands under the Resources Planning Act of 1974 (Table 12–1). Picnicking, nature study, pleasure driving, and sightseeing lead the land-based activities. These involve a great many American households, and represent favored activities repeated at least four times a year. Outdoor swimming stands out as the most favored water-related recreation, and many people appear to enjoy boating as well. Relatively few use the outdoors during winter for snow- and ice-dependent activities, except for local sledding and skating.[22] Probably a good many people engage in more than one of these pursuits, and most people seek out some type of outdoor recreation during a year. They might not venture far from their neighborhood, but somewhere in an outdoor setting most Americans find a way to rest and relax through outdoor recreation.

Figure 12–1.
Outdoor recreation includes all those leisure activities that take place in the outdoors and depend upon or become enhanced by the setting and associated natural resources.

TABLE 12–1
OUTDOOR RECREATION ACTIVITY IN THE UNITED STATES, 1977

Activity	Percentage of American households that participate in outdoor recreation each year: At least once	More than four times
Land-based activities		
Developed camping	30	12
Dispersed camping	21	9
Off-road driving	26	20
Hiking	28	16
Horseback riding	15	8
Nature study	50	36
Picnicking	72	49
Pleasure driving	69	57
Sightseeing	62	36
Water-based activities		
Canoeing	16	5
Sailing	11	5
Other boating	34	20
Outdoor swimming	61	47
Water skiing	16	8
Snow and ice activities		
Cross-country skiing	2	1
Downhill skiing	7	4
Iceskating	16	0
Sledding	21	12
Snowmobiling	8	5

Source: Based U.S. Forest Service [22]. Used by permission.

Statistics for nonconsumptive use of wildlife also demonstrate the popularity of outdoor recreation. Such activities involve about 27 percent of the country's population who observe wildlife in an outdoor setting, and 22 percent who call themselves bird watchers. Between 20 and 30 percent of American families purchase seeds for feeding birds, presumably to enhance viewing in their own yards or in their patios. Many others visit zoos and participate in organizations interested in animals. These include people of varied backgrounds and socioeconomic status.[14] Further as with so many other leisure activities, those related to wild animals became enhanced by the forest or range setting, or by the urban trees and shrubs that provide the essential habitat for the creatures to live in.

Growth of Recreation

Beginning in the mid-1950s, outdoor recreation activities began to grow. This came about in part from the increased leisure time available. Workdays dropped to eight hours, work weeks to five days, and employers adopted more flexible work schedules. From 1960 to 1969, vacation time benefits rose almost 50 percent, and reached an average of 2.2 weeks by 1970.[20] This trend continued. The number of leisure hours for a married working male rose from about 33.7 a week in 1965 to 36.1 in 1975. Employed working women increased their weekly leisure from 26.7 to 31.7 hours during the same period. At the same time, disposable income nationwide rose from about $244.3 billion in 1940, to $612.4 billion in 1965, and $966.0 billion in 1978 (based upon the value of 1972 dollars).[22] All these changes encouraged a rapid increase in the interest in recreation, and gave people the capacity to implement it.

During those days, automobiles and abundant fuel for them could be purchased at relatively low cost. Nationwide highway construction, including the interstate system, gave people better access to recreational facilities. Entrepreneurs, communities, and governmental agencies began publicizing the opportunities available. This encouraged people to look for exciting ways to spend their leisure time, and the outdoors seemed a logical place. In response, manufacturers developed new types of equipment. The improved materials and designs made sport and recreation easier and more pleasurable. All this encouraged Americans to spend their leisure time out-of-doors.

At that time, nearly all of the forests and rangelands in the United States had some potential for recreation. However, as of 1980, only about 29 percent of the noncorporate private land and 54 percent of that held by corporations has remained open to public use for hiking, hunting, fishing, horseback riding, off-road vehicle driving, picnicking, and the like. Another 54 percent of the noncorporate and 15 percent of the corporate lands are restricted, or available only to special groups who pay a fee for the privilege. As an alternative, most people can turn to the 290.7 million hectares of federal forest land, the 10.6 million hectares held by the states, and additional areas controlled by county and local governments as a major base of land for recreational activities.

On all types of ownerships combined, the nation has about 15,850 developed campgrounds, with nearly 58 percent operated privately. About 3.5 million families own vacation homes, with nearly 2.4 million hectares of land subdivided for this type of recreational use. In addition, nearly 5.2 million kilometers of rivers and streams remain open to public access, along with scores of lakes, ponds, and reservoirs. Downhill ski enthusiasts can find more than 2240 lifts across the United States, while opportunities for cross-country skiing remain almost limitless on millions of hectares of private and public land. Nevertheless, national statistics show a great need to expand the developed facilities to relieve overcrowding, to improve conditions, and to make them more readily accessible at lower cost.[22]

For the most part, studies of developed and dispersed recreation suggest continued growth well into the future. Land-based activities may increase by 20 percent over the 1977 level by the year 2000, and by as much as 60 percent by 2030. Water recreation could rise about one third by 2000, and more than double by 2030. Winter sport use may grow by nearly one quarter by 2000, and as much as 2.4 times by 2030.[22] These projections from the 1980 Resources Planning Act assessment forecast a growth due to an increase in the population, coupled with a continued capacity to take advantage of opportunities offered by our forests and ranges. However, if supplies of low-cost fuel diminish, or if costs inflate more than wages rise, the growth might not materialize. Although recreation close to home then may increase, activities at remote locations could decline.

Economic Worth

The growth of outdoor recreation has triggered a rise in economic activity as well. Sales of recreational equipment and associated items have reached into the billions of dollars. Estimates for 1978 indicate that Americans spent about $2 billion on organized camping in that year. And two years earlier, in 1976, they had spent $1 billion on recreational vehicles and trailers, and $5.6 billion on boats. Other expenditures for 1978 included $3 billion on hunting and fishing, $4 billion on swimming, $7 billion for golf, and $800 million on bicycles. Cities also reported spending more than $965 million on recreational services, and states another $670 million. The federal government alone listed expenditures of about $5 billion for park and recreation programs in 1976. Altogether, as of 1978, Americans were contributing about $180 billion annually for recreation to the U.S. economy.[21]

These expenditures obviously support a major industry that provides a wide range of services and materials, such as food and lodging, guides, equipment, and gasoline. These revenues go to individuals and companies who work for government or operate the service industries and various enterprises catering to the needs of recreationists. The spending has a multiplying effect in communities through sales, added income, and more employment. However, little goes directly to the forest landowner, except in the form of entrance and use fees for developed recreation sites or restricted lands. In the case of public lands, the government rarely realizes enough revenue from use fees to pay the costs of maintaining the recreation program. Legislative appropriations or timber revenues carry most of the cost and subsidize the recreation.

Consumptive use of wildlife for fishing and hunting also has great economic value. In 1978, expenditures for these activities totaled about $3 billion nationwide.[21] Surveys indicate that about 29 percent of Americans engage in some type of fishing, and spend an average of $160–$210 each per year in pursuit of their sport. Those who hunt game and migratory birds spend an average of $104–$196 per year each. In both cases, costs include equipment, license fees, ammunition and bait, travel, meals, and special clothing. Most of the money goes to a variety of vendors, but even state governments benefit. Of the $135 million available for administration of inland sport fisheries and wildlife resources in 1971, about 33 percent came from hunting license fees, and 29 percent from sales for fishing.[22] In addition, the Pitman-Robertson Act levies a manufacturer's excise tax on sporting arms and ammunition to go to state agencies for wildlife management. The Dingle-Johnson Act provides funds to states for fisheries programs.

These statistics point up the importance and economic worth of outdoor recreation in the United States. Clearly the qualities of forests and rangelands that attract people for leisure pursuits have economic relevance. They stimulate local economies, as well as businesses within and near the major population centers. Outdoor recreation also affects land prices, influences vacation homesite development, and promotes the sale of goods and services. It creates jobs in construction, service industries, administration, and maintenance.[9] On the other hand, many people spend little on recreation. They visit local parks, derive pleasure from greenbelts along roads and highways, and enjoy watching birds and small animals that live in the shade trees and shrubs of their neighborhood. The nation's urban forest of street and yard trees, the parks of cities, and the islands of forest-like vegetation in the expanding suburbs all enhance the quality of life for those who live there.

DEMANDS UPON FORESTRY

The growth of recreation has placed at least two demands upon traditional forestry programs, especially on public lands. First, as people crowded the campgrounds and day-use areas and taxed the limits of facilities available, they began to ask for more and better ones. This brought pressures to locate new sites and develop additional facilities to accommodate the demand. It also closed off some types of resource use in and around the facilities. Second, growth in hunting, hiking, pleasure driving, picnicking, and other dispersed activities brought more people onto the forest roads and access trails. Off-road vehicles of all types made previously remote areas more accessible. This compounded maintenance problems and fostered a demand for more trails and roads for public use. People asked for more parking facilities near the popular fishing, picnicking, and hunting spots. Foresters had to devote energy and funds to accommodate the needs, sometimes at the expense of their timber management programs.

As people used the forest more and more, they began to see timber harvesting and management at first hand. They were vocal in their criticism of some forestry practices which they felt degraded the esthetic values of the forest and interfered with their recreation. Foresters found themselves embroiled in conflict and faced with mounting pressure to forego some traditional methods, or to implement more costly alternative practices to enhance landscape values. It was just such a situation that helped to trigger public reaction against the use of clearcutting on federal lands. That led to court actions to prevent it in the Bitterroot, Medicine Bow, and Monongahela National Forests in the early 1970s. The National Forest Management Act eventually resolved the issue. But even in the private industrial forests in some regions, people continued to pressure for greater attention to their recreational interests. Many companies have responded as goodwill gestures, and as a part of their public relations programs.

Similar demands have impacted rangelands. In fact, no place seems exempt, no matter how remote. The result is conflict. For example, to maintain huntable populations of big game, ranchers must divide forage use between cattle and wild animals. Then hunters and others with off-road vehicles may damage the range vegetation and cause its deterioration. Vacation homesite developments confound planning and rangeland administration. Less direct impact comes from hiking, cross-country skiing, relic hunting, and other activities that do not disturb the range—but growth of any sort does increase the likelihood of range fires and vandalism. In contrast to these problems, sale of hunting privileges provides a major source of income to many western ranchers, and development of tourism attracts business to local communities and establishments. Clearly recreation managers and ranchers must work together to find ways to blend the alternatives so that they can capitalize upon the recreational potentials without threatening grazing and other commodity uses people demand from rangelands.[19]

In response to these increased recreational pressures upon forests and range, managers have looked for ways to adjust standard practices to enhance the recreational experiences of people who have little direct interest in commodity values of the land. These efforts have been most intense on public lands, and the U.S. Forest Service has implemented many changes in response to the multiple-use planning requirements of the National Forest Management Act of 1976. First, they tried to determine the kinds of things about which recreationists felt dissatisfaction, and then looked for ways to alter normal management to accommodate these interests. For example, one study of public reaction to scenes in western conifer forests revealed that people prefer landscapes that appear natural and orderly, rather than disturbed and disorderly. Also, revegetation and tree growth seem to temper a viewer's reaction to the same scene. In practice, this would suggest use of partial cutting compared with clearcutting, reducing logging debris after a harvest, correcting evidence of soil and litter disturbance, and promoting rapid revegetation along roadsides.[1]

Besides trying these practices, the Forest Service has also zoned the National Forests and identified areas where recreational and esthetic values are very important. They found that about 12 percent of the National Forest area fits into this special category. For these areas, the management must enhance visual qualities by using special logging methods, planting large-sized seedlings after cutting, cleaning up slash, lengthening the rotation for a stand, altering the shapes and sizes of cutting areas, and staggering the sequence of stands cut. While such measures have helped to maintain scenery that recreationists prefer, they also cost more to implement. In the Mt. Hood National Forest in Oregon, use of these practices has increased the cost of timber production by 14 percent over that by standard methods.[5] Since the Congress has not appropriated funds to offset this increase, the U.S. Forest Service has had to pay the costs out of its other timber management funds and to cut back on other work to make up the difference. The experience has shown that balancing timber and noncommodity opportunities is not easy. Benefits gained by giving attention to one program often results in offsetting losses among others. It will take time for foresters and planners to find workable solutions they can implement at acceptable costs in satisfying the multiple-use provisions of the National Forest Management Act, while they continue to operate within the limits of the annual appropriations from Congress.

PLANNING AND DEVELOPING RECREATION OPPORTUNITIES

In response to demands for outdoor recreation, resource managers can help landowners capitalize upon opportunities appropriate to the area and the land. They must first identify what the landowner hopes for, and then design a management scheme to accomplish it to the degree possible. This will depend to a major degree upon the nature of the property and its physical attributes. In addition, the manager must look for approaches that prevent possible damage to the forest resources during use, and find ways to protect users from any harm or danger. To accomplish the task, the manager must know what people want from the area. This applies both where a private landowner plans a commercial development or seeks only personal recreational use, and to governmental holdings open to the public.

While the manager may not find it feasible to satisfy all the wishes, the plan should be designed to capitalize upon the best features of the land by matching the opportunities to demand. Thus, after understanding what the landowner seeks, the recreational specialist and forest manager can inventory the physical resources and determine the degree to which an area will accommodate the requested uses. They can then move ahead to plan a development and management program that concentrates upon what the land can support in relation to what is asked of it.

Identifying the Potentials

While forested lands generally offer opportunities for many types of dispersed activities, careful determination of probable user demand would need to precede any investment. What people do with their leisure time depends greatly upon the options open to them, and having available a particular type of recreational facility might stimulate use. Yet success of a recreational development is contingent upon how easily people can get to the site, and whether they come in sufficient numbers to justify the cost. Planning, therefore, must take into account people's preferences for various types of activities, what types of settings they want to visit, accessibility and distance from population centers, the existence of comparable and competing facilities, the degree to which people use them, the types of people whom the development might attract, what conveniences they are likely to require, the cost of providing these, and the potentials for recovering investments and operating costs through fees and other charges.[3,17] While many publicly owned recreation developments may operate at a net cost to taxpayers, private enterprise must show a profit or terminate business. Thus a careful choice of the types of opportunities to provide and good management of the facilities after development are key elements in ensuring the success of a recreational enterprise, whether public or private.

Recreation developers and planners can use different devices to ascertain public interest and demand. In many cases, they will rely upon several sources of information, rather than only one. As a start, they can get much help from national statistics on user preferences and what types of activities people actually do engage in. National surveys conducted periodically from 1960 to 1977, for example, show a high preference for picnicking, swimming, fishing, biking, camping, and tennis. Further, these did not change in relative ranking or popularity throughout the survey period.[21] Consequently developers logically would expect that recreation developments catering to these interests might prove more popular than facilities that did not.

Table 12–2 gives a fuller listing of preferences in 1977, plus an indication of activities that seem to offer the greatest promise of rapid future growth. Persons planning recreational developments would want to scrutinize this list and determine whether the resources and facilities available would support them, including the high-growth activities. In addition, planners might undertake local market

TABLE 12–2
LEADING OUTDOOR RECREATION ACTIVITIES IN THE UNITED STATES, AND THEIR POTENTIAL FOR FUTURE GROWTH IN POPULARITY

Most popular activities*	Percent	Fastest growth activities†	Percent	Highest potential growth activities‡	Percent
Visiting zoos, aquariums, fairs, or carnivals	73	Cross-country skiing	25	Downhill skiing	6
Picnicking	72	Downhill skiing	17	Tennis	6
Driving for pleasure	69	Tennis	13	Waterskiing	5
Walking or jogging	68	Sailing	11	Horseback riding	4
Pool swimming	63	Snowmobiling	11	Cross-country skiing	4
Sightseeing	62	Waterskiing	10	Primitive area camping	3
Attending sport events	61	Canoeing or kayaking	9	Sailing	3
Other sports or games	56	Golf	9	Golf	3
Fishing	53	Off-road vehicles	5	Snowmobiling	3
Nature walks	50	Horseback riding	4	Canoeing or kayaking	2

*Percentage of total population participating in an activity at least once during a 12-month period.
†Percentage of participants just starting activity for the first time during previous 12 months.
‡Percentage of nonparticipants that would like to begin participating during "next year or two."
Source: U.S. Department of the Interior [22]. Used by permission.

surveys to measure preferences of people who would use the facility. In deciding to expand an existing facility, they could also interview users and ask them about the plans. Through these means, the developer can judge in advance whether an investment will pay off, and if the resources will support the types of uses people seek and would pay to gain access to.

Especially in public land management, foresters often must identify and protect areas with outstanding natural beauty as an important early step in providing recreational opportunities. Then they can fit the use to the characteristics of the place (Figure 12–2). They also must identify the carrying capacity of the site, and keep development and use within the area's capabilities to sustain the activity without destroying features that made the setting attractive in the first place. This depends upon both biologic and physical resources of the ecosystem, the number of persons using the site at any one time, and management policies and practices for maintaining the facilities and controlling use. Design of the facilities, limiting the number of users, spreading pressures over daily and seasonal periods, influencing attendance by fees and permits, rotating its use, offering alternatives to users, and similar administrative and management devices can help to maintain the quality of an area and safeguard its recreational attractiveness.[11,20]

Figure 12–2.
During early planning recreation specialists will inventory a site to identify the potentials it offers, and then fit the use to characteristics of the place and its capacity to sustain the activity. *(U.S. Bureau of Outdoor Recreation.)*

Wilderness Recreation

Wilderness provides a special type of recreational resource that has received increasing attention in public land management since the 1950s. It often serves as a source of controversy between those who would use the forests and rangelands for commodities or develop it to enhance recreation uses, and those who want wild areas set apart from human manipulation and intervention. The National Park System offers a unique approach to wilderness. The management combines development to provide services and intensive-use opportunities on a limited portion of the land, plus more remote dispersed use in unaltered settings over larger areas. In addition, the Wilderness Act of 1964 and actions taken subsequently to implement it resulted in the creation of a national wilderness system encompassing 7,391,538 ha of forest and rangeland as of 1979.[22] These have been withdrawn from the National Forests and other public lands outside of the National Parks. Later additions, such as those in Alaska, have pushed the total upward considerably. Other areas presently under study should add to these totals.

The U.S. Forest Service administers about 80 percent of the nation's designated wilderness. The Park Service, the Bureau of Land Management, and the Fish and Wildlife Service have considerably less. These set-aside areas do not support timber production or related activities. They provide recreation, offer scientific and educational opportunities, help to preserve several species of plants and animals in an undisturbed setting, support livestock grazing, and serve as watersheds and for water storage. And while these areas are not managed in the sense of development and major modification, federal agencies do support active programs to protect the wilderness, control its use, provide minimum services and trails, monitor the health and status of the ecosystems, and otherwise safeguard the wilderness from deterioration through careless human intervention.[11] All cost fairly large sums of money, and agency budgets based upon legislative appropriations subsidize much of the expense associated with the wilderness programs.

Fitting Use to the Resources

Clearly planning for investments in a recreation site or development of an available area depends upon the intended uses. But not all types of activity will demand the same degree of development.[2] For example, camping, picnicking, skiing, and boating all take specialized development and considerable investment. They require rather intensive use of a limited amount of space. On the other hand, hunting, sightseeing, and hiking may attract as many people, can take place in wild lands, necessitate far less development, and the activity per unit of area often remains low. In both instances, the sites must satisfy the users' needs, prove convenient and accessible, and provide an enjoyable setting for the activities in which people engage. Climate, topography, soils, water resources, and the general environment all influence the quality of the area. Thus these elements require careful attention, beginning with an inventory to ascertain whether the place offers promise for development to satisfy the projected degree of use and the types of facilities needed.[2,3,17]

Camping, picnicking, and hiking are major forest-based recreation activities. Though one can picnic almost anywhere in a forest area, a developed site will add convenience and provide a place for group activities. Trails may be used for either short hikes or longer trips, depending upon their length and the opportunities afforded for overnight camping along the way. Campgrounds probably attract people by their setting rather than their proximity to home. Generally all of these activities are enhanced by some type of construction that makes the site safer and more convenient. Lakes, ponds, and streams add to the attractiveness of an area by offering the possibility for swimming, boating, and fishing. Such developments usually require a large parking area, safe and abundant drinking water, good roads to accommodate cars and trailers, terrain suited to game and athletic fields, well-drained soils and suitable vegetation, well-planned sanitation facilities, fireplaces and tables, and freedom from annoying insects and harmful animals. Electricity, beach facilities, shelters, docks, and telephones may be other desirable improvements. Needs for different combinations of these will depend upon the intended use, and the types of people who frequent an area.[3,17]

Trails may serve a variety of activities, including hiking, horseback riding, cross-country skiing, off-road vehicle driving, and strolling. They need not go any place in particular, though they usually do lead to some objective. A trail should provide a safe and adequate passageway, give users the opportunity to enjoy scenic vistas, guide people to unique spots and diverse environments, or take them to a particular place. The kind of construction and maintenance will vary with the intended use. Some trails may resemble roads, while ones designed solely for pedestrians may be only narrow paths with sufficient clearing to allow people to pass safely. The surface, steepness, drainage, and overall design should reflect the objective, and must withstand the pressure of use without causing deterioration of the land and vegetation along the way. Parking lots at the beginning and end of trails must be able to accommodate the number of vehicles expected at any one time. On some trails, well-placed signs will assist in guiding the travelers, interpreting wayside features, pointing out hazards, and providing other types of information that may be useful. However, for all types of trails, the planner will need to consider safety, topography, and alignment in developing the design. In some cases, different types of trails must even be kept apart from each other to prevent infringement by one kind of activity upon the requirements demanded for another.[3]

Design of Facilities

In planning, forest recreation specialists may call upon the skills of many other professionals. Within planning departments of many public agencies, staff specialists provide standard designs for smaller structures, such as tables, lean-tos, foot bridges, fireplaces, pathways, picnic shelters, or other rustic-style facilities. Sanitation and water supply systems will usually require on-site engineering, as will major roads. Specialized buildings and larger structures take detailed plans by architects and should fit into the surroundings to maintain the area's natural beauty. For some structures, the developer will need to obtain local or state permits to assure compliance with rules, regulations, and laws governing development and operations. These will include requirements of the state health department relative to water supply, sanitary disposal, swimming areas, design of areas and buildings, fire safety, food handling, pest control, and supervision of users. Local zoning ordinances may also control the types of uses permitted.[3,17] All require careful scrutiny prior to any construction, and should receive attention during the early phases of planning. Intensively used sites will usually require the greatest degree of planning, while recreation facilities designated for dispersed uses normally demand fewer consulting services and permits.

MANAGING RECREATION AREAS AND PUBLIC USE

The management of forests for recreation should provide a maximum variety of opportunities for the most people at the least cost. Further it should safeguard the resources that make the area attractive for recreational uses.[2] Positive results come from properly integrating advance planning, adequate service to users, and wise management of the resources and facilities.

Management Responsibilities

In addition to looking after these broad concerns, the public manager will also have overall responsibilities for maintaining trails and parking spaces, providing signs and general interpretive information, and controlling use to keep it consistent with the area and its capacity to absorb the pressure. Yearly operations may include improvement of facilities, repair work, and some new development. Within these public agencies, the resource manager may have access to staff specialists who can give advice and in many ways assist with administration. For private corporations, or landowners who engage in recreation management, the forester would normally rely upon books, publications, and manuals for guidance. In cases where either the public or private landowner collects a fee for using a property, the manager may also have responsibility for controlling access and ensuring the rights of the paying users. In most public holdings that support dispersed uses, the management becomes integrated with other activities, adding diversity to the daily routines of the forester who has overall responsibility for planning and fostering resource use and management.

As mentioned earlier, where the landowner chooses to emphasize both recreation and commodity uses, the management plans for timber may need to include special practices to preserve the visual qualities of the landscape. At least, the forester will need to use techniques that do not destroy potentials for recreation (Figure 12–3). These may address the effects seen over broad vistas, or those immediately adjacent to places that people frequent. The silvicultural prescriptions and methods for implementing them may need to reflect sensitivity to recreation interests. Practices such as leaving buffer strips along major travel corridors and trails, minimizing noisy or infringing activities during seasons of peak recreational use, applying alternate methods that have a better appearance, reducing logging slash, screening landings and skid trails, and even foregoing the harvest of some timber may seem warranted.[18]

Serving User Needs

Aside from these considerations, management of a recreation area involves planning, maintenance, protection, and proper administration.[3] The user's enjoyment represents the end product. Consequently the management must serve those interests, especially for intensively used areas. In many instances, some higher administrative authority sets the policies and determines overall approaches and techniques for running and maintaining a recreation facility. Nevertheless the manager has many ways in which to influence daily operations and conditions of the development. As these affect the quality of the place and the experiences of its users, they influence the success of the operation. For private enterprises, this often means the difference between profit and loss.[3,12,17]

Customer relations and promotion also are important to success, especially for intensive-use areas. Good hospitality, convenient services, publicizing of the opportunities offered, and supervising the activities are as important as the nature of the facilities themselves. Employees must provide maximum service to the guests, and enforce the ground rules with understanding and tact. The success of an operation often depends upon the personalities, characteristics, and actions of the people who run it. Further, users tend to respect those facilities that the caretaker keeps clean, in good repair, and suitable for the use intended.[3,17] Thus, in many different ways, the attitudes and diligence of the manager in running the facility will determine how well it succeeds, whether the owner makes a profit, or if demand reaches the full potential offered.

Administrative Considerations

For intensive recreation areas and some day-use ones where the owner charges a fee, quality of the facilities alone may not be enough to provide users with an attractive experience. Instead the manager may need to arrange special activities to attract visitors, hold them for a longer stay, and encourage their return. These might include interpretative programs, organized group activities and entertainment, schemes that encourage interaction among users, or other techniques to lend interest to the stay.[12] Even in public campgrounds, guided nature walks, campfire

Figure 12–3.
In forested areas access roads for timber management may serve a dual role as hiking trails as well, if developed with concern for the esthetic and recreational values and the ways people will use them.

programs, concessions for leasing horses and bicycles, rental of canoes and boats, organized swimming beaches and pools, publicity about local attractions for campers to visit, and many other forms of informal programming often attract people to a site and help make the experience memorable. Herein the skills of recreation specialists and environmental educators may serve as a vital supplement to those of the caretakers or manager. Often users will gladly pay extra fees for these privileges, providing supplemental income to help pay the costs of the operation.

Forest-based recreation programs of federal, state, county, and some local governments represent enterprises of considerable magnitude. These provide opportunities for persons skilled in administration to organize the programs of several individual facilities into an integrated effort that serves a larger clientele. At this level, an administrator will supervise the expenditure and collection of large sums, and have the responsibility for planning, maintenance, development, and supervision on a rather large scale. Progressive site deterioration of heavily used public facilities also represents a growing problem. Rehabilitation programs could greatly improve conditions and make the facilities more attractive. Also, public managers need to encourage more uniform use over the range of sites available, thereby taking pressures off the few that have proved most popular in the past. Providing improved information about potentials of underused areas could help promote them. In areas where public facilities cannot accommodate more use, governments might even try to encourage private enterprise through such devices as technical advice on development and operation, tax credits, and other measures.[22] Innovative approaches to administering large-scale recreation programs could contribute importantly toward satisfying the growing demand for recreational services from the nation's forestlands.

VALUES OF THE URBAN FOREST

The urban forest comprises the trees that line the streets, open spaces, and yards of the nation's cities, towns, and villages. These provide a valuable resource offering many amenity values. In fact, while most people do not consider a city's vegetation a forest, the trees, associated plants, and animals make up a rather complex ecosystem. Tall plants furnish shade and cooling, act as sound barriers, and screen sights. They function as habitat for a host of small mammals and birds, add landscaping to help beautify city streets and parks, and offer many other benefits to home lots and neighborhoods. But without deliberate management, these urban forest resources can deteriorate and even prove a nuisance for some purposes.

The 1960's social revolution, spurred by added freedom, leisure, and affluence, renewed an awareness of the importance of the nation's cities. The movement also attracted forestry-trained professionals. They realized that the knowledge, skills, and experience of forest resource specialists would prove valuable in developing innovative methods for safeguarding the health of a city's trees and enhancing their contributions to the citizenry. They recognized that forestry had traditionally dealt with other complex ecosystems to satisfy specific goals. And it has a perspective of managing space resources to enhance human use and benefit.[6,15] Moreover people schooled in ecology, soils, wildlife, recreation, botany, pathology, entomology, silviculture, management, and many other aspects of forestry could develop innovative management techniques for capitalizing upon urban tree populations. They could help the cities to enhance the environment of these otherwise artificial ecosystems. This idea grew into a specialized discipline known as urban forestry. It has spread throughout the country and contributed to programs designed to keep our cities attractive, thriving places offering quality lifestyles to residents and visitors.

Urban forestry, as it has evolved today, involves the management of vegetation in urban and suburban areas. It includes arboriculture techniques for breeding, selecting, propagating, protecting, repairing, and shaping trees and shrubs. In addition, urban forestry looks to the urban vegetation as an ecosystem that warrants management to satisfy the ecologic requirements of the trees, plant associations, wildlife, and society's objectives as well.[13] This requires use of a system approach that addresses both the needs of individual trees and of the urban forest as a whole. When implemented, this system yields many of the social values that any forest can provide.[8]

Much like those in more remote areas, the urban forests contain mature trees in established neighborhoods, fringe forests along the periphery of the city, and relatively young trees around new developments and where planting programs have reforested the streets and plazas. The fringe areas may provide special challenges, since these forests continually suffer depletion by new construction. There the urban forester can work jointly with landscape architects and planners to leave uncleared greenspace as landscaping, as screening and sound barriers, for protection of soils at fragile places, and for other values they offer from just being there to help enhance the quality of life in a neighborhood (Figure 12–4). In more established parts of the city, urban foresters make inventories of street-side trees to keep track of their health, condition, and usefulness. These inventories indicate needs for removals to protect public safety, for insect and disease control, or for pruning and other forms of maintenance to keep the trees healthy and to repair damages. The inventories also show requirements for new plantings to replace or supplement the growing stock. Such elements parallel ones common to forestry in remote areas, but may operate on a more intense scale in the urban places. They will address needs for individual trees as well as for the entire stands of vegetation that populate a block or neighborhood. However, with the urban forests, the management attends to both the health and vigor of the forest itself, and to ways the trees and associated vegetation might directly enhance or interfere with the functions of society.[8] Further, the primary product of the forest lies in its existence rather than in any commodity the owners can derive by cutting the trees or selling different by-products from them.[16]

Figure 12–4.
Greenspace within and around populated areas provides a multitude of values, including settings for nearby recreational uses of many types.

Tree planting provides special challenges in urban areas. Besides the normal need to match a species to the soil and general climate, plans for urban plantings must also take cognizance of any effect the vegetation will have upon the way people function in the area. Layout plans must carefully map individual tree locations with reference to above- and below-ground utility lines, traffic flow for pedestrians and vehicles, possibilities for obstructing views and activities, safety, visual characteristics, and other features of importance to urban lifestyles and activities. Species selected for use must also withstand the special microclimate, drainage, soil features, air pollution, impervious surface coverings, compaction from construction, and other ways that human development has altered the potentials for tree growth and survival. Also, the species choices and design for a planting should consider later maintenance and clean-up problems, and attempt to minimize these. When the plan takes these factors into consideration, the addition of trees and other vegetation can greatly enhance a city environment.[8,23]

Healthy and well-managed urban forests offer several important benefits to the people who live in them. Quite important, they help to ameliorate temperature, wind and air movement, humidity, runoff, noise, glare, reflection, and the harshness of urban scenery. To illustrate, around homes the trees offer shade in summer and help keep the structure cool. In winter, the conifers act as windbreaks. Deciduous trees lose their leaves, so the house intercepts the solar energy directly and is warmed by it. Shrubs and trees between the house and street also absorb sound waves and can reduce noise several decibels. As landscaping, trees may add considerable resale value to a home. The urban forest also attracts wildlife and helps to provide essential habitat elements to sustain many small mammals and a host of songbird species. The arrangement of trees and shrubs around a place can influence where people walk and so channel pedestrial traffic. It also provides privacy and can lend a sense of intimacy even to places surrounded by the activity of scores of people.[8,13]

In these and other ways, the urban forest provides a wide range of amenity values that contribute to the lives of people. By capitalizing upon their understanding of ecologic principles, an appreciation of human behavior, and the skills of several related disciplines, foresters can help to keep the urban forest healthy, viable, and well adapted to serve the needs of the people who live and work within the area. Through their management, they contribute to the recreational and social potentials of a city.

CAREER OPPORTUNITIES IN RECREATION AND AMENITY MANAGEMENT

Foresters and other forestry-trained professionals become involved with recreation and amenity values of forested ecosystems from the most remote wilderness to the heart of a thriving city. No matter where they are, the presence of trees as the major vegetative cover demands the attention of foresters. They understand the forest as an ecosystem, appreciate its complexity, know how to maintain its health and vigor, and are experienced in manipulating it to satisfy specified objectives. Foresters also are responsible for administering millions of hectares of land that contribute amenity values, and in many ways plan and supervise the activities that provide opportunities for recreation and other noncommodity uses of forestland. Foresters become involved because people use forested ecosystems for recreation.

Such involvement has characterized forestry from its early days in North America, though with less emphasis than today. Early attention focused upon providing relatively primitive facilities for camping, hiking, picnicking, swimming, hunting, and other common forms of dispersed recreational use. Then, beginning in the 1950s, and especially after 1965, popular interest in outdoor recreation suddenly began to mushroom to the level we know it today. With that turn of events and the subsequent outflowing of people to spend their leisure time in rural areas, foresters found themselves faced with a greater demand for recreational use than they had experienced in earlier times. In fact, the importance of outdoor recreation and other amenity values of forests and trees intensified to the degree that it stimulated new areas of specialization within forestry as professionals responded to the need and altered their management to enhance the noncommodity values people sought.

The growing use of forests and rangelands for recreation of all types also has increased the chances for more frequent contact between professional resource managers and people who would use the land for leisure pursuit. On public lands, this interaction has precipitated innovative programs for integrating recreation with management of resources for other purposes. Therefore it behooves students of forestry to develop an understanding of recreation management and the ways people perceive forested lands as settings for outdoor recreation.

Many schools of forestry have formulated programs to prepare students for careers in recreation management as a unique field of forestry. Some have set up this program as an option within a regular forestry curriculum, while others may offer a baccalaureate degree in recreation, park management, wild land recreation management, recreation resource management, or natural resource recreation management. Programs in forest recreation emphasize area and facility planning, administration, and programming aspects of recreation. Background courses include history, philosophy, sociology, communication skills, economics, political science, plus the natural sciences. Special studies may involve environmental interpretation, environmental impact assessment, recreation planning and administration, land use and soils, park management, landscape and site planning, and law enforcement. Programs in schools of forestry will also include core courses in resources management.[7,10] Exact options depend upon the school and availability of recreation-oriented courses outside the forestry program.

Preparation for careers in urban forestry follow more traditional approaches. As with other kinds of land resources management, the tending of tree vegetation depends upon a strong foundation in ecology and biologic sciences. It also uses skills learned through study of arboriculture, pathology, entomology, soil science, plant materials, biometrics, wildlife management, recreation, and other basic technical subjects. To appreciate the social implications and aspects of urban forestry, some students may choose to strengthen their understanding in sociology, psychology, government, anthropology, and other behavioral and social sciences. Basic skills in landscape design, small structures, urban planning, communication, public administration, and related areas could prove useful, depending upon what aspect of urban forestry an individual selects to emphasize. Here, especially, individual students should seek advice from a faculty member who also has specialized in this discipline, and profit from experiences of the professor.

Many universities offer graduate programs in recreation, though not necessarily limited to forestry departments or schools. Programs in physical education, landscape architecture, environmental education, park management, and others offer appropriate graduate-level courses as well as their undergraduate counterparts. These deal with social and economic subjects in addition to biologic ones, and can lead to either a master of science or the Ph.D. degree. Several institutions also provide programs leading to comparable degrees, but with specialization in urban forestry. Again, these may center in the forestry and silviculture departments, or in landscape architecture and environmental sciences. Requirements for degrees in these disciplines will differ among institutions, and may require course work plus a thesis or other nonthesis options. Programs with internships or project options may provide chances for students to work directly with community parks and recreation departments, city forestry groups, or other government agencies that deal directly with providing and enhancing the many amenity values people derive from trees and forests.

Government agencies would provide most of the career opportunities in recreation, urban forestry, and amenity management. These include offices from federal through local levels. In addition, many large private corporations, land management and real estate firms, and even smaller businesses hire professionals. Responsibilities may include on-site operation and management of existing facilities, planning of new ones, or administration of programs. Many beginning professionals also find employment with interpretative programs in local parks or state and federal recreation areas. For the National Forests and in some states, forestry staffs will have specialists to assist in overall planning of management programs, and to assist other foresters in accommodating amenity values as a part of timber and range management activities. Individuals holding advanced degrees also find employment with federal, state, and university research programs, or teaching in schools of forestry and forest technology. Growing demand for recreation services and increased awareness of the value of noncommodity resources portend an important role for specialists who prepare themselves to address the nation's recreation needs and work to provide them, from the city to the wilderness.

REFERENCES

1. R. E. Benson and J. R. Ullrich, *Visual Impacts of Forest Management Activities; Findings on Public References,* U.S. Forest Service Research Paper INT-262, 1981.
2. C. F. Brockman, L. C. Merrian, Jr., W. R. Catton, Jr., and B. Dowdle, *Recreational Use of Wild Lands,* 2d ed., McGraw-Hill, New York, 1973.
3. R. W. Douglass, *Forest Recreation,* 2d ed., Pergamon Press, Elmsford, N.Y., 1975.
4. B. L. Driver and P. J. Brown, "The Opportunity Spectrum Concept and Behavioral Information in Outdoor Recreation and Resource Supply Inventories. A Rationale," in H. G. Lund, V. J. LeBau, P. F. Ffolliott, and D. W. Robinson (eds.), *Integrated Inventories of Renewable Natural Resources: Proceedings of the Workshop*. U.S. Forest Service General Technical Report RM-55, 1978.
5. R. D. Fight and R. M. Randall, "Visual Quality and the Cost of Growing Timber," *Journal of Forestry* **78**(9):546–548 (1980).
6. Florida Department of Agriculture and Consumer Service, *Urban Forestry in Florida,* Pamphlet Florida Department of Agriculture and Consumer Service, Division of Forests (undated).
7. R. D. Greenleaf and W. F. LaPage, "Recreation in the Forestry Curriculum—Observations," in *Recreation Resource Management and the Professional Forester, Proceedings of Recreation Working Group Technical Session, Joint Convention Society of American Foresters and Canadian Institution of Foresters,* October 24, 1978, St. Louis, Mo., Society of American Foresters, Washington, 1978.
8. J. Grey and F. J. Deneke, *Urban Forestry,* Wiley, New York, 1978.

9. C. R. Jensen, *Outdoor Recreation in America. Trends, Problems, and Opportunities,* 2d ed., Burgess, Minneapolis, 1973.

10. D. M. Knudson, "Recreation Education—Environmental Dimensions," in *Recreation Resource Management and the Professional Forester, Proceedings of the Recreation Working Group Technical Session Joint Convention Society of American Foresters and Canadian Institution of Foresters,* October 24, 1978, St. Louis, Mo., Society of American Foresters, Washington, 1978.

11. Ibid., *Outdoor Recreation*, Macmillan, New York, 1980.

12. W. F. LaPage, *The Role of Customer Satisfaction in Managing Commercial Campgrounds,* U.S. Forest Service Research Report NE-105, 1968.

13. S. Little (ed.), *Urban Foresters Notebook,* U.S. Forest Service General Technical Report NE-49, 1978.

14. T. A. More, *The Demand for Non-consumptive Wildlife Uses: A Review of Literature,* U.S. Forest Service General Technical Report NE-52, 1979.

15. N. A. Richards, "Forestry in an Urbanizing Society," *Journal of Forestry* **72**(8):458–461 (1974).

16. Ibid., *Greenspace Silviculture,* SUNY College of Environmental Science and Forestry, Syracuse, N.Y., unpublished manuscript.

17. C. R. Smith, L. E. Partain, and J. R. Champlin, *Rural Recreation for Profit,* 2d ed., Interstate, Dansville, Ohio, 1968.

18. Society of American Foresters, *Timber Harvesting Guidelines for New York,* Pamphlet, Society of American Foresters, New York Section, 1974.

19. L. A. Stoddard, A. D. Smith, and T. W. Box, *Range Mangement,* 3d ed., McGraw-Hill, New York, 1975.

20. U. S. Bureau of Outdoor Recreation, *Outdoor Recreation, A Legacy For America,* U.S. Department of the Interior, Bureau of Outdoor Recreation, Washington, 1973.

21. U.S. Department of the Interior, *The Third Nationwide Outdoor Recreation Plan. The Assessment*. U.S. Department of the Interior, Heritage Conservation and Recreation Service, Washington, 1979.

22. U.S. Forest Service, *An Assessment of the Forest and Rangeland Situation in the United States,* U.S. Department of Agriculture, Forest Service, FS-345, 1980.

23. R. F. Watt, "Shade-Tree Selection," in S. Little (ed.), *Urban Foresters Notebook,* U.S. Forest Service General Technical Report NE-49, 1978.

CHAPTER 13

MANAGEMENT PLANNING AND ADMINISTRATION

Landowners can derive timber and other plant products, minerals, water, animals, and recreational opportunities from their forests. The way that they use these goods and the degree to which they use them depend upon personal interests, and upon an array of ecologic and economic factors that influence management decisions. The economic or institutional factors include pressures to generate income, requirements of law, mandates of governmental rules and regulations, needs to supply products to support other operations of an enterprise, conditions of available markets, costs imposed by taxes and administration, and a host of other business-type matters. In many cases, the basic interests of the landowner in holding the property will influence the way an individual, firm, or agency reacts to each economic and institutional opportunity. Not all owners will respond in the same way. Their interests show up in the approach taken to management.

In addition to the economic factors, foresters must take into account many biologic and ecologic ones as well. These include conditions of the physical site, and the interactions of living organisms with that environment. The species present, growth rates attainable, timber quality, types of animals, amounts of water yielded, and other tangible attributes reflect the nature of the physical environment. Foresters can influence stand composition and development by silvicultural practices. They can also introduce a species that would not normally grow at the site, and arrange the trees in rows and at spacings fitted to special landowner interests. However, foresters cannot overcome the inherent capacity of the site to grow trees unless they spend a lot of money to do it. Thus they must learn to recognize what a site will produce, and gear their management to fit those limitations and opportunities.

The economic and institutional factors must work within these biologic or ecologic limits. Owners can choose to manage or not. They can decide to invest either relatively large amounts or relatively few resources in the management. They can elect to exercise various silvicultural options that seem possible under the conditions present. The choices will depend upon many factors, such as the availability of markets for various potential products, the costs of producing the goods and services buyers want, the accessibility of the resources, and the owner's general reasons for owning the land.

OWNERSHIP OBJECTIVES AND MANAGEMENT PLANNING

Approaches to Management

Relatively little forest and rangeland has a single-use restriction. Where this attitude prevails, the owners or persons controlling the property have decided to serve one overriding interest and feel that to encourage other uses would detract from the primary purpose of ownership and management. Therefore they exclude those other options. In some cases, the owners will not necessarily bypass all supplemental benefits, but they take advantage of the extra uses as an incidental spinoff of ownership rather than as a deliberate part of management. Thus the single-use interest governs activity on the property to the exclusion of other potentially compatible benefits. Such a philosophy often characterizes the management of government and private wilderness areas, wildlife refuges, recreation parks, and other types of preserves. In addition, many private individuals who control relatively small parcels of forestland own them primarily for personal recreational uses and reject other possibilities in order to preserve the conditions they deem important to their primary interest. Here the forest manager will devise a program to protect the requisite conditions and to enhance the uses the owner seeks. Timber or mineral extraction may have no place in these schemes. Instead the owners will measure the output of management in less tangible ways or through indirect benefits deemed important.

Foresters call the second approach dominant use. In this case, management may aim to promote more than one value, but the forester gives primary attention to one preferred use. The landowner would not necessarily prohibit other uses and might even encourage them as long as they help to enhance or do not deter from the primary purpose of management. This philosophy might apply to industrial forests where the firm grows products to supply a mill. Efforts would focus upon growing wood products, but the corporation would also hope to capitalize upon the wildlife, recreation, and water values if the opportunity presented itself. Yet the bulk of energy and attention will go toward growing timber and harvesting it to supply a mill. Interestingly this same philosophy can apply to many nonindustrial private forests. There the landowner might have primary interest in recreational values, but would use timber management to help pay costs of ownership or to enhance the recreational opportunities on the property. Overall the recreation interests will dominate management, and that may limit some of the options available to the forester in managing the timber. However, this approach to management provides more flexibility. It probably characterizes much of the private forest ownerships in the United States.

The third approach of multiple use can apply to any type of forest ownership. It means managing the forest and its associated resources deliberately to satisfy two or more objectives at the same time.[9] This differs from dominant use in that no single value gets primary attention. Instead, the forester would try to identify several compatible uses and integrate the management for them to the fullest degree possible. In most instances, the management would not attempt to accommodate all possible uses. Rather the forester would seek a harmonious combination of several in which the owner is interested. Such a philosophy governs the management of most public lands, and governmental foresters will strive to serve a broader range of interests than with private forests.[7] Yet many private forest owners also deliberately manage their land for multiple use, and ask the forester to sustain the yields of these integrated values over the long run.

Interests of Ownership

The objectives for private ownerships rest soley with the individual or corporation holding title to the property. Historically such owners also have had exclusive rights to control access to these properties, and the sole privilege of enjoying or disposing of economic goods derived from them.[2] A private owner may also sell the property to someone else at any time, and at a price negotiated between the two parties. The private owner pays all the costs of maintaining and owning the land. This includes property taxes assessed by local governments, and state or federal levies on income derived from the sale of products or the leasing of use to others.

Various governmental rules and regulations or zoning ordinances may limit some management options on private lands. But these will not impose a particular type of management upon private ownerships within the United States. Rather they tend to govern activity that might trigger pollution or impinge upon the rights of others. The Federal Water Pollution Control Act is a good illustration. These regulations require landowners to use practices that limit or control sources of pollution. They do not prohibit activities, but control the way in which a landowner may do certain kinds of work that might affect water quality. As long as a management scheme can be devised that does not cause a deterioration in water quality, the owner may proceed with the intended uses.

State or local land ordinances will usually apply to large regions and create zones or areas where the agency controls or prohibits some possible uses or management practices. Rather than preclude a use, the regulation may require the landowner to obtain a permit that specifies conditions the manager must satisfy in carrying out the activity intended. The limitations over timber harvesting along wild, scenic, and recreation rivers in some states illustrate one type of such land-use restrictions. Establishment of forest districts where forest clearing and many types of land development cannot take place exemplifies another. In some states, the regulations may require deliberate efforts to gain prompt restocking of a forest stand following harvesting operations. However, these regulations do not mandate that the landowner must apply some particular management. The owner can choose to let the land sit idle. Or the manager can devise an appropriate approach that fits the constraints of the regulations and attempt to capitalize upon the goods and services obtainable from the forest to the degree permitted.

The diversity of landowner interests reaches its extreme among the nonindustrial private group. Though they are lumped into a single category for most statistical purposes, these people have various reasons for purchasing forestland. Some simply hope to derive pleasure from owning the land and visiting it occasionally. Others use the forest as a vacation retreat and as a place in which to pursue recreational activities. Some hold the land as an investment and speculate upon earning money from its resale after its value increases. Still others seek to capitalize upon timber growth for profit. In contrast, industrial owners have a commercial interest in their holdings. These may include nontimber values such as minerals, or the potential to sell recreational opportunities. The corporate interests, however, usually involve commercial enterprises that depend upon the land for materials, or as a land base to use in generating revenues for the firm.

Across the United States, private owners of various types control about 54 percent of the total forests and rangelands.[14] Collectively they hold 71 percent of the commercial forestlands, and the nonindustrial owners alone about 59 percent. East of the Great Plains, private ownership includes 82 percent of the commercial-quality timberlands, and in the South about 91 percent. By contrast, west of the prairie, private owners control only 32 percent of the commercial forest, including 25 percent in the Rocky Mountains and 37 percent along the Pacific Coast.[13] Altogether nonindustrial owners hold about 59 percent of the commercial forest timberlands and 83 percent of the private area.

Despite their diverse interests, these owners sell substantial amounts of timber annually to the forest industry. They control a significant proportion of the nation's timber reserves and play a major role in supplying the raw material needs of industry. In 1976, for example, private nonindustrial owners supplied about 48 percent of the 408,476.4 thousands of cubic meters of wood products removed from U.S. forestlands. The forest industry took almost 29 percent off their lands, and the remainder came from public lands controlled by various levels of government.[14] At least among the private sector, the decision to capitalize upon these timber values or any other ones of the forest remains soley with the landowner.

Government agencies usually have less freedom in setting a course for management. In most cases, the foresters must operate under legislative mandates that specify the goals. These usually espouse a multiple-use philosophy, except on lands specifically dedicated to a single or predominating use. The management must normally balance alternative uses to serve the overall good and interests of taxpayers. For example, the Multiple Use and Sustained Yield Management Act of 1960 requires that federal forestlands provide for several uses at the same time, and that planning and management must deliberately accommodate these. What uses will prevail in any forest depends upon local needs as well as national interests. State and local governments have similar laws for operating lands under their jurisdiction, and the foresters must comply with general provisions of these statutes. Each agency will have its own procedures for identifying and accommodating the mix of values to emphasize and integrate on various properties. The exact approach will recognize local needs and the different opportunities available from place to place.

Most of the public lands lie west of the Great Plains, and the federal government controls the majority of these. East of the prairie, federal, state, and local governments hold 14 percent of the commercial timberlands. The federal government has only 7 percent, and the proportion diminishes considerably east and north of Ohio. In contrast, public agencies control 68 percent of the commercial forests west of the Great Plains, and the federal government owns 58 percent alone.[13] This includes 94 percent of Alaska. Eventual settlement of the land law claims in that state, however, will substantially reduce federal control there.

The federal government has a fairly consistent policy guiding the management of its holdings, as do state and local governments for lands under their jurisdiction. The policies for private ownerships can differ widely, even within a given locality. Despite these differences, all owners need to manage their properties deliberately if they expect to benefit fully from them. Forest values do not accrue without effort. It takes harvesting and using of the wood, animals, water, and any other conceivable produce from the plants and animals that occupy the land (Figure 13–1). In addition, foresters must actively protect the resources from destruction and harm. They must control insects, diseases, fire, and other things that could reduce the benefits. Further, active management implies a commitment to perpetuate the forest over the long run, and to safeguard its productive capacity for future generations. This requires a planned approach to determine the needs and arrange for systematic attention to them in the most efficient manner possible.

Management Planning

The discipline of forest management includes the overall administrative, economic, legal, social, technical, and business methods and principles applied to the

Figure 13–1.
Use of a forest involves harvesting its products, whether in the form of timber, recreational activity, water, animals, or any other benefit the owner hopes to derive from the land.

operation of a forest property for a specific objective.[9] Through these, the forester will develop a plan to keep the forestlands productive. The management may range from relatively simple to fairly complex, depending upon the owner's objectives and the nature of the forest resources. The incentive may come as financial or less tangible benefits, and business properties must usually show a profit on the investment of ownership and management. However, in most cases, no owner is likely to assume the expense, effort, and risk of management without the prospect of satisfying some specific goal that provides the primary motivation for ownership.[12]

The management and working plans for an ownership will serve to direct efforts and expenditures toward those activities that promise adequate returns for the owner. They will provide guidance to the forester in helping to economize on the use of funds and time, to focus the efforts toward a common purpose, and to gain continuity over the long run.[5] The plan will guide decisions for a reasonable period into the future. Consequently it should include a concise statement of the owner's objectives, rank these by priority, list the resources available on the property, delineate how to capitalize upon them in serving the objectives, and outline a schedule of operations to achieve the goals. Also, the plan would provide a statement of investments and expected returns so the owner could prepare an appropriate financial plan.[12] The forester must often include a variety of options for the owner to consider, and explain how each will provide specific benefits. In effect, this helps to make the advantages of management clear. Then the landowner can decide how and with what intensity to embark upon a management program.

In essence, planning means identifying and weighing alternatives in order to reach a deliberate decision about what actions to take. The process serves to assure fulfillment of the firm's or an individual's objectives as nearly as possible. It may not necessarily involve highly complex steps, though some firms engage in far more elaborate planning than others. But all need to plan their future if they hope to achieve any particular set of goals. Some may plan ahead for relatively long periods, while the planning horizon for others will remain relatively short.

Despite the differences in complexity or time span, all planning begins by identifying the objectives. The forester then will look at the alternatives available, assemble information about each, analyze the prospects, and decide what combinations to pursue. In addition, the forester must make a plan for keeping the information up to date so that the managers have a current picture of the progress.[8] These steps aid the decision making. They require skill and imagination to assure the wise use of capital and resources in helping to satisfy the goals of management.

Implementing the Plan

Once the forester develops a plan, the landowner, supervisor, or corporate officer in charge must decide to implement it. If done well, the plan will provide them with assurances that daily activities will fulfil the overall objectives for the property. This applies to both simple and complex approaches toward forest management. With the largest multiple-use holdings of a governmental agency or a large corporation, the plan may guide the expenditure of large sums of money and control use of important quantities of resources. Even with single-purpose or limited acreage ownerships, however, the management must proceed according to a deliberate plan before the forester can give the landowner reasonable assurance that the purposes of ownership will be met.

The forester must also devise a harvesting plan that results in a balance between growth or additions and the amounts removed periodically. The approach must provide for prompt restocking to replenish the harvested crop, and for tending those portions of the resource not scheduled for immediate harvest. This will assure a sustained yield over the long time. Further, for multiple-use properties, the harvest, tending, and restocking will consider not just the trees, but also the continuing use and replenishment of water, wildlife, recreational opportunities, and whatever combination of values the landowner seeks. Only by knowing the amounts available, the degree to which people use the resources, and the rate at which they are replenished will the forest manager be likely to succeed. As foresters can balance these, they will provide a sustained yield of multiple benefits to capitalize on both now and in the future.

ORGANIZING THE FOREST PROPERTY

In conjunction with efforts to plan the long-term management of a forest property, the forest manager must become familiar with the resources available and general conditions that might affect various operations and uses. This involves the mapping of boundaries, inventorying of resources, subdividing the area into relatively homogeneous management units, determining the losses to mortality and the additions by new trees, ascertaining the growth and removals, and quantifying accessibility and suitability for use in available markets.[1,7] The process will begin with a land survey to locate and mark the property boundaries. This step often accompanies the purchase of a property. If it does, the forester can begin directly to inventory the resources and accumulate data for management planning.

Modern forest inventory usually begins with a study of aerial photographs. Foresters can map the boundaries of forest stands on them, and these smaller areas will serve as the basis for later management and record keeping. The photo maps will show the forester where to locate different stands, how various resources relate to each other spatially, and the amounts of space covered by different stands or features of the landscape. In some cases, a property may have essentially one condition over a relatively large area, and the forester will subdivide it into several smaller operating units or compartments for convenience of management and record keeping.

After mapping the forest from photographs, the forester or crews of forest technicians will visit the stands and gather information about the species, numbers, sizes, and condition of trees. They may also classify these for value and suitability for use in local markets. With ownerships interested in managing more than just the timber, the inventory may also assess habitat conditions, available food, status of a watershed, grazing capacities, potentials for recreational use, existing mineral resources, and other conditions of the ecosystem that might contribute to the ownership interests.[1] Field checks also often include assessments of such physical conditions as topography, accessibility, priority for management and use, or other kinds of information that might help the forester to interpret the findings and balance the needs for one use with those of others.

The local forester will usually have responsibility for planning, undertaking, and summarizing results from fairly simple inventories for limited-acreage properties. Inventory specialists and biometricians often participate in developing the system for larger holdings, and for integrated surveys to answer multiple information needs. These may include fairly elaborate integrated sampling schemes to reduce costs without sacrificing data essential to management planning and control. The plan will also include ways to accumulate and summarize the data, probably employing modern high-speed computers to provide the firm or agency with a data management system to generate timely information in a variety of forms. Ultimately the design and approach to multi-resource inventories for a particular situation will depend upon the information needed by administrators for decisions, and how the mensurationist can translate the measured data into information useful in the management process.[3,4,10]

Besides helping the forest manager to develop plans to coordinate future activities, the inventory also provides periodic assessments of the status of resources and the progress of the management. By using remeasurements from two time periods, the forester can estimate the change in the resources. This shows the effects of growth, renewal, and harvesting on the condition and volume of trees on the ownership. Results will also reflect the conditions of wildlife habitat and food supply, the nature of recreational opportunities, and the status of other resources the management centers upon. Managers can use these data in their routine planning to accommodate changes in markets and use potentials, unexpected events or catastrophes, changing interests of the landowner, or new opportunities resulting from a variety of different developments. Potentials to input new information and efficiently generate new assessments from the data give the forester the capability to monitor changes in the forest resource, and to adjust the management as needed to keep the program focused on satisfying the targets set by the landowner.

Overall management endeavors to produce goods or values sought, and to arrange their timely harvest or use for the continuing benefit of the landowner. To realize this goal, the forester must have access to markets that will absorb the amounts and kinds of goods produced, and at acceptable prices. This applies both to hard goods such as timber or minerals, and to less tangible values such as hunting and other recreational opportunities. Ultimately the available markets will control management activities and influence the investments a forestry operation can sustain.

Even though a buyer may pay hgh prices for some materials, the forester must find outlets for all the products and at prices that pay the cost of management. When the markets take many different types of products and pay good prices for them, the landowner will have more options open. Then the forester can generally implement a more diverse and more intensive type of management. For multiple-use properties, the marketing will prove more complex. The forester must balance potentials for capitalizing upon one source of income against the effects upon other values. Some times, it may be necessary to forego one option to leave other opportunities open for the landowner. For example, some situations may call for reducing the output of timber to improve potentials for recreation. In others, the forester may need to limit access for general hunting to protect other resources or to generate income through leasing programs. In these cases, the owner will need to cut back on the kind and intensity of one use to improve chances for some other benefit. To reach the proper decision, the forester must understand the objectives for management and blend potential uses through a well-conceived management plan. On the basis of these goals, the forest manager can outline the various opportunities and determine the best way to balance uses for the overall good of the ownership.

MANAGING THE BUSINESS ENTERPRISE

Aside from the management of the various resources of the forest ecosystem, forestry also includes the administrative, economic, legal, and other activities needed to organize a business enterprise and keep it functioning effectively. For either corporate or individual businesses, this will mean gathering together individuals with various skills needed to arrange and carry out the many types of management activities. These people will supervise operations to permit prompt and efficient fulfillment of the firm's goals.

The Business Team

In many large enterprises, the management team will include field foresters to implement the silvicultural, harvesting, inventory, and other on-the-ground programs. In addition, it will have staff specialists who will attend to the record keeping, planning, data management, personnel management, and other types of administrative services. An enterprise may also hire individuals for marketing, sales, promotion, and procurement activities. It will have a group to set policy, decide about overall program goals, and generally guide the course of activity. The complexity of this organization will depend upon the size and nature of the enterprise.

For small firms, individual professional foresters may serve in more than one capacity and the organizational structure will remain simple. Certainly many businesses would benefit from having a team of specialists available to deal with different aspects of the operations, but unless the landowner can pay the costs, the business cannot accommodate them. Such an owner can hire accountants, lawyers, bookkeepers, and other business specialists part time. But in many cases, the forester will provide some of the business management services in addition to caring for the property, growing the timber, and arranging for the harvest of products. Large firms generally are able to employ specialists and fit them into a rather complex organization. Their staff also will include teams of technicians and others who assist in various management tasks.

Generating Revenues for the Enterprise

The growing and sale of timber products are the backbone of most forestry enterprises, although many individuals and businesses also capitalize upon other products. They may rent or lease hunting and recreation rights, sell gravel and other road-surfacing materials, market secondary products such as Christmas trees and maple syrup, or find other ways to generate income as a supplement to timber revenues. Actually most corporations will try to take advantage of a wide array of opportunities in order to maximize profits, commensurate with good sustained yield management.

Most of the forest products industry depends upon wood purchased from private individuals who supply at least part of the raw material for a mill. The firms suffer occasional setbacks related to fluctuations in the demand for goods, and the way market opportunities influence prices. During periods of low demand, corporations will usually continue to cut wood from their own properties, but will reduce purchases from other landowners to keep the raw material inventory at the mill in balance with production. Then private landowners may find their income diminished, and need to adjust the cutting and other operations of the enterprise. On the other hand, if a firm finds timber in short supply, it can increase the cutting of its own timber reserves, or raise prices for purchased wood. This will encourage harvests by other landowners who derive their income by providing the raw material to keep the corporate mill operating. Generally, then, the activity on noncorporate holdings reflects the health of the wood products industry that purchases logs and pulpwood in the area.

In addition to finding and maintaining steady, well-paying markets, the forester as business manager must also consider the operability of available timber. This depends upon the anticipated cost of harvesting it, compared with its market value.[8] Cost factors include its location, accessibility, the quantity per hectare and across the sale area, and the difficulty of logging as influenced by topography and terrain. Tree sizes, quality, species, usable lengths, and soundness all influence value. Where the value exceeds the operating cost, the manager can consider a sale.

In many areas where good roads come reasonably close to harvestable tracts, accessibility usually does not affect marketability as much as does the condition of the timber itself.[11] Constraints imposed by demands for alternate uses as set forth under multiple-use philosophy may also reduce the availability of timber. They may necessitate modes of harvesting that can increase the cost sufficiently to make the timber unprofitable to market. This basic concept also would apply to the marketing of other forest uses. Managers will need to evaluate the demand, accessibility, quantity, quality, and similar factors that make use of the resource desirable and influence the amounts people will pay for the privilege.

Managing the Risks

One important business management aspect of forestry involves minimizing the risk and chances of financial loss. In part, this requires active programs to protect the timber against losses to insects, diseases, and other destructive agents. It may also include protecting the land and its resources against trespass and safeguarding other assets such as roads, machinery, and structures from harm and deterioration. In addition, the business manager must also anticipate consequences of widely fluctuating values for timber and other marketable resources. Income from sales and leases may fluctuate widely from year to year, depending upon the amounts available for harvest and the prices they attract.

The manager must keep costs in balance with income over the long run. It takes many years to grow timber, and repeated investments to protect and manage it through extended rotations. This tends to reduce the rate of interest owners can realize from investments in forestry. Commonly they do not rise much above 3–10 percent. Further, firms cannot easily get insurance against losses to windstorms, fire, and other catastrophic events that might threaten their growing stock or destroy the products and use opportunities they depend upon for income. When these strike, the owner must salvage any usable products and absorb the loss for any values not recovered. Hence success of most forestry operations will depend upon the business acumen of its managers. They must learn to skillfully manipulate the forest, available labor forces, marketing and sales, use of credit and insurance, and other business aspects to maximize financial returns and minimize operating costs.

With vertically integrated firms that both grow the timber and convert it into some type of manufactured product, management decisions often reflect a need to provide a steady influx of raw materials to a mill. Procurement foresters will satisfy this demand by purchasing wood from other landowners, and scheduling harvests off corporate lands. Companies may have their own logging crews who go out onto private lands and cut, skid, and haul the purchased timber. In other cases, they simply purchase the wood from independent contractors who deliver the pieces to the mill ready for use.

These procurement activities demand careful coordination with the mill manager. The forester must anticipate increases or slowdowns in production, and stockpile adequate reserves to cover short periods in which logging and hauling might be difficult. Many firms also hire foresters to work with the private landowners and provide timber management advice and services as an incentive for the landowner to sell timber to the firm. Overall the foresters in charge of these procurement programs must skillfully plan and schedule the acquisition, harvest, and delivery of the raw materials to provide a steady input to the mill, at the lowest possible cost.

Foresters and Business

The foregoing activities represent some important concerns for foresters involved with the business aspects of forestry. Others parallel the activities and elements of all business enterprises, such as accounting and financial management, recruitment and supervision of personnel, public relations and promotion, marketing and sales, and acquisition and expansion of facilities and improvements. As with any other business, those of forestry employ capital to realize a profit. The firm strives to become self-sustaining and to expand. It must have sales of goods and services that exceed costs of production. In addition, it must provide an adequate interest payment on the capital invested and risked on the enterprise. The managers need to develop an understanding of markets and their trends, an appreciation of the technology of production and ways to control it, and a capacity to make wise choices about the kinds of goods to produce and how to regulate their output. They must also be skilled in selecting and supervising a staff capable of carrying out the activities of the firm, and be able to develop and manage budgets to control outlays and prices.[5] These represent important concerns for the business of forestry where the owner requires a profit in terms of goods, services, or values realized in return for the investment in owning and managing forest resources.

PUBLIC LAND MANAGEMENT

Public forests supply both timber and many noncommodity benefits. Further, they remain open to public entry with few constraints over access or use potentials. Interestingly, to a large degree these are lands no one else wanted. Many eastern forests held by states once supported farming, but failed to compete with more productive lands with richer soils in the Middle West. Some were too remote or had too short a growing season to offer potentials for alternative uses. Thus states acquired them after former owners failed to pay taxes. The states then worked to convert them to tree cover as part of the great conservation movement that was popular early in the 1900s. Most federal forests represent the residual of the great public domain. And while private individuals and corporations would gladly purchase the better and more productive lands if available, many hectares of federal forests and rangeland have limited productive potential. In fact, they represent only 28 percent of the commercial forestland in the United States.[13] Still, federal lands provide a multitude of benefits to the nation.

Uses of Public Lands

The legislation authorizing acquisition by or granting title to a governmental agency usually mandates a different management and use than for private lands. For the federal forests, legislative provisions require multiple use except where land is specifically set aside for wilderness, parks, refuges, or other special single purposes (Figure 13–2). For the remainder, management must provide integrated benefits to meet local and national needs. State and other governmental lands also have similar purposes specified in the enabling legislation that allowed their purchase or provides funds for their management. While these laws may not dictate management for all possible uses, they normally call for an integration of several in a plan that takes cognizance of the timber, range, water, wildlife, and amenity values.

Many states did not begin to acquire forestlands until early in the 1900s, following the depression. In eastern regions, the former owners often cut over the

Figure 13–2.
Legislation authorizing ownership and management of public lands usually requires approaches that integrate concerns for timber, forage, water, wildlife, amenities, and other values to realize multiple benefits from the resources.

land before abandoning it for taxes, or selling it to state or local government. Consequently many state and locally-owned forests had little timber left. Early stages of management primarily involved acquiring the land, surveying it, and marking the boundaries for easy identification. Many agencies planted the open fields to conifers, creating important areas of softwood plantations. By the 1950s, the management had shifted away from caretakership activities such as tree planting and protection and into inventory, management planning, intermediate stand treatments, and improvement cuttings. Then, by the late 1960s, the cutover stands had once again regrown substantial volumes and could support active timber sales programs. Growth of public interest in outdoor recreation beginning at about the same time increased pressures on these lands. This necessitated careful planning to integrate uses in a way that provides the most public good. Timber sales will continue to grow, but many of these public lands will also be in demand as sites for recreation, wildlife habitats, scenic enjoyment, and many other purposes.

The vastness of the National Forests makes them especially important as suppliers of timber. In 1977, those lands contained about 51 percent of the entire softwood sawtimber volume in the United States, and almost 93 percent of the softwood growing stock volume in smaller sized trees.[13] The importance of these species for construction purposes and many fiber products makes the National Forest System critical as a source of material to supply the needs of the country. Under the National Forest Management Act of 1976, the Forest Service must regulate timber harvests of each National Forest to a nondeclining sustained yield level, subject only to minor short-term deviations. It cannot cut more volume in any period of time than growth off the forest will sustain in perpetuity. However, since much timber on the National Forests has reached a state of maturity or beyond, the forests probably accumulate substantially less volume than younger and more thrifty stands would produce.

Paying for Programs

Government agencies operate under the limits of appropriations. In part, these funds may carry a mandate to implement particular rules and regulations, or to enforce laws. The remainder goes to a general program with a broader edict to manage the resources wisely. In few cases will public foresters face the economic pressures that govern industrial firms or private investments. They face no mandate to earn a profit or realize a specific rate of return on the investment of management or on the capital worth of the resources. Actually, the enabling legislation or agency policies may require expenditures for activities that generate no revenues or do not return sufficient amounts to pay the budgeted costs. Data in Table 13–1 illustrate this point by figures for the National Forests. Here revenues from timber far exceeded the appropriation for managing it. Yet recreation, wildlife, protection, and other functions required considerable outlay with no or only limited revenues in return. The deficit represents the commitment by govern-

TABLE 13–1
TOTAL EXPENDITURES AND RECEIPTS ON NATIONAL FOREST SYSTEM LANDS AND BY SELECTED FUNCTIONS, FISCAL YEAR 1980

	Millions of dollars	
Function	Expenditures	Receipts
Total National Forest System	$1709.4	$1288.5
Selected functions		
Timber management and sales	329.1	625.4
Range management and improvement	35.4	15.8
Recreation, wilderness, and land uses	101.8	18.0
Mineral and power leases and permits	15.9	219.3
Road and trail construction	392.6	164.2
Wildlife and fish habitat	38.8	2.6
Soil and water	47.3	0.5

Source: Adapted from U.S. Forest Service, *Report of the Forest Service. Fiscal Year 1980.* U.S. Department of Agriculture, Forest Service, Washington, D.C., 1981.

ment to provide services to taxpayers, and to subsidize those programs for public benefit.

This does not imply that governmental foresters do not insist upon operating in the most efficient and cost-effective way. They still must assess the alternatives and determine how to spend the available funds effectively. In some cases, they may elect to support activities that bring no revenues in return, but that will enhance a noncommodity value that people seek. Hence government agencies usually evaluate the wisdom of an expenditure on the basis of the benefit expected relative to its cost, and opt for the package of activities offering the best cost-benefit ratio among the alternatives considered.

Resolving Demand for Alternative Uses

Government foresters also deal with another situation not common among private operations. They must resolve conflicts in demand for use of resources in a way that balances opportunities for the overall good. This situation grows out of the multiple-use philosophy mandated by the enabling legislation or administrative orders that guide government forestry programs. These usually require active efforts by the agency to seek public opinion concerning goals for a property, and to delineate efforts for integrating them in management. The process often uncovers conflicts over what values to emphasize, even to the point that addressing one demand might exclude others. As a result, the agency must resolve disagreements over alternative uses in a way that effectively serves local needs.

To ensure adequate results, many government agencies maintain elaborate planning programs. These involve teams of specialists who can identify needs and opportunities for the different available resources and devise ways for addressing them. The specialists will have backgrounds in timber, water, wildlife, recreation, protection, and other disciplines related to forestry. By working together, they can identify areas of potential conflict, devise ways to integrate uses to resolve the differences, and translate these into plans for management. They will also work with field foresters to find avenues for addressing alternate needs in the routine of their timber and land management activities. By anticipating needs and opportunities through this integrated use planning, the agencies hope to identify potential areas of conflict and resolve them in a way that balances the management to provide the most good to the people who use and depend upon the resources of the forest.

FOREST ADMINISTRATION

The planning, budgeting, scheduling, supervising, and associated activities needed to keep a private or public forestry organization running fall under the purview of the administrator. That person must channel the knowledge and talents of staff specialists to plan and put into practice a management program that best meets the wishes of an owner. Further, administrators must accomplish this task within the constraints of the financial and technical resources available. In this role, the administrator will serve as liaison between the owner, board of directors, elected officials, and technical specialists. This liaison will involve setting of goals based upon the objectives and instructions of the owner, and deciding how to allocate resources to accomplish them most efficiently.

Budgeting and Operations

The operating budget is one of the principal tools available to the administrator for influencing activities. It will include items for capital expenditures, such as for land acquisition, purchase of major equipment, and investments in development of roads and structures. Such a capital budget will represent the end product of a long period of preparation to identify needs, determine their justifications, assess their payoff, and develop specifications for the expenditure. In addition, the administrator will prepare an annual operations budget to cover salaries, materials and supplies, travel expenses, maintenance costs, taxes and overhead, and other items necessary for the smooth functioning of the organization. The annual budget will also show anticipated revenues from sales and fees, and compare these with costs. A balance sheet will reflect the expected profit or return on the investment in maintaining the program and paying off ongoing capital expenditures.

In addition to a budget, the administrative staff will also prepare annual operations plans to show how budgeted items will translate into program activity. These plans might include such items as the annual cutting schedule, a listing of stands scheduled for harvest and regeneration, and details about road and access construction. The plans will allocate staff time for organizing and maintaining recreation programs, scheduling intermediate stand treatments, maintaining various facilities, and carrying out whatever other activities the firm or agency will engage in each year. Items included will reflect long-range plans developed for the property and show how supervisors will channel the year's activities to assure fulfillment of the goals set forth in the overall management plan.

Personnel and Training

The forest administrator must identify the types of people needed to accomplish the tasks of the organization, and recruit them for the staff. In large organizations, this function may rest with a full-time personnel manager who will also serve as the liaison between the employees and the firm's management. In addition, that person will oversee the implementation of administrative guidelines for public employees. Another member of the administrative staff may have the responsibility for developing training programs. These will introduce new employees to work assignments, organizational policies, and methods for undertaking various tasks. This staff person will encourage and arrange continuing education for older employees, and help train key people in new techniques adopted from research results or administrative innovation (Figure 13–3). Together the training officer and personnel manager will often have responsibility for employee evaluation and for setting performance standards for the different work tasks. This assessment will serve as a means for evaluating people for promotion and assigning new responsibilities as they develop. Altogether it may take 8–10 years of experience to prepare a person to assume major administrative responsibility. The ongoing training program will play a key role in this preparation.

Record Keeping

The forest administrator will be responsible for developing and maintaining a program of reporting and record keeping to keep track of the organization's activities. Some will include records of silvicultural activities, recreational use intensity, maintenance accomplishments, resource inventories, and other kinds of information the firm requires to assess accomplishments of the organization and plan its future efforts. Other reporting will account for income and expenditures and accumulate data needed in planning future budgets, or for charting efforts to maintain financial solvency. All these report and record-keeping efforts will utilize a variety of means, including computer files and data banks that the administrator can manipulate in various ways to facilitate information dissemination and evaluation.

Figure 13–3.
Forest administrators will provide continued training for the staff as a means of introducing new techniques, improving employee skills, and preparing them for promotion to positions of increased responsibility.

Directing Operations

Through this complex of activities involved in goal setting, planning, budgeting, establishing an organization, recruiting and training, making work assignments, supervising, and reporting the forest administrator directs the organization's work. All these activities help in making many routine as well as complex decisions required to sustain the functioning of an organization.

The administrators must encourage eagerness and devotion and foster the organization's growth and success. They must effectively delegate responsibility to subordinates throughout the organization, and furnish them with information needed to make decisions, for which supervisors must provide the support to implement them. In directing these efforts, administrators must maintain contact with employees in order to identify problems that might develop and help them find suitable solutions. In essence, an administrator's skill in utilizing the employees' talents will ensure fulfillment of the program's goals.

Top executives must operate with a considerable element of uncertainty about the future. They can organize planning and information gathering. They can assure effective evaluation and analyses of available data. But they must depend upon people within the organization to remain attentive to changing situations and new opportunities, and to react creatively and effectively in carrying out their responsibilities. The success of an administrator depends upon that individual's ability to work with people and to challenge them into effective action. Administrators must remain ready to accept this responsibility, to devote themselves to the objectives of the organization for which they work, and to lead others in striving together for that purpose.

CAREER OPPORTUNITIES IN FOREST MANAGEMENT AND ADMINISTRATION

All organizations need managers and administrators. In small firms or agencies, the structure may remain relatively simple and the means of controlling activities fairly informal. Larger organizations develop a staff with more built-in levels of authority to accomplish their tasks. If the programs cover a wide geographical area, even more complicated structures may develop, with subdivisions handling programs in different regions or as various major tasks of the organization. Still, at all levels of complexity, the forester as administrator will be responsible for identifying the purposes of ownership, surveying the resources available, and planning a program to use those resources effectively to serve the interests of an owner.

The general forestry education provides students with at least a basic preparation for fulfilling these tasks. Above all, managers must have the capability to reason, to integrate various pieces of information, to decide what actions to take, and to plan how to implement them. These skills depend upon such courses as communication skills, psychology, sociology, economics, political science, mathematics, and others that help stimulate thinking and build general intellectual capabilities. Later, courses in business management, forest regulation and finance, management of the business enterprise, forestry economics, public administration, personnel management, accounting, land use planning, and others will introduce students to specific skills useful in administration and management.

Some individuals may elect to focus their careers on timber management planning. They can take electives such as biometrics and mensuration, forest inventory, computer applications, resource planning and development, photogrammetry, and other courses that strengthen skills for planning and controlling timber management programs. Comparable sets of courses for recreation, water, and wildlife would prepare individuals for similar planning efforts with those resources. Other students will find challenges in business management and take electives in accounting, finance, law, personnel management, and principles of management. For students who might seek careers with government agencies, elective courses such as public administration, political science, budget planning, and government should prove useful. The faculty adviser can provide valuable assistance in identifying particular courses that the college might offer to help a student prepare for a specific type of career.

Schools of forestry or other departments at the university also offer graduate programs related to business management and administration. Individuals who seek careers in corporations might well consider a program leading to a master of business management degree rather than to strengthening their technical skills in forestry per se. Others will find offerings at both the master of science and doctor of philosophy levels in resource management and policy, forestry and resource economics, regional resource development, business administration, public administration, and many related fields. The master's programs often do not require a thesis, but instead may offer internships and projects as a mechanism for giving students practical experience or allowing them to apply basic skills in a working situation. These often prove valuable in helping individuals to experience the dynamics of administration in large organizations, and to see how successful administrators accomplish their tasks.

Graduates from both undergraduate and advanced studies will find employment in a wide range of public and private agencies and firms. These may include positions in field offices or at the main headquarters of the organization. Early assignments may move the individual about to experience many different aspects of the organization's functioning. That gives new employees a feel for the overall structure of the agency or corporation, and helps them and their supervisors to find an area where the individual might be best suited to work. For individuals who have personal and technical talents, promotion may come rapidly with commensurate increases in responsibilities and salaries.

Persons interested in pursuing careers in management and administration in private industry need not limit themselves to large corporations. Many smaller firms hire professionals to help run the business, and give to these individuals a broad range of responsibilities. While the administrative structure and function may be simple, the individual will find ample challenges in having to serve in different capacities. In that way, a smaller staff can fulfil all the needs of an enterprise. Some may also find a challenge in starting their own company or joining into partnership with others. Also many nonforestry companies will seek out graduates who can provide technical liaison with forestry organizations with which the firm does business.

Many foresters who do not aspire to administrative work will draw frequently upon techniques of management and contribute toward the overall management of the organization that employs them. Professional people usually have at least a small staff of technicians and others who assist in the forestry work. Organizing these people, scheduling their work, overseeing their activities, and developing the budgets to support them will parallel the general programs common to broader administrative activity, except that they work on a smaller and simpler scale. Many individuals who initially seek careers conducting and supervising field operations eventually gain supervisory responsibility. They win promotion into management positions, assume increasing responsibilities for administrative work, and become less directly involved with activities in the woods. Experience usually brings increased responsibility to most professional foresters, along with greater involvement in supervision and management. After 8–10 years, many forestry professionals end up in administrative careers rather than working directly in the forest manipulating the resources. This seems the inevitable course for those professionals interested in such challenges.

REFERENCES

1. T. E. Avery, *Natural Resource Measurements,* McGraw-Hill, New York, 1975.
2. R. Barlow, *Land Resource Economics,* Prentice-Hall, Englewood Cliffs, N.J., 1958.
3. L. E. Beeman, "Computer-Assisted Resource Management," in H. G. Lund, V. J. LaBau, P. F. Ffilliott, and D. W. Robinson (eds.), *Integrated Inventories of Renewable Natural Resources: Proceedings of the Workshop,* U.S. Forest Service General Technical Report RM-55, 1978.
4. D. D. Chapman, "Design of Integrated Natural Resource Surveys," in H. G. Lund, V. J. LaBau, P. F. Ffilliott, and D. W. Robinson (eds.), *Integrated Inventories of Renewable Natural Resources: Proceedings of the Workshop.* U.S. Forest Service General Technical Report RM-55, 1978.
5. H. H. Chapman, *Forest Management,* Hildreth, Bristol, Conn., 1950.
6. M. Clawson, *Forests for Whom and for What,* John Hopkins Press, Baltimore, 1975.
7. K. P. Davis, *Forest Management: Regulation and Valuation,* 2d ed., McGraw-Hill, New York, 1966.
8. W. A. Duerr, *Fundamentals of Forestry Economics,* McGraw-Hill, New York, 1960.
9. F. C. Ford-Robertson, *Terminology of Forest Science, Technology, Practice and Products,* The Multilingual Forestry Terminology Series, no. 1, Society of American Foresters, Washington, 1971.
10. W. E. Frayer, "Objectives of Multiple Resource Inventories in Relation to Design Specifications," in H. G. Lund, V. J. LaBau, P. F. Ffilliott, and D. W. Robinson (eds.), *Integrated Inventories of Renewable Natural Resources: Proceedings of the Workshop,* U.S. Forest Service General Technical Report RM-55, 1978.
11. T. C. Nelson and R. N. Stone, "Supply and Forest Owners," chapter 6, in W. A. Duerr (ed.), *Timber! Problems, Prospects, and Policies.* Iowa State University Press, Ames, 1973.
12. H. L. Shirley and P. F. Graves, *Forest Ownership for Pleasure and Profit,* Syracuse University Press, Syracuse, New York, 1967.
13. U.S. Forest Service, *Forest Statistics of the U.S., 1977.* U.S. Department of Agriculture, Forest Service, Washington, 1978.
14. Ibid., *An Assessment of the Forest and Rangeland Situation in the United States,* U.S. Department of Agriculture, Forest Service, FS-345, 1980.

CHAPTER **14**

THE NATURE AND PROPERTIES OF WOOD

Wood and various by-products of trees have helped to sustain human life since prehistoric times. Despite the availability of many synthetic substances, wood remains in high demand, primarily because of its properties. For example, wood has great strength and resilience in relation to its weight and makes a superb building material for many uses. Wood has durability and resists oxidation, salts, acid, and other corrosive agents. It has high shock value, and when dry provides good insulation against heat, electricity, and sound. Its fibers have value for use in paper manufacture and several other types of products. People can subject it to various chemical reactions and extract basic constituent materials, and then reuse these in other processes to make several manufactured products. Wood also burns readily and serves as a vital fuel in many parts of the world. Above all, it has intrinsic beauty and good grain patterns that skilled craftspeople can enhance with stains and finishes. It shapes easily, and fastens with simple devices. It can be combined with other materials, and repaired readily. And wooden structures can be altered or remodeled without great difficulty.[11,18] Because of these qualities, wood has great value as a universal material for both functional and esthetic uses.

HOW TREES GROW

The stems, branches, and twigs of trees support the food-producing leaves. They also provide the structure for conducting the flow of water, nutrients, and food substances between the foliage and other parts of the tree. Roots anchor its great mass and serve as the conduits for water and minerals from the soil. Chemical processes taking place in the leaves synthesize these basic materials plus carbon dioxide into products such as cellulose and lignin that form the structural parts of the tree. These must give its parts great strength to support the weight of the leaves and upper branches. The main stem must have great flexibility to sway with forces of the wind without breaking. These same properties enhance the value of wood for many domestic uses.

Trees grow in two basic ways. First, shoots and roots become longer, and then increase in diameter or girth. Elongation takes place only at the root, stem, and branch tips. There, behind a protective covering of outer tissues, actively dividing cells compose an area called the apical meristem. These new cells develop at the tip and increase the shoot or root length. This apical meristem gives rise to all cells that make up the primary stem tissues, plus the buds that develop into branches, leaves, and flowers. A short distance back along the stem or root, the primary xylem (wood) and phloem (bark) develop from a thin layer of cells called the cambium. Cells produced on the inside of the cambium become woody. These add strength and diameter to the stem. Cells produced on the outside develop into bark that protects the wood and conducts fluids. Near the bark's outer surface, another bark-forming region called cork cambium forms a corky tissue. It serves as a barrier to water loss. As the stem grows in diameter and forces the cork cambium outward, it ruptures and dies. A new cork cambium then forms inside from young cells, only eventually to rupture, die, and slough off the outer face of the tree. This process recurs continually as the tree grows.[3,9,11,18]

Each year, a new sheath of wood forms over the tree's entire surface, from root tips to the ends of each branch. This sheath completely encases the stem, becomes lignified (woody), and develops into permanent tissue. Once laid down, the annual sheath never increases in length or diameter. Instead an entirely new sheath covers it the next year. And as one sheath forms over the last one, the tree diameter gradually increases. Also, each year the actively growing apical meristem tissues extend the tip of branches and roots.[11] As a result, the annual sheath of wood must cover a progressively larger surface area. To maintain this growth, food production must increase proportionately. If it does not, diameter growth will slow and the thickness of the new woody sheath will diminish. If drought, defoliation, or other factors interrupt growth and it starts up again, more than one sheath of wood may form in a single season. Trees with small suppressed crowns may have an incomplete woody sheath.[18]

Wood closest to the cambium and extending only a short distance toward the center of a tree contains living cells. As new cells form over the older ones, the inner tissues die, and the dead cells undergo chemical and physical changes. These inactive cells may fill with various materials that increase the wood's weight, though not always enough to make an important difference.[1,18] The deposited materials may also improve the resistance to crushing, and the durability.[1,18] For some species, the chemical changes darken the wood, while other species remain bright in color. Injuries to the tree also trigger oxidation and other chemical changes in wood cells and cause discoloration of the wood present at the time of the injury.[13] Consequently trees that normally do not develop darkened inner wood might discolor as the result of a wound.

If a tree is cut down, the annual sheaths of wood appear as concentric rings on the top surface of the stump. Foresters call them annual rings, or growth rings. In temperate regions with distinct seasons, trees develop separate growth rings each year. Thus, from a cross section of these, the forester can see a history of the tree's growth. Periods of good growth and favorable site conditions show up as wide rings. If the tree suffers gradual suppression as it matures, the growth rings become increasingly narrow and more difficult to separate. Trees growing on good fertile sites and free of serious competition grow in diameter at faster rates than do trees on impoverished soils. Years of drought make for poor growing conditions and the trees develop narrow rings sandwiched between wider ones from better years. Incidence of fire that damaged the cambium, mechanical injuries to a side of the tree, increases in growth due to release from crowding, and similar events also might influence the growth or condition of a tree. These cause distinct patterns of tree rings that are easily visible.[1,18] For regions that undergo periodic droughts, scientists can study these patterns and find common combinations of growth rings among many trees. Then, by counting back among very old trees, they can establish a type of visual calendar that charts the ring patterns against a baseline of time. Later they can match these against the growth patterns in boards, posts, and other objects and determine the age of those pieces of wood. This science of dendrochronology has value to archaeologists, and has traced pieces of wood back as far as prehistoric times.[1,5]

THE STRUCTURE OF WOOD

If you use a saw to cut a length of wood from a log, and then split it into pie-shaped pieces with an axe, you will see three basic faces on each piece: (1) a cross section opened by the saw cut; (2) a radial section exposed when you split the piece lengthwise; and (3) a section tangential to the bark (Figure 14–1). The cross section on the end displays concentric growth rings as laid down each year by the cambium. By counting the number of rings from the outside to the pith at the very center of the tree, you can determine the age. These rings form annually among trees in climates with a distinct summer and winter, or with alternating wet and dry seasons. Widths of the annual rings reflect the tree's vigor, how rapidly it grew in diameter, and whether it ever suffered suppression.

The tangential face has a bold grain pattern caused by exposing segments of the inner and outer surfaces of the annual rings. These give abstract images of irregular shapes and contrasts. The radial face displays a more uniform and finer grain, and shows a side view of the growth rings. Hence lumber from the log's tangential face may have a conspicuous flat-sawn grain pattern. A radial cut through a log produces quarter-sawn lumber with grain appearing as long, uniform

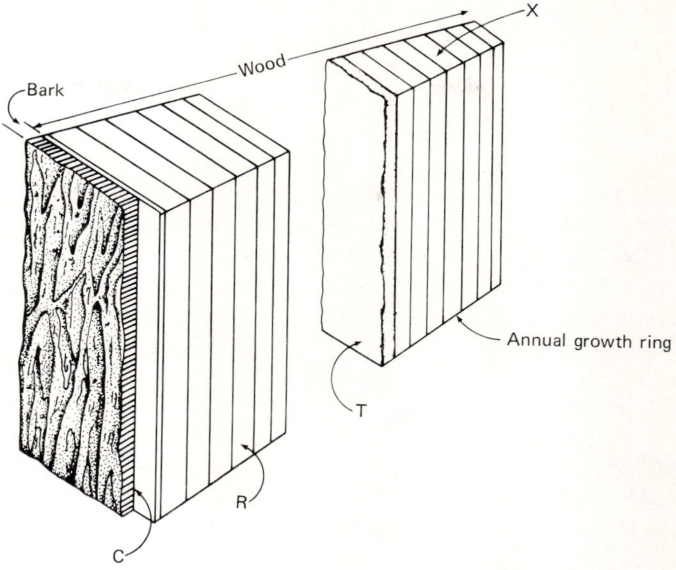

Figure 14–1.
The three basic faces of a piece of wood have distinct characteristics, as shown here for the cross section (X), radial section (R), and tangential face (T). The cambium (C) lies between the bark (phloem) and wood (xylem).

lines. In both cases, the ends of the boards will show the distinct growth rings of the cross section. When you look at furniture, lumber, and other items of wood and see these basic features, you can tell the original orientation of the board in the tree.

Wood has an unseen molecular and submicroscopic structure as well. When you view it through a light or electron microscope, you see a highly structured cellular material far more complex in construction than homogeneous materials such as steel, aluminum, and glass. Yet a tree synthesizes this complex structure from simple and readily available carbon dioxide and water. The formation of these into wood begins with production of glucose into the long-chain polymer called cellulose. Cellulose has the empircal formula $(C_6H_{10}O_5)n$, where n represents the number of anhydroglucose units in the molecule.[15] These long-chain molecules form into strands of microfibrils (Figure 14–2) that give wood its tensile strength. They account for 40–50 percent of the wood substance. The hemicellulose and lignin fill voids around the microfibrils and hold them in place. Hemicellulose accounts for about 20 percent of softwood (conifer) substance, and 30 percent of hardwoods. Lignin also forms an amorphous material around and within the woody cell walls, binding the other components together to give wood

Figure 14–2.
Lumen lining of Douglas-fir tracheid showing helical thickening and crisscross arrangement of microfibrils. *(Courtesy Wilfred Côté and University of Washington Press.)*

its stiffness. It makes up 15–35 percent of the substance in wood, with its highest levels in conifers.[11]

The cellulose, hemicellulose, and lignin together account for 71–98 percent of dry wood's substance. The remainder includes various gums, resins, and other miscellaneous materials. In living trees, however, water accounts for about one half or more of the total weight. This water carries basic minerals into and through the tree. As the water flows through a tree, it also transports sugars down from the leaves to the growing branches, trunk, and roots. It moves foodstuffs into and out of storage in cells, and carries the surplus mineral substances for deposit in the leaves and eventual discard if not needed in the tree. More than this, water is a basic constituent of sugars and cellulose.

A look at the cell wall structure of wood shows the long-chain molecules bunched into microfibrils and reinforced with hemicellulose and lignin (Figure 14–2). Aggregations of the microfibrils form cell wall tissues.[4] This complex cell wall has three major layers displaying a cross-laminated structure (Figure 14–3). In the thickest layer, the microfibrils align in the vertical direction. This helps to support the tree's weight, and provides the major resistance to stress parallel to the cell's length. Figures 14–4 and 14–5 show portions of the tree stem on an enlarged scale for a hardwood and a softwood respectively. These illustrate the alignment and diversity of cells as viewed in cross section, tangentially, and radially. Open cells in the wood conduct water and other fluid materials. In dried wood, the empty cells also form dead air space that affords wood its insulating qualities against transmission of sound and heat. Such a pipelike structure also provides economy in the distribution of wood substance, giving it great strength for its mass. Some of the cells are living and function to transform plant foods, changing sugars to starches, and later back to sugars. In this way, food stored in late summer becomes available the following spring to support rapid growth in the new plant tissues.

Figure 14–4 gives a more detailed view typical of hardwood trees with their long heavy-walled and pitted wood fibers. These lend strength and stiffness to the wood. Large-celled vessels conduct water, and other thin-walled cells serve for transportation and storage of foods, or as gum canals. Ray cells seen from the tangential face vary from broad and conspicuous to narrow and inconspicuous.[4] They transport materials across the tree stem. Both the radial and tangential faces show pits in the sidewalls of the different cells. In some species, called ring porous, water-conducting vessels laid down early in the growing season have large diameters, while those formed later in the year have smaller ones. This gives the cross section a distinctive appearance of concentric light- and dark-colored circles. Diffuse porous woods have more uniform vessel sizes.

Closer examination of conifer wood shows somewhat different construction (Figure 14–5). In temperate zones, they commonly exhibit a conspicuous variation in cell size between spring wood and summer wood. Some species contain scattered large channels lined with thin-walled cells. These form ducts for transport of resins up, down, and across the stem. The water-conducting cells, called

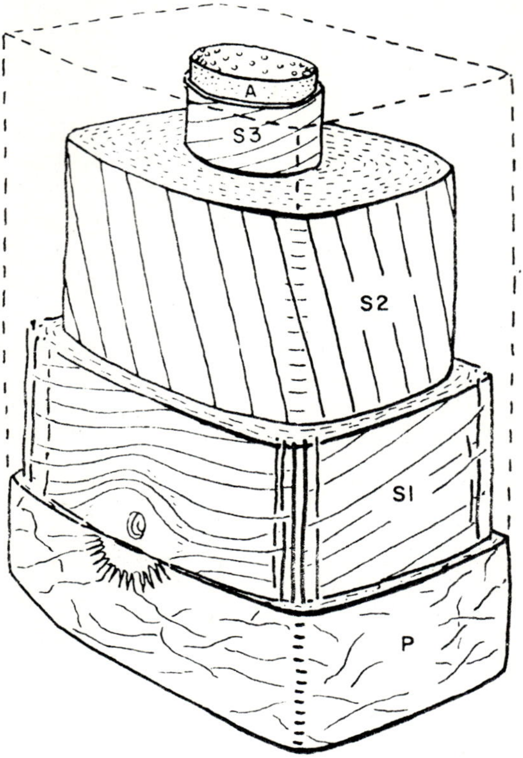

Figure 14–3.
Schematic drawing showing the structure of the wall of a woody fiber, with (A) the inner layer; (S1, S2, S3) the three layers of the secondary wall, distinguished from each other by the different orientation of microfibrils; and (P) the thin primary outer wall. *(Courtesy Modern Materials, Henry H. Hausner, ed., Academic, New York, 1958.)*

tracheids, also help support the tree's weight. These small- to medium-diameter cells have small circular pits on the sidewalls. If a conifer tracheid becomes plugged with resin, water will pass from it to an adjacent cell via the pits. Ray cells, seen in the tangential face, run at right angles to the tracheids. The horizontal ray cells primarily transport fluids from the center of the tree to the outside, and in the other direction. Movement of materials through the rays equalizes pressures between growth rings. The rays also serve as locations for food storage.

WOOD FEATURES AND DISTINCTIVE CHARACTERISTICS

Prospective users in the United States have available more than 100 species of woods to choose from, though only a relatively small number may grow at any

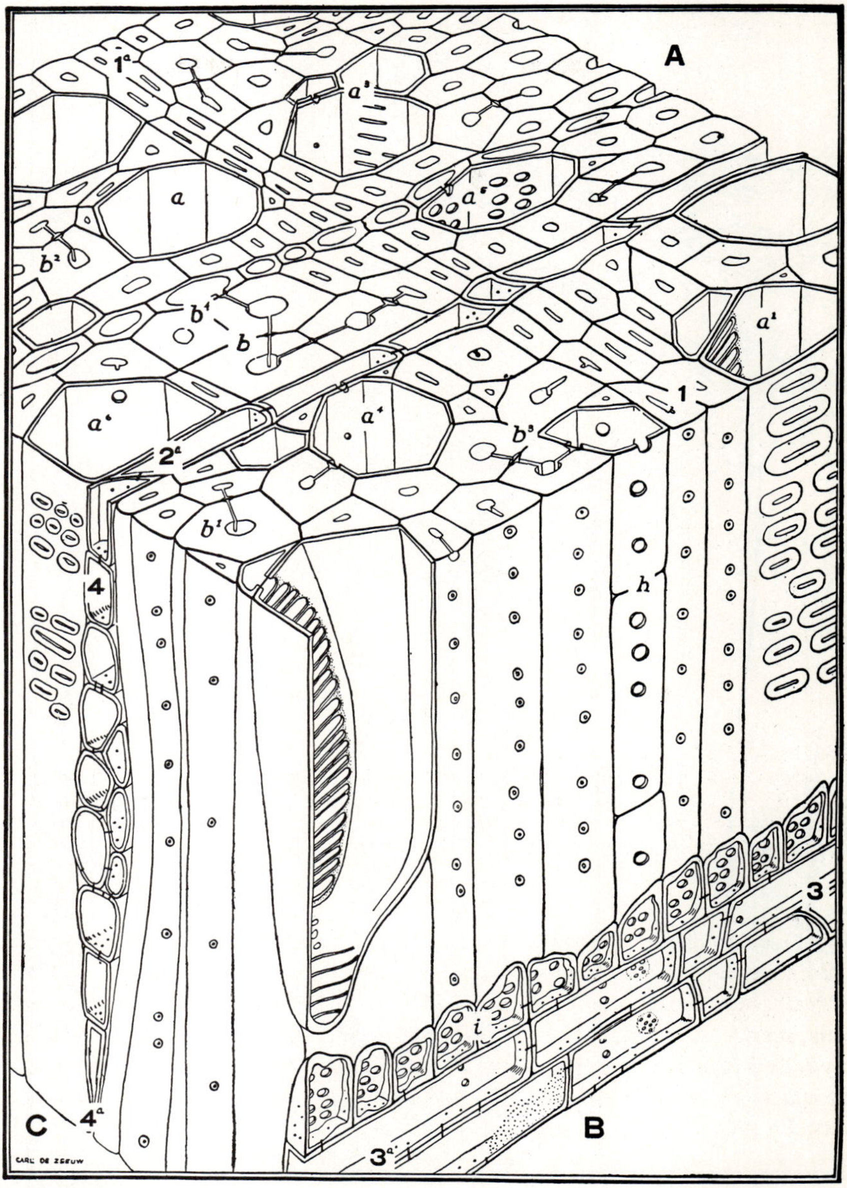

Figure 14–4.
Schematic drawing showing cellular structure of red gum. Face A: Cross section showing vessels (a) and fiber tracheids (b). Face B: Radial section showing ray (3) and wood parenchyma (h). Face C: Tangential section showing ray cells (4) and fiber tracheid (1). (*From Panshin and deZeeuw, Textbook of Wood Technology* (ed 4), *McGraw-Hill,* New York, 1980.)

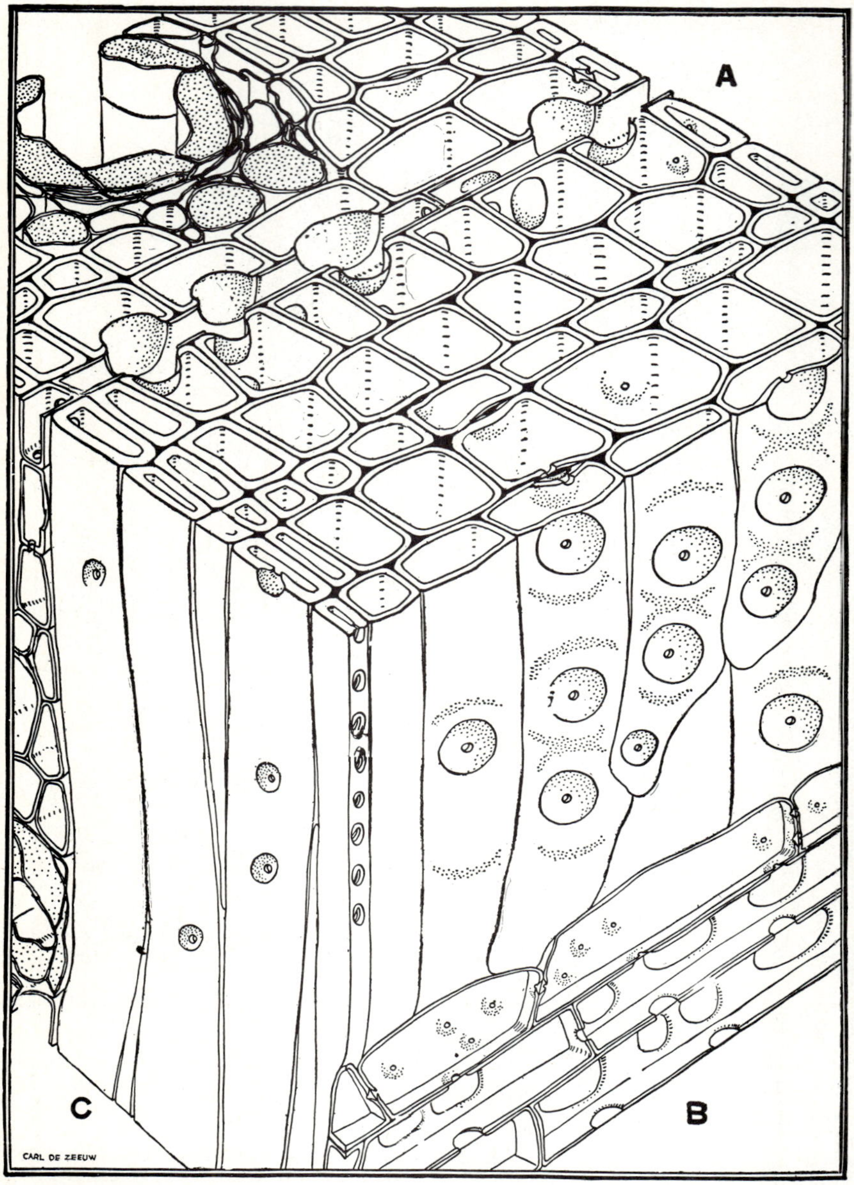

Figure 14–5.
Schematic diagram showing cellular structure of eastern white pine. Face A: Cross section showing resin canal. Face B: Radial section. Face C: Tangential section. *(From Panshin and deZeeuw. Textbook of Wood Technology (ed 4), McGraw-Hill, New York, 1980.)*

single locality. Commercially about 60 native species have importance, with another 30 imported from various parts of the world in the form of logs, cants, lumber, and veneer.[18] Tropical species far outnumber those in the temperate regions, and several dozen may occur on a single hectare within a tropical rain forest. Still most tropical species have little commercial value, although advances in wood technology and manufacturing and the demand for increasing supplies will certainly result in a broadening of the number accepted for commercial uses—as has occurred with temperate species. This diversity provides people with woods that have a wide range of properties and potentials.

Many species have distinctive characteristics that make them easily identifiable, even by casual observation. Readily recognized features include color, odor, luster, taste, grain, figure, density, ray size, vessel size and arrangement, and hardness.[11] The absence of vessels, for example, distinguishes the conifers from hardwood species. The hardwoods occur in groups as either ring or diffuse porous wood. More specialized features help separate the species within these groups, such as the lightness and softness of balsa wood versus the heavy, dense wood of ebony. Incense cedar gives off the distinctive odor we associate with wooden pencils. Oaks show obvious light-colored rays that easily distinguish them from most other hardwoods. In northern white oak, bubblelike outgrowths (tyloses) of cells plug the vessels, but these remain open in many red oak species. Features such as these help to distinguish one wood from another, and people who work regularly with them have manuals and keys with descriptions of the different species.[3,4]

With all the distinguishing features of the various species, wood has several general characteristics that make it as good as or better than other materials for structural purposes. These include the following:[10,11]

1. It is easily fashioned into a variety of shapes by using simple hand tools and machines.

2. It is readily joined with nails, screws, and other connectors, or fastened with adhesives to yield a joint as strong as the wood itself.

3. It has high dimensional stability with regard to changes in temperature compared with metal, and in the direction of the grain with regard to moisture changes.

4. It is resistant to corrosion by saltwater or weak solutions of acids or alkalies, and if it is kept dry it endures indefinitely.

5. In large sizes, it heats slowly, and loses its strength only gradually when exposed to fire.

6. When dry, it insulates against heat, sound, and electricity.

7. It has excellent rigidity and strength for its weight, compared with other construction materials.

8. It displays an intrinsic beauty in its infinite variety of grains, colors, and textures that enhances the objects made from it.

These properties make wood a superb structural material for many uses, especially for applications where it is protected from wetness and high moisture.

Wood also has several features that makes it useful for a variety of nonstructural purposes as well.[5] Above all, the distribution of forests around the earth makes wood abundant and relatively cheap, especially in moist temperate and tropical regions. The absorbant surface of wood holds glue, stain, and paint readily. This facilitates its joining, decoration, and preservation. The character of wood makes it suitable for bending, and thus for use in musical instruments. Its high shock resistance sets wood apart as an excellent material for railroad ties, flooring, hammer and tool handles, sports equipment, and many other objects that require resiliency. The tough, energy-absorbing nature of wood means that when stressed it warns of potential breaking by first sagging, cracking, and splintering. This quality makes wood highly valued for supporting loads, as with mine props, beams, and stringers. And finally, when disposed of it decays or has value as fuel.

Despite these advantages, wood does have shortcomings. These limit its use for some applications, or interfere with its durability. To illustrate, knots, spiral grain, tension and compression wood, checks, and similar characteristics weaken it. Consequently designers often specify bulkier pieces and members than would be needed in wood free of such defects. Wood will shrink and swell across the grain as moisture changes, limiting its use for applications that require greater dimensional stability. When struck or forced along the grain, a wooden piece will split. And under prolonged pressure, wood tends to collapse. However, not all species exhibit these features to the same degree, so proper choice of the one to use will help to minimize the disadvantages. Also, some wood properties result from poor or unusual tree growth, and proper silvicultural practices will often improve its qualities. Best results often come by maintaining uniform growth rates and favoring high-density wood.[3,7]

Cellulose and lignin do not differ much among woods. Rather their mass and distribution in the cells largely determine the wood's properties. These substances form in developing cell walls in response to local stress. Wood laid down in the main stem tends to have greater density than that from upper branches and roots. Also, wood produced early in the growing season when trees use considerable moisture and food for developing leaves and new stem tissues tends to have thinner walled cells with large cavaties, or lumens. Later in the summer, the new cells have thicker walls, and the wood a greater density. In the southern pines, gums and resins impregnate the wood and give it heavy weight, add stiffness and brittleness, and result in other special properties.

A tree's growing conditions also influence properties of the wood produced. Youthful trees grow with greater vigor and form wide growth rings. As they mature and become crowded by other trees, the growth rate slows and the rings become narrower. Further, within a single tree the wood properties may vary by position in the tree, season of the year, and age. These depend upon the size of cells as well as the supplies of light, nutrients, and moisture available to support photosynthesis and growth.[5,7] Resultant differences in wood properties among trees of a single species, and between species, may profoundly affect the suitability for some uses.[5,11,15,17]

Stiffness, strength in relation to weight, and ease of fabrication represent the three most desirable qualities of wood for structural purposes. Hardness makes some species such as oak and beech particularly valuable for flooring. Poles and pilings require the full strength of the tree trunk and tallness, strength, and minimum taper found in several conifers. Sawn wood should retain its straightness and not twist and warp when dried. Some species perform better than others in this regard. For construction use, a wood must take nails easily, thereby precluding use of high density species like the hickories. Most conifers serve extremely well for framing and boards in construction. The toughness of hardwoods makes them better suited for pallets and furniture. Wood used in contact with the ground must have natural tannins, oils, or resins that make the pieces resistant to wood-destroying fungi. Otherwise it must be treated with chemical preservatives. Thus, choice of a species depends upon the properties of its wood and understanding of what characteristics will best serve the intended use.[11,17]

CHEMICAL COMPONENTS OF WOOD

Wood tissue forms from a group of large molecules that make up the bulk of the cell walls, plus several inorganic substances. Some remains after combustion, and this ash contains mostly the inorganic elements. A second type of substance, called wood extractives, can be removed by steam or neutral solvents. These include a wide range of organic compounds. The wood itself is made of cellulose, hemicellulose, and lignin. These remain insoluble in neutral solvents, and do not volatilize with steam.[10,19,20] Variations in these component substances and differences in the cellular structure among species make some woods heavier than others, some stiff and others more flexible, and some harder. However, within a species the wood composition has similarity from tree to tree, giving it properties a user can learn to depend upon.[18]

Chemicals in wood ash come by deposition from the tree's sap stream. Some enter into live tissue in the wood rays, cambium, and bark. As a result, wood ash contains all the elements essential to plant growth, except for the carbon, hydrogen, oxygen, nitrogen, and sulfer lost during combustion. Some of the common components of ash include calcium and potassium. These various chemicals add little strength to the wood, but those dissolved in the moisture in the wood increase its electrical conductivity. Several tropical species and a few others also have a high silica content that will rapidly dull tools during machining.[15]

Extractive substances include tannins, waxes, oils, resins, gums, sugars, and pigments. Some may occur in sufficient amounts to add to compressive strength, and others affect decay resistance. In a living tree, these may seal over wounds, repel insects, and serve other functions. In contrast, the major components of wood remain insoluble and nonvolatile, though combustible. These substances include lignin and the polysaccharide system termed holocelluloses, made up of cellulose and the alkali-soluble hemicellulose, together with some peptic materials. Lignin, hemicelluloses, and cellulose are found in all portions of cell walls in wood. And since cellulose and lignin do not react to most chemicals, wood does not easily corrode in air or saltwater. Wood can be decomposed by fungi in a process known as decay, or by other wood-destroying organisms such as termites and marine borers.

Cellulose from wood has many uses, either as the major constituent of fibers or as reconstituted from solutions. In a reconstituted form, it goes into many products. For example, purified cellulose can be dissolved with strong alkali and reformed by acid coagulation into fine strands for rayon production, or into sheets to make cellophane. Treatment with nitric and sulfuric acids forms nitrocellulose known as guncotton, with highly explosive properties.[20] Nitrocellulose with less nitrogen than found in guncotton goes into the manufacture of lacquers and plastics. Cellulose forms esters with acetic acid to yield cellulose acetate that will soften with heat.[16] These have use in various plastic materials, including acetate fibers for textiles and films for photographic work. Plasticized ethyl cellulose has value as an insulating coating for wire, or when extruded in strips can be used as a material for furniture seats. It has flexibility, toughness, stability, low moisture absorption, takes printing, and fabricates well. Thus, when treated through various chemical and manufacturing processes, cellulose from wood can serve as a raw material for a variety of products that no longer look like or have properties of the original wood itself.

Cellulose and lignin belong to the family of chemical compounds known as polymers. These comprise a number of simple units, called monomers, chemically linked to form long chains or networks. In general, the longer the cellulose chain, the stronger is the fiber formed from it. Cellulose also has a high molecular weight made up of units of anhydroglucose, or glucose lacking a molecule of water. In a highly organized parallel condition, cellulose molecules form microcrystals that show up with x-ray diffraction. The linear nature of cellulose molecules, together with their packed parallel organization, gives cellulose fibers their high tensile strength. Cellulose chemically has the same empirical formula as starch, but in starch the cellular units form a spirally coiled chain. In cellulose, they form a linear chain. This small alteration in structure gives them vastly different physical properties and chemical behavior.

Lignin does not form a linear polymer as cellulose does, but rather a three-dimensional macromolecule. Phenylpropane, $C_9H_9O_5{:}OCH_3$, makes up its basic structural units. This lignin is deposited between cells and throughout the cell walls of wood. The molecular weights of soluble lignin derivatives cover an immense range from below 1000 to values greater than 10^6 for lignosulfonates.[12] It may comprise a mixture of two or more compounds, and when broken down chemically yields a variety of substances.[15] Among these, vanillin is valuable for flavorings, pharmaceuticals, and perfumes. Lignin tends to show color reactions with various chemicals—an undesirable property for papermaking. Like coal, it will yield petroleum-type derivatives when hydrogenated at elevated temperatures and pressures. Its complex structure differs between hardwoods and conifers, and hardwoods have higher methoxyl content in the aromatic rings.[2,20]

Several commercial processes will yield a variety of useful by-products from wood, and many of these date back for centuries. New techniques improve the potentials, but the traditional ones illustrate this value of wood. For example, the destructive distillation of wood yields charcoal, wood alcohol, acetic acid, turpentine, and tars. Extractive processes with water or petroleum solvents can remove such materials as tannins, turpentine, rosin, and essential oils. Hydrolysis can change the cellulose and other carbohydrates into sugar. The resulting sugars will yield molasses upon evaporation, serve as a medium for growing yeast, crystalize into glucose sugar, ferment into ethyl alcohol and other products, and serve as a feedstock for furfural and hydroxymethyl furfural and a variety of alcohols and acids. Another process, called hydrogenation, causes hydrogen gas to react with wood components at high temperatures and pressures to form a liquid comprised of complex cyclic alcohols, phenolics, and neutral oils, plus a tarlike residue. Hydrogenating wood in an aqueous alkaline suspension can leave the cellulose as a pulp residue, or under severe conditions will break the cellulose into sugars and glycerine.[11,14] Through such modes of processing, industrial firms can derive a variety of products essential to daily life, but not normally readily recognized as a by-product of our forests.

PROTECTING WOOD AGAINST DAMAGE AND DETERIORATION

Dry wood retains its properties and remains strong and intact for centuries. But it may suffer damage from wood decay organisms and insects if the moisture content, oxygen, and temperature remain at suitable levels for a prolonged period. Principally degradation of wood results from fungi, bacteria, insects, and marine borers. The fungi cause molds, stains, and decay. Bacteria primarily make the wood more absorptive. Some affect its strength. Insects and marine borers may eat or bore holes into the wood, affecting its strength and beauty. However, users can circumvent problems with most of these by taking precautions in the way they use and protect the wood.[18]

The likelihood of damage and degrading by fungi molds and many insects depends upon the wood's moisture content and the temperature. These organisms need a humid, warm environment, and cannot thrive in dry wood. For most of the United States, woods with a moisture content below 20 percent will not provide a good environment for them. Thus users can get good protection by properly seasoning or drying the wood following primary processing, by covering it during storage, and by ensuring adequate preparation and finishing of exposed surfaces. These steps will usually suffice for keeping woods at an appropriate moisture content. Wood used in contact with moist surfaces or exposed regularly to wetting, however, will need special treatment with preservatives to make it unsuitable for invasion by the wood-damaging agents.[18]

For products such as poles, posts, pilings, railroad ties, mine props, and others exposed to moist soil or in humid environments, some type of preservative treatment will extend the usable life considerably. This involves chemical treatment following primary processing, using one of two basic types of preservative. The preservative oils such as creosote and solvent solutions of pentachlorophenol provide considerable protection from weathering outdoors. But these may give wood an odor, change its color, reduce its paintability, and make it less fire resistant. Water-soluble preservatives such as acid copper chromate, ammoniacal copper arsenite, chromated copper arsenate, chromated zinc chloride, and fluor chrome arsenate phenol take paint better and give the treated wood a cleaner surface. They also have usefulness for outdoor applications, and some resist leaching better than do the preservative oils.[10,18]

The effectiveness of preservative treatments depends, in some measure, upon the species and the method of application. Round pieces scheduled for treatment must first be peeled and sometimes dried. A few species absorb the preservative better if first mechanically incised along the surface. Lumber will often be dried to reduce the moisture content, and should not have strips of bark on the edges. Some species readily take preservatives by a nonpressure soaking treatment, while other woods require pressure treatment. For some products such as telephone poles made from decay-resistent species, treatment may include only the lower part that will contact the ground. Other species may require full-length preservation. Lumber for frames and other structures should first undergo shaping, boring, and other processing so the preservatives will enter through the cut surfaces. Given such preparation and treatment, wood will last for long periods even under rather adverse conditions.[10,18]

Common wood-damaging insects include the ambrosia and powderpost beetles, termites, and carpenter ants. Marine borers include shipworms, pholads, *Limnora,* and *Sphaeroma*. Ambrosia beetles attack freshly cut lumber and some rustic products. Preventive measures against these insects include spraying the logs to keep them wet and then rapidly drying the lumber to reduce the danger of damage. In some cases, control may involve spraying and dipping the lumber in insecticides. Powderpost beetles attack both fresh and seasoned lumber. Control involves applying insecticides, or finishing the lumber on each surface with a suitable finish.[18]

Subterranean termites require a constant source of moisture and maintain their colonies in the ground. From there, they build tunnels up into wooden structures to get at the wood they use for food. Protection involves treatment of the wood, preventing access by using solid or capped foundations, destroying tunnels, or treating the soil with insecticides. Protection against dry wood termites requires pressure treatment with preservatives for posts, poles, and timbers. Painting protects the surfaces in structures. The measures to prevent decay and termites usually work against carpenter ants as well. In addition, control of moisture and provision of good ventilation usually afford considerable protection against carpenter ants. Protection of structures, boats, barges, and pilings against the various marine borers requires full treatment with creosote, coal tar, or comparable preservatives. Attention to these rather simple precautions will extend the life of wood and protect it against damage under most circumstances.[18]

WOOD AS A FUEL

The abundance of low-cost coal, petroleum, and natural gas in North America encouraged many families and industrial users to adopt these more convenient fuels beginning in the 1920s. Consequently a major industry built upon cutting of firewood and making of charcoal largely vanished across the continent, except for nonessential uses and some specialized industrial applications. By midcentury, most people considered firewood a luxury for burning in living room fireplaces, and charcoal had value only for backyard barbecues. Few families depended upon wood as a source of heat or for cooking. Then shortages of oil and gas in the late 1970s pushed the nation into a crisis. During a series of especially cold winters, many families ran short of fuel, schools closed, and factories shut down to conserve limited supplies of natural gas and oil. The situation worsened as oil-producing countries limited their production as a means of increasing price and gaining some political leverage on industrialized nations. Prices rose sharply and nationwide programs of conservation sprang up in response.

These events triggered activity to find alternate fuels, more efficient ways to burn them, and effective means to insulate homes and other buildings against heat losses. In response, many homeowners purchased wood-burning stoves and furnaces and turned once again to wood as a major or supplemental fuel. Industries began exploring the use of wood for power and steam generation, and many wood-using plants and sawmills began converting wood wastes into energy. Schemes even arose for electric power generation based upon wood fuels, and conversion of coal-fired boilers to facilitate wood began to increase the demands for using it as an energy source.

Where wood occurs in surplus amounts, suppliers who use modern harvesting equipment and efficient handling methods can deliver wood chips for industrial burning at costs competitive with some traditional fuels. Waste materials from sawmills and wood-processing plants especially offer great potential as a reasonably economical fuel. As a result, many sawmills, paper mills, and wood-processing plants have converted their own boilers to use waste wood and bark for fuel, thereby cutting back on their purchases of oil and gas for heating and steam generation. This development has helped many companies to cut fuel costs and gain considerable energy self-sufficiency. Expanded use of mill wastes or of fresh wood chips for other types of commercial-scale heating and power generation could become important as the nation searches for alternative energy sources.

For homes, wood seems to offer more limited potential as a widespread fuel. Many houses lack an adequate design for efficient distribution of heated air or the strategic placement of wood-burning stoves. Wood also requires a fairly large storage space, does not handle well, and does not lend itself to the convenience of automatic stoking. Consequently most stoves or furnaces require frequent refueling. The newer slow-combustion stoves have improved efficiency, and some require attention only at 8–12-hour intervals.[11] If left untended for longer intervals, the homeowner must provide a supplemental backup system that uses conventional fuel. Burning wood also emits considerable smoke and could pose potential pollution problems in congested areas, if enough homeowners there utilize it. Yet, with efficient combustion in well-designed stoves, wood burns quite cleanly and leaves little ash. The ash even has value as a fertilizer. And by converting to wood burning, many families have reduced their fuel costs. That has made wood attractive as fuel in many rural and suburban areas.

Although wood will not replace other fuels for most purposes, it may prove valuable as a supplemental energy source. Where sawmills and processing plants have converted to wood burning, they have saved considerably on fuel costs, thereby holding down the final price of their end products. Many homes do not lend themselves to wood-fired heating as a primary source, but adding space-heating stoves can supplement the oil and natural gas furnaces and reduce fuel bills. Eventually, if electrical generation from wood-fired boilers is found economical and practical, especially for the smaller generating plants, electric heat may offer advantages over the more conventional sources in some areas. In all these cases, proper design and maintenance of the heating system are essential for safe operation. In particular, homeowners must clean their chimneys and flues regularly and adhere strictly to fire codes for proper installation and use of their stoves and furnaces.

CAREER OPPORTUNITIES IN WOOD TECHNOLOGY AND CHEMISTRY

The widespread uses and importance of wood and its by-products provide many career opportunities for persons interested in wood technology, materials science, wood products chemistry, wood products engineering, and several related disciplines that contribute to our understanding of wood as a material and of how to use it effectively for serving human needs. Allied fields include those related to ways insects and fungi affect wood in products, architectural design, and engineering. An understanding of the properties of wood has application in many aspects of its manufacture into solid wood products, papers, fiberboards, and other wood-based goods.

Academic preparation in the several fields related to wood technology and chemistry draws heavily upon basic understanding of physics, mathematics, chemistry, and engineering. A program in natural products chemistry would also have a major concentration in several branches of chemistry and biology. Elective courses might include botany, zoology, entomology, and physiology. Some of these courses also are important to students interested in paper science and chemical engineering. Emphasis in materials science and wood technology would rely more heavily upon physics and engineering aspects of wood, with less concentration in chemistry. However, courses in biology also have value in developing appreciation for the growth processes that produce the wood. Additional study in wood structure and physical properties, especially relative to engineering and manufacturing, round out the education in wood products technology.

Not all schools of forestry offer programs for these careers, and some schools concentrate upon them rather than traditional forestry. Nevertheless most provide courses to give forestry students an introduction to wood as a material, and to the different processes for converting it into useful products. Individuals from these schools can strengthen their preparation in the physical sciences and mathematics and transfer to wood products and chemistry programs at the graduate level. These advanced studies would lead to a degree as master of science or doctor of philosophy, and prepare students for careers in research, testing, product development, and teaching.

Graduates from natural products chemistry, wood technology, and materials science programs will find employment within the forest products industry, and in research and testing laboratories. They may serve as staff specialists to aid engineers and manufacturing personnel in quality control, problem solving, and product development. Others may work as technical representatives providing liaison between a manufacturer and customers. In these many capacities, they would provide technical services to others who need to understand the qualities of wood and wood by-products better, and to learn how to capitalize upon those qualities effectively in the manufacture and use of different forest products.

REFERENCES

1. M. Bramwell (ed.), *International Book of Wood,* Simon & Schuster, New York, 1976.
2. F. E. Brauns, "The Proven Chemistry of Lignin," in C. J. West (ed.), *Nature of the Chemical Components of Wood,* TAPPI Monograph Series 6, 1948, pp. 108–132.
3. H. A. Core, W. A. Cote, Jr., and A. C. Day, *Wood Structure and Identification,* Syracuse Wood Science Series 6, Syracuse University Press, Syracuse, N.Y., 1976.
4. W. A. Cote, Jr., *Cellular Ultrastructure of Woody Plants,* Syracuse University Press, Syracuse, N.Y., 1965.

5. W. M. Harlow, *Inside Wood,* American Forestry Association, Washington, 1970.

6. W. M. Harlow and E. S. Harrar, *Textbook of Dendrology,* McGraw-Hill, New York, 1969.

7. F. F. P. Kollmann and W. A. Cote, Jr., *Principles of Wood Science and Technology,* Springer, New York, 1969.

8. J. E. Langwig, J. A. Meyer, and R. W. Davidson, "New Monomers Used in Making Wood Plastics," *Forests Products Journal* **19**(11):57–61 (1970).

9. P. R. Larson, *Wood Formation and the Concept of Wood Quality,* Yale University, School of Forestry, Bulletin no. 74, 1969.

10. A. J. Panshin, C. deZeeuw, and H. D. Brown, *Textbook of Wood Technology,* vol. 1, *Structure, Identification, Uses and Properties of the Commercial Woods of the United States,* McGraw-Hill, New York, 1964.

11. A. J. Panshin and C. deZeeuw, *Textbook of Wood Technology,* 4th ed., McGraw-Hill, New York, 1980.

12. K. V. Sarkanen and C. H. Ludwig (eds.), *Lignins, Occurrence, Formation, Structure, and Reaction,* Wiley-Interscience, New York, 1971.

13. A. L. Shigo and E. H. Larson, *A Photo Guide to the Patterns of Discoloration and Decay in Living Northern Hardwood Trees,* U.S. Forest Service Research Paper NE-127, 1969.

14. A. J. Stamm, "Chemicals from Wood," in *Trees, Yearbook of Agriculture,* U.S. Department of Agriculture, Washington, 1969.

15. Ibid., *Wood and Cellulose Science,* Ronald Press, New York, 1964.

16. SUNY College of Forestry, *Proceedings: Conference on Tropical Hardwoods,* August 18–21, 1969, SUNY College of Forestry, Syracuse University, Syracuse, N.Y., 1969.

17. H. Tarkow, "Properties and Behavior of Wood," *Journal of Forestry* **68**:408–410 (1970).

18. U.S. Forest Products Laboratory, *Wood Handbook: Wood as an Engineering Material,* U.S. Department of Agriculture, Forest Service, Agricultural Handbook no. 72, 1974.

19. C. J. West (ed.), *Nature of the Chemical Components of Wood,* TAPPI Monograph Series 6, 1948.

20. L. E. Wise and E. C. Jahn, *Wood Chemistry,* 2d ed., vol. 1, Reinhold, New York, 1962.

CHAPTER 15

WOOD PRODUCTS AND THEIR MANUFACTURE

The wood-using industry includes three basic types of manufacturing or industrial activity: primary processing, secondary manufacturing, and construction. Primary processing takes wood in the form of logs and bolts and converts them into such products as lumber, veneer, and pulp. From this stage, secondary manufacturing firms convert the basic materials into a variety of consumer goods. In some cases, a firm may integrate both primary and secondary activities, whereas other companies are involved with only one aspect of wood use. Many primary products go directly to the construction industry, or to homeowners and others for do-it-yourself projects. Construction also uses many finished goods that include wood materials. At each stage, the processing or other activity adds value to the initial economic worth of the harvested timber. This comes from the sale of goods, and from wages paid to workers.

USE OF WOOD IN THE UNITED STATES

As of 1977, about 388 million cubic meters of wood were used annually in these industries, or directly by consumers.[16] Most went to primary processing plants such as sawmills, planning mills, veneer and plywood plants, pulp mills, paper mills, and paperboard plants. In addition, a wide variety of other operations converted some timber into such products as wood shingles, cooperage stock, excelsior, particle board, and gum and wood chemicals. In 1972, these uses combined included about 28,375 firms across the nation.[16] They employed approximately 487,900 people and manufactured products worth about $23.1

billion. Secondary manufacturing took this output and converted it into some type of more refined product—wearing apparel, containers, furniture, millwork, prefabricated products, turned and shaped goods, paper and paperboard products, fibers, plastics, textiles, and many other consumer items. In 1972, this group of firms turned out goods worth nearly $34 billion, and employed almost 2.74 million people. The construction industry also employed in the vicinity of 795,250 with jobs directly attributable to timber or the use of wood products.[13]

Today, in sheer volume of material used, residential construction takes most of the nation's solid wood products (Figure 15–1). This includes softwood lumber and plywood, hardwood plywood, particle board, and insulation board. These products go into the upkeep, renovation, and building of housing. Another 10 percent goes for nonresidential construction. Manufacturing uses about 40 percent of the hardboard and particle board, and about 10 percent of the lumber, veneer, and plywood. They are utilized in a wide range of products such as household furniture, consumer goods, and commercial and industrial equipment. About 16 percent of lumber and 4 percent of plywood is allotted to wooden pallets, shipping containers, and other products for the shipping industry. Additionally about 13 percent of the wood produced from the nation's forests is exported, including 5 percent of the total lumber production.[16]

Table 15–1 shows how U.S. consumers used about 388 million cubic meters of wood for various purposes in 1977, and the projected demand for the year 2030.

Figure 15–1.
Residential construction, upkeep, renovations, and furnishings use most of the nation's solid wood products. *(U.S. Forest Service.)*

These data also reflect the relative importance of different segments of the wood-using industry. For example, pulp and paper manufacture took almost 44 percent of all wood used, including a considerable volume of mill residues. Projections suggest that the amount may rise to 394 million m^3 by 2030, and account for 49 percent of the total domestic wood consumption. Other miscellaneous industrial products took 10.7 million m^3 of roundwood in 1977, and that volume should increase to 25.5 million by 2030. These industries also used 14.6 million m^3 of mill residues, with that amount projected to hold fairly steady into the future. In addition, combined uses of wood for fuel by homeowners and industrial firms equaled 35.8 million m^3 in 1977. This amount is likely to rise, but the future demand for wood as fuel is unpredictable. Many complex factors can influence demand. The remaining 60 percent of wood used in 1977 went into lumber, plywood, panel products, and for other items required mostly in construction. Their use should also increase substantially and push total demand for all products to just over 800 million m^3 by the year 2030.[16]

TABLE 15–1
APPROXIMATE LEVELS OF WOOD USE IN THE UNITED STATES, AND PROJECTIONS OF DEMAND TO THE YEAR 2030

Type of use	Approximate consumption of wood in the United States,* millions of cubic meters 1977	2030
Pulp and paper products:		
Roundwood	100.9	194.6
Mill residues	70.5	98.6
Total	171.4	394.1
Other industrial products:		
Roundwood	10.7	25.2
Mill residues	14.6	14.6
Total	25.3	39.8
Fuelwood:		
Roundwood	13.2	57.4
Mill residues	22.6	unknown
Total	35.8	unknown
All uses combined—including lumber and panel products	388.0	801.5

*These estimates of the breakdown between roundwood and mill residues were developed from various available statistics, and may represent only approximate levels of use rather than exact amounts. They serve primarily to illustrate the general levels of wood consumption.
Source: Adapted from U.S. Forest Service [16]. Used by permission.

The several products mentioned above and many others not listed require wood either in the form of round pieces, as lumber cut to a specified widths and lengths, as mechanically shaped parts, as sheets of veneer, as chips or ground particles, or as pulp made by separating the wood into its component fibers. Uses that retain the integrity of wood in its structure as it comes from the tree qualify as solid wood products. These depend greatly upon properties of the wood as found in a tree, and how these vary among species. Pulping reconstitutes the wood into a form with properties quite different from those of solid wood. This may make additional uses feasible, yet render the wood unsuitable for more traditional applications. Available technology for processing wood into solid or reconstituted products, and techniques for reusing them in final manufactured goods, also influence the uses to which a species may be put.

MANUFACTURING AND PROCESSING OF SOLID WOOD PRODUCTS

The principal primary solid wood products include round timbers, sawn mine timbers, railroad ties, lumber, veneers, shingles, shakes, and several other forms of wood resulting from the processing of trees and logs. These items are used in structures and other construction, and in furniture, plywood, containers, and numerous consumer items. Such diversity results from wood's suitability for hewing, sawing, planing, turning, carving, and joining. The choice of a species to use for these purposes depends upon the structural and intrinsic properties required, as well as on consumer preferences and tradition. In many instances, more than one species would be suitable for the same application. However, the softwoods generally go into construction. The hardwoods, in their solid form, provide much of the material for cabinets, furniture, blocking, ties, props, and other uses not accommodated by the conifers.

Lumber Manufacturing

Lumber manufacturing is a series of steps performed with several different types of equipment. In most cases, the sequence starts by removing the bark from a log. The debarked log then goes through saws that cut slabs from the outer surfaces and reduce the squared piece into lumber and timbers. Next the lumber goes through machines to square the edges and trim it to length. Following this processing, the manufacturer will generally dry the lumber to reduce its moisture content. Some firms sell the green lumber directly to others, which dry it before using the pieces in their operations, or reselling it. After drying it, the producer will plane the surfaces to smooth them and reduce the thickness to the buyer's specifications. Finally, a sawmill will package the lumber for shipment to a secondary manufacturing plant or some other user. Each mill will have different means for accomplishing these tasks. However, modern sawmills have become highly mechanized and often control different steps in the processing by connecting the equipment to computers to monitor and regulate the operations.

The Headrig. The machine used to convert a log into lumber, called the headrig, comes in several different designs. One of the oldest has a large circular saw that can be up to 1.5 m in diameter. This saw turns on a stationary arbor at high speeds, often driven by electric motors at up to 2743 r/min, depending upon the type of wood and its condition. The log rests on a movable platform called the carriage, pulled or pushed along a set of tracks by cables or other mechanical devices. The operator, called the sawyer, controls a series of push-buttons or switches hooked up to several mechanical levers and holding devices. By operating these, the sawyer can push a log onto the carriage, turn it into position, and move the carriage back and forth. The sawyer can also move the log closer to or away from the saw to get lumber of different thicknesses. Further, by turning the log from time to time, the sawyer can control the quality, yield, and width of lumber that comes out of a log.

Another type of headrig incorporates a bandsaw that turns over wheels up to 2.7 m in diameter. These look much like the bandsaws used by carpenters and cabinet makers, but are much larger. They also have a movable carriage and control devices as do the circular mills, but the bandsaws use thinner blades that convert less of the log into sawdust. They often cut with greater accuracy and produce lumber more regular in thickness. Most bandmills cut on one side of the saw blade. Some have cutting teeth on both edges, where the sawyer cuts off a board on the forward stroke of the carriage, and another as it returns to the starting position. Some headrigs have multiple bandsaws set up in series so the sawyer can cut off more than a single piece of lumber with each pass of the log. Some types of multiple-saw headrigs have a series of circular saws or other arrangements that can even reduce a log to lumber in a single pass through the machine (Figure 15–2). One type of headrig, called a chipping saw, employs knives to cut chips off the outer surfaces and square the log into a cant. The cant then passes directly through a series of multiple saws that reduce the squared piece into lumber during a single operation.[11,17] The mill can feed the chips directly into a boiler as fuel, or sell them to a pulp mill or secondary wood-using plant.

Mill Residues. While sawmills primarily produce lumber, all generate large quantities of waste materials called residues. These may amount to 35–40 percent of the total volume brought to the mill in the form of logs.[11] Some mills even lose as much as one half as waste materials of various sorts. The residues include the bark taken off the logs, and sawdust and shavings from the different machines. They also get waste from the slabs and from poor-quality lumber in the center of many logs. Sawmill owners try to merchandise the residues to supplement income derived from the lumber sales. For example, bark is used for mulching, as a soil supplement in landscaping and gardening, and as fuel. Sawdust and shavings go

Figure 15–2.
Modern sawmills equipped with automated multiple-saw headrigs can convert large volumes of logs into lumber at high rates of production and with considerable economy as compared with the older circle or band mills. *(Weyerhaeuser Co.)*

into animal bedding, or for remanufacture into some secondary products such as particle board. Chips made from slabs and poor lumber make good raw material for pulping, and then go into paper or fiberboard. Also, in attempts to reduce fuel costs, many sawmills and processing plants have installed wood fuel boilers to generate steam and power to operate the machinery and equipment. This use of bark and wood residues has grown rapidly since the late 1970s and has resulted in a sharp reduction in energy costs for many companies.

Controlling Quality. The sawyer plays an important role in the sawmill and can influence the quality and amount of lumber obtained from a log. When first pushing the log onto the carriage, the sawyer must decide how to position it for the most profitable first cut. This should set the stage for getting optimum yields of the types of products sought. In sawing for volume alone, the sawyer may not even turn the log more than once. But hardwoods and the old-growth conifers usually have outer wood free of knots. The center of the logs will contain old branch stubs and perhaps other types of defects. If a mill hopes to produce and sell lumber to firms that require knotfree material, the sawyer may need to turn the log several times to realize the greatest volumes of quality boards off the outer surfaces.[11,17] In some highly mechanized mills, electronic scanners linked to computers feed information onto a display panel to which the sawyer can refer in deciding the optimum way to cut up a log.[17]

Mills carry out a second step in quality control by edging and trimming the lumber to meet users' product specifications. This includes reducing the pieces to standardized widths and lengths. In addition, some boards that come from the headrig may have rounded edges, fragments of bark, cracks and splits, and other imperfections. By carefully setting the edging and trim saws, the workers can cut these defects out of the lumber and improve its quality. They usually follow special rules and guidelines in making these decisions, and the owner will tell each operator what to look for and how best to edge and trim the lumber to fit available markets. These operations also lend themselves to electronic scanning techniques and automated control assisted by computers.

Lumber Drying and Processing. Once sawn, the lumber must dry to a reduced water content before remanufacture into secondary products, or use for construction. Sawmills can employ two approaches to bring this about. For some species and purposes, they simply stack the lumber into piles, place spacers between the boards to provide good air circulation, and let natural air drying occur. Some mills also place a roof or other covering over the piles to protect them from precipitation. The duration and degree of drying differ with the species and locality. In hot, dry climates the moisture may drop to as low as 6 percent, while in more temperate localities the content may drop only to 12–18 percent. Drying can take between 20 and 300 days, depending upon conditions and the amount of moisture loss sought.[17] For some situations, this will suffice. However, to speed the drying time and reduce the moisture content, many modern mills force-dry the lumber in specially constructed kilns. By controlling the drying conditions, they can turn out a consistent product that many secondary users prefer. Many mills use their residues to generate heat and steam to run these dry kilns.

Following drying, the lumber may go through a planer to surface the boards and reduce them to a standard thickness. Lumber scheduled for remanufacture into items such as furniture might go directly from the kilns to the secondary mill without further processing. Softwood lumber used in construction will be surfaced before packaging for shipment. Some sawmills also will take lumber from the dry kilns, cut it into smaller pieces of specified dimensions, and sell these to furniture and other manufacturing plants. In other cases, the mill may run dried lumber through shapers to cut in tongues and grooves, convert it to moldings, or otherwise process it into special products for resale. In fact, the basic lumber produced at the sawmill will go to a variety of processing operations that convert it into the many types of products used in homes and offices, and for score of other essential purposes.

Veneer and Plywood

Instead of cutting thick boards from the logs, a mill can peel, slice, or saw off thin sheets into a product called veneer. Veneers are used as a surface covering on furniture and paneling, and when laminated together with glue they form plywood. Generally, to manufacture decorative veneers from hardwoods, the producer will slice or peel the logs. Veneers for most plywoods and for use as panel stock are peeled from logs by using large lathes. These rotary veneers constitute more than 90 percent of the amount produced today.[11]

Rotary veneer goes into many kinds of products. In addition to plywood, these include baskets, tongue depressors, ice cream sticks, and a variety of novelty goods. To produce it, the manufacturer must first soak or steam the logs to soften the wood, and then mount them between the chucks of a lathe. As the lathe spins the log, an operator lowers a knife against the surface. The knife extends the full length of the log and cuts off a thin layer of wood from the outer surface of the spinning bolt. The initial cuts shape the log into a perfectly round cylinder, and then the knife peels off a long continuous sheet of wood much like paper toweling off a roll. After the long sheet pays off the lathe, it passes through trimming knives that slice it into standard lengths. In other operations, the manufacturer will cut out any defects and plug the holes with wooden patches. Once trimmed, the veneer passes through drying machines that rapidly reduce moisture content. After that, it can go to gluing and pressing machines for manufacture into plywood by laminating the plys with the grain alternating at right angles between successive layers. For specialty paneling, the manufacturer can add an outer layer of decorative wood and then finish it with a protective coating to seal the surface and give luster to the appearance.

For sliced veneers, the manufacturer uses an entirely different process. After soaking or steaming, the logs go through a sawmill for processing into squared pieces called flitches. From the saw, the flitches go to a slicing machine. Once clamped into place, the flitches pass in an oblique direction past a large sharp knife that slices off a thin veneer the length and width of the flitch. Normally the manufacturer piles the sheets from each flitch separately, keeping the pieces in the same order in which they came off the slicing machine. The sheets from a flitch are dried and sold to furniture and other manufacturers as a unit. The separate sheets cut from a single flitch will have a repetitive grain pattern and color. By matching these, skilled craftspeople can create spectacular designs on furniture and decorative paneling. Because sliced veneers are used mostly in decoration, they come primarily from fine hardwoods that have distinctive grain and color. These decorative veneers may also be produced by sawing the flitches using thin band or circular saws, but few companies produce sawn veneers today. Potentials for reducing waste resulting from the saw kerf have increased the relative popularity of sliced veneers.

Plywood has three, four, five, seven, or some other odd number of plys, with the grain of successive plys running in alternate directions. The core sheets of plywood and panels need not have the same quality as the outer surfaces. Thus a firm may use lower grade veneer for the inner plys, and save the better material for surfacing sheets. Adhesives for binding them together include quick-setting synthetic resins such as phenol-formaldehydes, ureaformaldehydes, melamines, and resorcinols. Or they may be animal or vegetable glues of the cold setting type. Because of the crossbanding of alternate layers, the glued plywood has great strength in all directions and does not change much dimensionally with fluctuations in moisture content. It has proved superior to conventional lumber for many construction and allied uses.

Laminated Wood

Heavy veneers laminated into thick materials may be used as substitutes for conventional lumber in studs and rafters. Also, by gluing together conventional lumber into laminated pieces, a manufacturer can make heavy-duty timbers and beams used as support members for construction. This laminated wood differs from plywood. It consists of successive laminea glued together with the grain of all layers running parallel. As a result, the layers can be bent and glued to make arched or curved pieces as well as straight ones. Large trusses, beams, and arches can span distances of 91.5m. They have application in constructing such buildings as factories, gymnasiums, churches, and other structures that need large amounts of clear floor space.[4] These laminated timbers add beauty to the structure, and are even popular for homes that have tall vaulted ceilings. The lamination gives the timber greater strength than solid wood pieces of comparable size, and makes it suitable for replacing steel for many purposes. Laminated-veneer lumber is also produced in sizes comparable to conventional sawed lumber for a variety of other construction purposes.

Particle Board

Sawmill residues and small-diameter trees have applications other than as fuel or for pulp. When converted to chips or flakes, they can be used as a raw material for various types of particle boards. The process involves screening the chips or flakes, coating them with adhesives, and spreading the coated materials onto platens in layers to form a board. In some processes, the manufacturer sorts the coarser material into the center of the board and puts the finer pieces in the outer layers. Once spread, the material must be pressed and heated to cure the adhesives and give the sheet a uniform thickness. With this process, a manufacturer can also turn out continuous sheets, or extrude the material into special shapes. The products lack the strength of plywood and may have considerable weight due to the adhesives, but can serve as a good core material for counter tops and table tops, for subflooring, some decorative paneling, for other interior uses protected against wetting, and for applications requiring a wide, smooth panel that need not bear heavy loads. Though originally developed to utilize wood wastes, particle board has better qualities when manufactured from engineered chips of uniform size and character. It provides a ready market for small-sized trees such as those removed during thinnings.[1]

Secondary Processing of Solid Wood Products

Secondary manufacturing of solid wood products includes numerous operations that process lumber or wood into some other form or item. It does not include the use of wood for home construction. Many of these operations involve only simple steps and basic types of machinery. For example, to make ice cream sticks, the manufacturer stamps the pieces from sheets of veneer, tumbles the sticks in a cylinder with talc to smooth the surfaces, and packages them for shipment. For more complex products, such as furniture, the wood must be formed, shaped, turned, joined, and finished in different ways. And the operations must be varied to make pieces of different designs, shapes, and characteristics. The exact steps differ according to each type of secondary product. In some cases, the firm may produce only a limited number of items and employ a few workers who prepare the goods by hand. In the larger and more modern secondary processing plants, the operations will include complex machines and assembly-line techniques. These will help to increase production, reduce costs, and standardize the end product. However, even with large manufacturing firms, the better grades of goods such as furniture still require considerable hand labor to capture the full beauty of wood and make the item esthetically appealing.

As a first step in secondary processing, a manufacturer must often condition the wood to suit the intended use. For many purposes, wood needs some drying before machining, especially if not purchased kiln dried. Most plants have a dry kiln for this purpose and give the wood a final treatment prior to cutting it into pieces and surfacing them to smooth the faces. For bent wood products, the manufacturer will first steam, soak, or treat the wood with plasticizing chemicals before shaping it. Prior to this step, the lumber may be cut into smaller pieces fitted to the intended use. Then after conditioning, the wood will go through a series of cutting and shaping operations to prepare it for final assembly.

Cutting operations include ripping and cross-cutting, planing the surfaces, shaping the wood with various high-speed cutting heads, turning the wood on lathes, boring holes in it, and carving the surfaces by hand or with mechanical routers to add special surface features. Each operation has its own technical problems related to the species of wood, the types of machines used, and the nature of the cuttings made. Proper design and the choice of wood going into it also influence the quality of the end product. Success depends upon having the correct blend of design, engineering, and quality control to ensure consistent output of finished goods with minimum losses and defects during manufacture.

Joints often are the weakest part of a manufactured item, and the weakness of wood across its grain accentuates the problem. The need to speed the manufacturing process and cut costs has fostered new joinery techniques that employ various mechanical fasteners to facilitate the process and improve the strength of a joint. These include wooden pegs or dowels, nails and screws, staples, bolts, rods, and other types of connectors. In most cases, a manufacturer will also use glue to help make the joint more durable.

Modern adhesives provide an excellent way to fasten two pieces of wood together. When properly glued, the joint will have a strength comparable to that of the wood itself. Some adhesives and glues will withstand repeated wetting and drying, while others have no resistance to moisture. Some tend to break down following a prolonged period of wetting. The choice of glues and selection and control of methods for their application require specialized skill. These procedures receive continuous attention from research, testing, and development groups. However, when properly matched with an appropriate machining process, modern adhesives have allowed manufacturers to join pieces along their edges, at the ends, and in a variety of other ways. These techniques have opened possibilities for using even relatively narrow and short pieces in the manufacture of high-quality products at reduced costs.

As a final step before packaging, the manufacturer will sand the surfaces and apply a combination of stains and finishes to give the item a desired appearance and to protect the wood. Rough surfaces will not cover well with paint or other finishes. With ordinary care, sanding by relatively simple techniques usually gives good results. Final finishing involves application of sealers, stains, lacquers, varnishes, paints, or other coatings. Some require rubbing or sanding between coats, and buffing to lend a final luster. One type of modern finish includes the penetrating oils, stains, varnishes, and resins. Another type does not penetrate but covers the surface, such as shellac, some varnishes, lacquers, paints, and related products. The penetrating seals, and especially those where the chemicals form a hard resin upon curing, give the surface maximum resistance against marring. However, few of the finishes applied to wood surfaces have the same resistance to wear as the wood itself.

MAKING PAPER, PULP, AND FIBERBOARD

The Chinese made paper as early as 105 A.D. using bast fibers from the inner bark of mulberry trees, old linen, and fishnets. Later, Arabs learned the art and improved and perfected the process using other fibers in place of the bark. However, not until early in the 11th century was paper made in Europe, by Spaniards. The first U.S. mill was built in 1690. During most of this early period, linen rags provided the fibers and the scarcity of raw materials kept prices high. Then in the middle of the 19th century, scientists learned to make paper successfully from wood fibers, thus reducing the cost of raw materials and of the final product. As a result, per capita consumption of paper and pulp products rose rapidly. Use grew from 9 kg per person in 1869 to 26 in 1899, 161 in 1948, and 278 in 1977.[5,6,16] Projections indicate that demand may be more than triple this level by 2030.[16]

At present, wood makes up more than 95 percent of the fibrous material in paper, with rags, straw, cane bagasse, bamboo, and marsh grasses still providing some raw material for limited production in a few places. Conversion of these materials into paper takes two steps. First, the manufacturer must reduce the wood or other material to pulp, and then process it into paper or paperboard. In the pulping process, the wood is treated either mechanically or chemically to separate the fibers into a pulpy mass. Afterward the fibers can be suspended into a slurry and spread out into the sheets that form the paper. The process of papermaking, then, essentially consists of separating the fibers and later rearranging them in a new configuration.

Pulping Processes

One principal method for mechanically producing pulp involves pressing the wood against the face of a large rotating grindstone to break the fibers apart. This groundwood process yields about 92 percent of the fiber material in the original wood. Spraying water on the stone keeps it cooled to about 75 °C, and the water also picks up and carries off the separated fibers in a slurry for screening to remove knots and slivers. The greatest cost for groundwood pulp comes from the power required to separate the fibers, and only such species as spruce, fir, some pines, and aspen have proved useful for groundwood pulp.

The groundwood process yields a pulp comprised largely of broken fibers, fiber fragments, and fiber bundles. It has low strength and produces a low-grade, low-cost paper or paperboard for uses that do not require permanency. Blending in pulps from other processes to get longer and stronger fibers will upgrade the quality. Typical newsprint, for example, contains about 85 percent groundwood fibers and 15 percent from other processes.

By modifying the process to defiber untreated or chemically pretreated chips in a disk refiner, a mill can use many additional species and also sawmill residues (Figure 15–3). These machines take in wood chips, soften them with steam, and separate the fibers between a stationary or a rotating metal disk. Another mechanical method called thermomechanical pulping (TMP) defibers the chips in a disk refiner at increased temperature and pressure (typically 125–135 °C and up to 2 atmospheres). This gives a stronger and longer-fibered pulp than either the groundwood process or by defibering the chips at air temperature without added pressure. The TMP method will make newsprint that does not need to be supplemented with chemical pulp to add strength.

Production of high-grade pulp from wood involves any one of several chemical methods that primarily remove the lignin, but also considerable amounts of hemicellulose. The chemical methods also cause some degradation and loss of

Figure 15–3.
A disk refiner such as the one shown can convert softened wood chips into pulp for subsequent remanufacture into paper and other products.

cellulose. One approach, called the chemical acid-sulfite pulping process, consists of cooking wood chips in a digester under pressure in a solution of calcium bisulfite and sulfurous acid that contains 6–7 percent sulfur dioxide. During cooking, the sulfite and hydrogen ions react with lignin to form water-soluble lignosulfonic acids. Then the cooked chips are blown into a large tank, where they disintegrate. Subsequent washing removes the dissolved lignosulfonic acids from the pulp fibers. The discharged liquor can undergo further processing into useful products such as alcohol, fodder yeast, and various lignin derivatives.[12,15]

The sulfite process can be used with hardwoods, and also with such conifers as hemlock and fir. The process yields a high-quality pulp used in catalogs, glassine, wrapping papers, paperboard, and newsprint. Bleaching the pulp with hypochlorite makes it suitable for white bond and ledger papers. By using sodium or magnesium instead of calcium as a base for the sulfite liquor, it is possible to reduce the discharge of spent sulfite liquor by a recovery process similar to those used with kraft pulping. The process can also be used to produce a pulp low in hemicellulose, termed dissolving pulp, needed by rayon manufacturers.

The sulfate or kraft process provides the strongest pulp for papermaking. It uses a mixture of sodium sulfide and sodium hydroxide as the active cooking liquor. Chips continuously fed into large digesters are converted by the liquor into pulp and later drawn off at the bottom. When drained and concentrated to 60–65 percent solids, the spent cooking liquor burns to yield sodium carbonate and sodium sulfide. Dissolving these in water and adding slaked lime cause a reaction that yields fresh cooking liquor. Among the chemical by-products of the process that also have commercial value are tall oil, sulfate turpentine, and alkali lignin. The recycling of the chemicals in the closed system minimizes pollution that might otherwise come from discharge of the spent cooking liquor. Use of improved furnaces and evaporators controls the release of mercaptans and other foul odors. Further, steam generated in the recovery furnace will supply almost all the heating needed in the pulping process.

The kraft process will work effectively with almost any wood, including resinous softwoods. It has become the dominant pulping system throughout the world. On the average, the process yields 45–48 percent of the dry weight for softwoods, and 45–55 percent for hardwoods. Sulfate pulp from softwoods has excellent strength, toughness, and durability. It is useful in manufacturing wrapping paper, high-strength paper bags, heavy paper for corrugated and solid-fiber paperboards, fiber drums, and cans. Bleached sulfate pulp also goes into envelopes and writing, printing, and onionskin papers, or for food containers, folding boxes, and strong white wrapping papers.[11]

Besides these fully chemical or mechanical means of defibering wood to make pulp, combination methods are useful for some woods or for making certain types of pulp. One approach with hardwoods, the neutral sulfite semichemical process, involves cooking wood chips with sodium sulfite buffered with sodium bisulfite, sodium carbonate, or sodium hydroxide to keep the liquor neutral or slightly alkaline. This process yields about 65–85 percent pulp, based upon the dry weight of wood, or up to 40 percent more than conventional chemical methods. A mill may have conventional or rotary digesters, or even a cooking process employing a combination of continuous digesting and refining under pressure. These processes make it possible to use low-grade hardwoods and realize a high yield from them. The resulting pulps can go into such products as corrugated board, liner board, insulation boards, and roofing felts. Other special semichemical processes include the cold-soda method, which consists of soaking hardwood chips in a sodium hydroxide solution at room temperature for one to two hours before fiberizing them in a disk mill.[11]

A relatively new and still experimental process called oxygen-alkali pulping uses oxygen as a major agent. It produces a pulp somewhat lower in strength than does the kraft process, but one that is easier to bleach. The method requires no sulfur in the processing,[8,9] and employs a much simpler recovery system. The resulting pulp has flatter fibers with improved flexibility and bonding. In this process, chips are cooked at 150–173°C in sodium hydroxide or sodium carbonate, and then defiberized in a disk refiner. The material undergoes a second cook with additional alkali, in the presence of gaseous oxygen under pressure. Although the process requires careful control so as not to over- or under-cook or improperly refine the pulp, it essentially frees the manufacturer of concern for atmospheric pollution.

Pulps from these various processes have a wide range of color, from light tan for sulfite pulps to dark brown for those from the kraft process. Before they are used in white papers, the pulps must undergo bleaching by chlorine, hydrogen peroxide, chlorine dioxide, or some other bleaching agent. This involves a multistage operation that also removes soluble impurities by alkaline extraction and washing. Such steps require large quantities of water, which is discharged. Newer methods use sodium hydroxide and gaseous oxygen at a pressure of about 6 kg/cm^2, followed by washing and treatment with chlorine dioxide in stages to achieve the desired brightness. The effluent from some of these steps can be recycled in the pulpmill, thus reducing pollution.[5] The amount of bleaching needed depends upon the intended use for the pulp, plus its initial color. Following bleaching, the pulp will undergo further processing to manufacture it into a final product.

Paper Manufacture

To make paper, the manufacturer runs the bleached or unbleached pulp through a beater or other refiner that subjects the fibers to rubbing and crushing to increase surface area and bonding strength. Removing the water by drying causes strong hydrogen bonds to form between the cellulose fibers. The reaction reverses when a paper is wetted, making it possible to recycle it later for reuse. Additives to the refined pulp give the paper desired properties. Some improve wet strength, others give it a smooth surface, some make the paper erasable, and others add opacity. Sizing the paper by adding rosin and alum improves the printing qualities. Titanium dioxide or clay adds to whiteness, and dyes impart color. Once refined and treated with additives, the pulp slurry consists of about 0.5 percent solids, and 99.5 percent water.

In continuing the process, the manufacturer flows the refined slurry onto the bed of a moving fine-mesh screen at the front end of the paper machine. The paper sheet forms on the screen as the water drains off while the sheet moves across the machine at speeds up to 1500 m/min. Gravity and suction remove the water. As the sheet gains strength, it moves onto a moving woolen or synthetic felt, and through press rolls that squeeze more water from the sheet. After going through the press rolls, the sheet moves over steam-heated rolls to dry thoroughly. It then may go through a series of highly polished rolls that press and smooth it before winding it into a roll at the end of the machine. Such a Fourdrinier paper machine produces newsprint at speeds that approach 900 m/min, and sanitary tissue at 1500 m/min (Figure 15–4).

A second type of paper machine has three to five or more hollow wire cylinders

Figure 15–4.
This Fourdrinier paper machine has a forming wire screen that is 120 cm wide and operates at 610 meters per minute. Its parts, from left to right, include the head box from which the pulp suspension flows, the forming screen, the press rolls, the enclosed drying rolls, and finally the rolls that wind up the finished paper. *(International Paper Co.)*

that serve as the forming wires. Each cylinder rotates partially submerged in a vat of pulp and forms a thin ply of paper on the wire surface as it turns in and out of the paper pulp. Separate sheets form on the different cylinders and they can be pressed together to form a multiply paper. This sheet then passes over drying rolls similar to the ones on a Fourdrinier machine. These are adaptations of the Fourdrinier machine such as the Inverformer, which will also produce multiply sheets at a much higher rate than the cylinder process. With these processes, the manufacturer can laminate sheets of different grades of paper, using coverings of high-grade paper over poorer filler layers sandwiched between them. A similar type of paper machine with a single cylinder can produce tissue paper, and ones that laminate several plies can make paperboards and bristols.[9]

The characteristics of the final paper manufactured by either of these processes may not be desirable for some uses, and so manufacturers may coat them with clay or other materials that brighten the color and fill in rough spots between the fibers. These papers are intended for use in printing that must reproduce fine detail. Also, coating of lower grades of paper may upgrade them and increase the value for other applications. Production of coated papers has greatly increased in the United States, showing a threefold increase for book papers between 1950 and 1966, whereas noncoated book paper increased only 50 percent during the same period.[3]

Pulping processes use large amounts of water to debark the bolts, grind the wood, cook the chips, and separate fibers, and to wash, screen, and further refine and bleach the pulp. Papermaking also requires water to suspend the fibers and transport them onto the paper machine. For chemical processes, the spent cooking liquors go through recovery processes that burn off vapors and pass them through a stack to volatilize the chemicals they contain. These steps in pulp and paper manufacturing all can produce useless by-products to be disposed of, into either the atmosphere or nearby waters. Consequently, growing concern over water quality and requirements to safeguard water from pollution have increased attention to appropriate means for handling these waste products.

Water from the paper machine, recycled many times before discharge, carries small particles of fiber and additives that pass through the machine's forming wire screen. This white water goes into settling tanks (Figure 15–5), along with coagulants to hasten settling of the solids. The treated water will then flow to a large lagoon for aeration to oxidize the organic compounds. However, even after these steps, the wastewater still contains some particulates and dissolved material. Requirements of federal air and water pollution regulations have forced considerable investments to reduce these discharges, with the greatest improvements needed for the older mills. Satisfying the goal will take increasingly complex monitoring and control systems to clarify effluent and atmospheric discharges, plus considerable research to find effective ways to do it inexpensively. Fulfillment of such requirements will be a major challenge to the industry in the years ahead.

Figure 15–5.
Primary wastewater clerifiers remove settable solids as part of the complex modern pollution control system at this Ticonderoga, N.Y., mill. The tall structure at the rear houses the company's continuous digester. *(International Paper Co.)*

Fiberboard

Using processes similar to those for making paper, manufacturers can form wood and other plant fibers into rigid and semirigid sheets of various densities to make fiberboard for insulation and construction. Thickness, density, and size of the sheets vary with the intended use. Wood pulped for this purpose retains nearly all the cell wall material to form fibers that are coarser than those used in paper. The pulping methods include mechanical, thermomechanical, semichemical, and explosion processes. These reduce wood to a fibrous state with the aid of few, or no, chemicals. The explosion or Masonite process first heats wood chips for a short time in a special "gun" with steam at high pressure. When discharged through a quick-opening valve, the chips explode due to the action of the entrapped steam, which reduces them to a fibrous state. A special semichemical method, called the Asplund process, forces wood chips into a cylinder reaction chamber and treats them with chemicals and steam for up to 20 minutes. The softened chips go into a disk refiner for further processing. This method yields about 90 percent of the wood dry weight as fiber.

Once in a slurry and treated with additives to give the product specific properties, the fibers can be spread on a Fourdrinier or cylinder machine to produce the board. The sheets then run through a tunnel drier before being trimmed to a size. Hardboard requires curing on a hot press to compress it to the desired thickness, and then later conditioning to bring its equilibrium moisture content to 5–12 percent. If not treated properly, it would warp and change dimensions. Fiberboard in thick, low-density sheets is of use for insulation and other purposes in building construction. Further, similar manufacturing processes will yield a flexible fiber insulation up to 5 cm thick. This material serves as blanket insulation to install between studs in building construction. Medium-density fiberboard also has many uses in furniture manufacturing, and fiberboard pressed into hardboard has structural value. These types of products have become popular for many construction purposes.[11]

Fiberboard manufacturing processes that turn out insulation and hardboard can use almost any moderate- to low-density wood. For many applications, the raw material can come from logging waste, cull trees, used lumber, cane bagasse, bamboo, and other fibrous materials. Potentials for marketing poor-quality trees for fiberboard manufacturing have opened many silvicultural options for improvement cuttings and use of thinnings that have little value for products requiring higher quality wood.

CHEMICAL TREATMENT OF WOOD

While wood has many values and uses, some properties limit its application unless it is modified chemically. These techniques do not change the outward physical structure of the wood or convert it to some new form the way pulping does. Rather, the chemical treatment modifies some characteristic of the wood, either temporarily or permanently. As a result, the treatment can improve the wood's versatility. For example, some species of wood bend fairly readily when softened by steam, making them useful for such items as tennis rackets and bentwood chairs. Also treatment with liquid or vapor ammonia plasticizes the wood. The ammonia seems to soften the lignin and supplant hydrogen bonds between cellulose molecules, permitting the cellulose crystals to slip over one another. Ammonia-treated wood thus bends more readily than steamed pieces and retains the shape better, even after wetting (Figure 15–6). The possibilities for industrial application appear considerable in wood processing and for artistic and ornamental uses.[2]

The steaming and softening of wood serves other purposes as well. If heated when moist and compressed, the wood forms into a densified and comparatively stable product called staypac. If impregnated with soluble resins and cured, the wood becomes hard, dimensionally more stable, and decay resistant. Or the manufacturer can substitute polymeric chemicals and polymerize them within the wood by using gamma radiation or a chemical means. Compression of the wood before curing the resins or polymers yields a high-density and stable product called compreg. It, too, has high stability and decay resistance, but the treatment makes the wood more brittle.

The chemical used for treating the wood will affect the properties imparted. This may alter the use. For example, methyl methacrylate-wood polymer has the

Figure 15–6.
Ammonia-softened wood will bend readily into decorative shapes, such as shown in the ornamental screen, and wood impregnated with chemicals may have use for many purposes, such as the folding chair with a wood-plastic hinge. *(State University of New York, College of Environmental Science and Forestry.)*

toughness of wood, and leaves it with a good finish, but does not prevent moisture-induced dimensional changes. On the other hand, using T-butyl styrene gives the wood good dimensional stability.[7] Also, treatment with polyethylene glycol so that the polymer forms 16 percent of the wood's weight reduces drying shrinkage considerably and helps the wood to resist checking and splitting.[15] Another process can replace the hydroxyl groups in wood with acetyl groups and reduce the tendency of wood to swell when wetted, and does not affect other properties. These and various other treatments offer promise for inhibiting swelling and shrinking, and afford a high degree of protection against wood-destroying organisms. Their use can improve the applicability and versatility of wood for many special purposes.

MARKETING FOREST PRODUCTS

The preparation of salable goods and products does not represent the end of activity in the forest products industry. Rather it begins an additional sphere of endeavor involving the marketing, distribution, and sale of the items. From the consumer's point of view, these activities bring needed goods to the marketplace and make them readily accessible. However, for the industry, the marketing and sale of a firm's output generate revenues to pay the costs of the operation, cover the salaries of the employees, and provide a profit to investors.

Marketing plays an essential role at all stages of forestry and product manufacturing. Landowners with timber to sell must market it effectively to realize the best return on their operations. Likewise, a sawmill or other primary processing plant must effectively market its lumber, veneers, pulp, or other products to firms that convert the rough materials into finished goods. Then, at the end of secondary manufacturing, those firms must arrange for the ditribution and sale of their products to retail outlets accessible to customers. These operations involve research, planning, promotion, selling, packaging, and distribution at all levels.

The effectiveness of the sales program often depends upon the planning that goes into developing the firm's marketing strategy. This includes attention to a series of matters involved with both the manufacturing and merchandising operations. It starts by making choices about the types, mix, amount, and quality of products to manufacture. In the process, a firm will often first conduct a marketing survey to assess interests and needs among potential customers. The results will indicate who might purchase the goods, how to reach those buyers effectively, and what methods might entice them into purchasing the goods the firm has to sell. In developing the strategy, the management must also decide what prices to charge based upon consumer demand, the degree of competition by other firms, and the costs of manufacturing. It must decide how and where to distribute the goods, what retailers or secondary manufacturing plants will handle its output, and how to ship the products efficiently. A firm may even contemplate whether it will gear up for these operations or depend upon an independent wholesaler to act as an intermediary. Finally, the marketing strategy will include a plan for advertising, personal selling, and special promotion to attract customers and secure their orders.[14]

These activities take place continually, not just when a firm is beginning operation. They provide an ongoing assessment of potentials for maintaining and increasing sales, or for adjusting the types and amounts of products offered. For small enterprises that handle a limited range of goods or deal with small-volume primary processing, the marketing program might consist of fairly simple steps used by the owner to keep tabs on a sales and distribution program. For large, vertically integrated corporations that purchase or grow the raw materials, convert them into primary products, and finally remanufacture them into retail goods, the marketing strategy will differ. It may represent the collective efforts of a special staff involving business, retailing, manufacturing, economics, and other specialists. They try to find ways to create a desirable public image and gain recognition among potential customers. In the final analysis, the ultimate success of a firm may depend as much or more upon its marketing program than upon the techniques it uses to manufacture its products.

Such products as lumber, pulp, veneer, poles, pallets, and plywood really differ little except in relation to the characteristics of the tree species used to make them. The highly competitive business operates within a narrow range of prices among firms. Improper manufacturing can result in faulty products, but firms that ignore quality control usually do not survive long. Hence companies that deal in these goods must win and keep their customers by the services they offer, the dependability of their supply, their fairness in pricing, and the general goodwill they create over the years in dealing with potential customers. These firms must also work constantly to find new buyers and to develop specialty markets offering better prices that will return greater profit to the enterprise. The owner must keep track of changing market demands and learn to anticipate ones that might affect the firm's operations. This ongoing research will also prove essential to managers in setting prices they will pay for the logs and bolts acquired from landowners. In fact, only by routinely studying the marketplace and the likely demands for future buying can the firm gain the insight needed to control many phases of its business.

Markets for forest products fluctuate with changes in the economy. During times of good business activity and much construction, demand may rise and available goods may command high prices. But downswings in the economy can seriously affect the health of the forest products industry. For example, the inflating of costs leading up to the early 1980s pushed home building costs rapidly upward. Coupled with a shortage of low-interest mortgage money, this decreased new home buying and sharply curtailed the demand for softwood lumber, plywood, fiberboard, and other wood-based products used in the construction industry. Tightness of spendable income plus the cutback in new home construction also impacted the firms that manufacture cabinets, flooring, furniture, and other furnishings. These developments had repercussions across the country, resulting in the shutdown of many softwood sawmills and plywood plants. While most firms could not control events, those with the most diverse markets and a capacity to shift production to alternate products better withstood the economic squeeze. Good marketing programs undoubtedly helped them to uncover new markets or to merchandise their output more effectively as the traditional markets declined.

As a part of their marketing strategy, many large firms also maintain a staff of technical representatives who work with their customers in solving problems related to the reuse or further processing of the product. Such a technical service program may get support from a research and development group within the corporation. The technical representatives will have frequent contact with customers, keep watch over problems and potentials, offer advice for overcoming difficulties, gather information about product specifications or changing needs, and share what they learn about new markets or additional uses for their products with their employers. These representatives thus facilitate a firm's promotion activities to obtain new customers and hold onto existing ones. They also help managers discover new ways the firm can market its goods or retool to provide additional products a customer might wish to purchase. In smaller firms, the owner or some key staff may also serve in a dual capacity as technical representative to service the needs of customers and maintain the goodwill to ensure a continuing business relationship with buyers. But no matter how large or how small the firm, the offering of effective service provides one important strategy for attracting and maintaining customers. This proves especially important for firms that sell goods commonly available from other companies competing for the same business in the same sales areas.

CAREER OPPORTUNITIES WITH FOREST PRODUCTS MANUFACTURING

The forest products industries with their diverse activities and the many types of products they manufacture provide a wide range of career opportunities. These include jobs for people who take responsibility for producing and delivering to the mill the roundwood, lumber, pulp, or other wood-based raw materials needed. They employ people to manage operations and oversee business aspects of the enterprise. Among these are accountants, economists, planners, business managers, design and engineering specialists, and personnel managers. They also require highly skilled engineering and technical specialists to work with design, functioning, and maintenance of the machines, and in solving technical problems associated with the manufacturing processes. These may include chemical engineers and chemists schooled in pulp and paper manufacture and wood material science, wood products engineers and wood technologists who understand the physical properties, mechanical engineers, and many others who have various types of engineering and chemical skills. Additionally the firms will need marketing and sales staffs and technical representatives. In some cases, individuals schooled in forest products manufacturing may form their own companies to produce primary or secondary products, or handle their distribution and sales.

Not all schools of forestry offer programs in wood products technology, pulp and paper manufacture, wood products engineering, or related fields. Some have curricula for several of these disciplines, while other schools concentrate on a few aspects of the broad field. Some colleges may offer programs such as paper science or wood-related engineering outside the forestry school, but may not maintain a forestry curriculum per se. Also, colleges will provide at least introductory courses for the students who specialize in forest resources management and related fields, and who need to develop a basic understanding of the forest products industry their efforts support.

Undergraduate programs in wood products engineering prepare students to work in a variety of professional occupations involving the use of wood as a material. These may emphasize manufacturing, building construction, marketing, systems engineering, and related areas. Basic preparations include strong emphasis on chemistry, physics, mathematics, biology, computer science, economics, accounting, English and communication skills, and engineering courses. Upon this foundation, students can build technical capabilities in wood structure and properties, engineering of materials, mechanics, adhesives and coatings, fluid mechanics, hydraulics, management, advanced accounting, structural design, and similar areas. Some may emphasize management through electives, while others use that time to strengthen technical skills. For careers in marketing of forest products, students may take such courses as marketing, money and banking, financial accounting, economics, management of operations, and technical courses related to wood as a material and its processing. Emphasis on production systems might include such courses as electrical science, time and motion study, thermodynamics, material handling, construction concepts, and quality control. While different schools may have fairly well-defined programs of study for these options, students will find time for elective courses and should discuss these with faculty advisers to determine the most useful ones to take to strengthen personal interests.

Programs in paper science and engineering will begin with basic courses in botany or biology, chemistry, physics, mathematics, computer science, economics, English, and communication skills, plus options to strengthen physical science capabilities. From this preparation, students progress into courses in advanced chemistry, wood structure and properties, fluid mechanics, papermaking, principles of mass and energy, pulp and paper processes, process control, pulping technology, paper properties, mass transfer, paper coating and converting, water pollution engineering, colloidal and surface chemistry, polymer chemistry, applied mathematics, principles of management, engineering design, mechanics, material science, and the like. For electives, students may take additional courses in the managerial sciences, such as accounting, finance, banking, personnel management, and others that facilitate preparation for employment in management. Or they can strengthen technical interests and build capabilities in chemical engineering as well as paper science. Students should discuss these possibilities with faculty advisers before selecting particular courses.

Study at the graduate level toward either the master of science or the doctor of philosophy degree allows specialization and preparation for careers in research and development, problem solving, teaching, and related fields requiring advanced study. These may involve such technical subjects as paper and wood chemistry, paper or mechanical engineering, wood properties science, and other disciplines related to the physical and chemical properties of wood and its derivatives. Also, individuals interested in the management of wood products companies can pursue graduate studies in business management, marketing and sales, and other areas related to business and management aspects of the enterprises. Those holding these advanced degrees can find employment in the research and development groups of corporations, on management and business staffs, and in overseeing the technical aspects of production. Others can find positions in research programs operated by universities, or in groups such as the Forest Products Laboratory or units of many Forest Experiment Stations maintained by the U.S. Forest Service.

REFERENCES

1. L. E. Akers, *Particle Board and Hardboard,* Pergamon Press, London, 1966.
2. R. W. Davidson, and W. Baumgardt, "Plasticizing Wood with Ammonia—A Progress Report," *Forest Products Journal* **20**(3):19–24 (1970).
3. D. Hair, *The Use of Regression Equations for Projecting Trends in Demand for Paper and Board,* U.S. Forest Service Forest Resource Report 18, 1967.
4. W. M. Harlow, *Inside Wood,* American Forestry Association, Washington, 1970.
5. E. C. Jahn and F. W. Lorey, "Industrial Chemistry of Wood," chap. 15, in J. A. Kent (ed.), *Riegel's Handbook of Industrial Chemistry,* Reinhold Publ. Corp., New York, 1974.
6. E. C. Jahn and S. B. Preston, 1976. "Timber: More Effective Utilization," *Science* 191:757–761 (1976).
7. J. E. Langwig, J. A. Meyer, and R. W. Davidson, "New Monomers Used in Making Wood Plastics," *Forest Products Journal* **19**(11):57–61 (1970).
8. R. Marton, A. Brown, and S. Granzow, "Oxygen Pulping of Thermomechanical Fiber," *TAPPI* **58**(3):64–67 (1975).
9. R. Marton and B. Leopold, "Oxygen-Alkali Pulping of Conifers," *APPITA* **27**(2): 112–117 (1977).
10. National Academy of Science, *Fibers in Renewable Resources for Industrial Materials,* National Academy of Science, Committee on Natural Resources., Washington, 1976.
11. A. J. Panshin, E. S. Harrar, J. S. Bethel, and W. J. Baker, *Forest Products, Their Sources, Production, and Utilization,* 2d ed., McGraw-Hill, New York, 1962.
12. I. A. Pearl, "Silvichemicals, Products of the Forest," *Journal of Forestry* **63**(3):163–167 (1965).
13. R. B. Phelps, *Timber in the United States Economy 1963, 1967, and 1972,* U. S. Forest Service General Technical Report WO-21, 1980.
14. S. U. Rich, *Marketing Forest Products: Text and Cases,* McGraw-Hill, New York, 1970.
15. A. J. Stamm, *Wood and Cellulose Science,* Ronald Press, New York, 1964.
16. U.S. Forest Service, *An Assessment of the Forest and Rangeland Situation in the United States,* U.S. Department of Agriculture, Forest Service, F.S.-345, 1980.
17. E. M. Williston, *Lumber Manufacturing. The Design and Operation of Sawmills and Planer Mills,* Miller Freeman, San Francisco, 1976.

CHAPTER **16**

FORESTRY AS A CAREER IN THE UNITED STATES

The Society of American Foresters defines forestry as the combination of science, art, and practice of managing and using forest resources for human benefit. As the preceding chapters have pointed out, it encompasses many diverse activities and disciplines involved in the creation, management, protection, and use of forestlands and their related resources for continued use.[2] Among principal uses are those of timber, soil, water, rangeland, wild animals, and a host of noncommodity values. These latter ones arise through the recreational use of forested lands and from the ways in which forests otherwise enhance our daily lives. Though frequently described as a single profession, forestry really involves many different specialties. All work toward a common goal of helping society to benefit from the rich resources of forested lands.

Forest activity developed slowly in the United States. Early approaches primarily considered the timber and sought efficient ways to extract and process it for human benefit. Slowly interest developed in forests as watersheds, as habitats for wild animals, and eventually as a source of many amenity values. This growth and broadening of forestry interests placed new challenges before the skilled professionals who seek to manage the many resources in a more integrated way.

As the expanse of forestland dwindled and the amount left undisturbed by human activity shrank, needs for more comprehensive approaches aimed at total ecosystem management, protection, and use became a necessity. Resource managers faced many difficult choices in resolving conflicts over what uses deserve greatest priorities. But with the broadening of demands, an approach has evolved that gives attention to a fuller range of forest values, not just to timber or any other single resource. Forecasts of an ever-shrinking forestland area, coupled with an anticipated rise in demand by a growing worldwide population, will make integrated management and use vital in the future.

Timber resources provide most of the revenues from forestland and offer to society much of the tangible benefit. One economic study based upon 1972 statistics indicated that timber-dependent activities across the nation involved about 3,058 thousand people. They worked in a variety of roles, from forest management through manufacturing, transportation, and marketing. About 20,000 of these, or 17 percent of all people employed in timber management activities, were professional foresters. Altogether one in 15 of all workers in the United States had some relationship to timber or its use. Their salaries and the value they added to the wood through their various activities helped to give timber a total worth of $26,079 million. For each $1 paid to a landowner for the standing timber, subsequent manufacturing, transportation, and marketing added about $21.[5]

Most of the economic contribution of timber comes from its manufacture into various products, its use in construction, and the costs of transportation and marketing. About 40 percent of the value added to timber through economic activity in 1972 came from functions centered in the northern states, east of the Great Plains. This comes from primarily and secondary manufacturing, construction, transportation, and marketing. The South generated about 35 percent of the economic benefits, and produced 47 percent of the total timber harvested in the country. Western states provided 40 percent of the timber, and generated about 25 percent of the economic worth attributable to timber in some way.[5] These data highlight the historic importance of the West and South for timber growing, harvesting, and primary processing. These regions have always provided many opportunities for forestry careers.

By far most foresters work in timber management programs, and the long-term prospects for employment in these fields appear good and to be growing. We cannot do without forest products, and foresters provide the necessary services for assuring sustained production to meet the demand. In addition to traditional uses and the projected rise in the use of paper and fiber products, requirements for wood as a fuel and as a feedstock for various chemicals may increase as well. Water and forage also are valuable, and attention to the watershed and rangeland characteristics will continue to receive high priority, especially in western regions. Both have important linkages to food production and should grow with the greater demand for food by the world's ever-increasing population. Requirements for these commodities, and others, are likely to continue to dominate forestry activity in the future. Concerns for wildlife and amenity values also will attract attention. As a result, the profession will face real challenges in finding better ways to serve the growing demand for forest-based commodities, while still enhancing other values and preserving the integrity of forests as fairly complex natural ecosystems.

THE WORLDWIDE CHALLENGE

In 1980, the U.S. Council on Environmental Quality and the U.S. Department of State estimated global changes in population, natural resources, and environment by the year 2000.[8] They drew several conclusions that have relevance to forestry. For one, their data indicate that the world population could increase by 55 percent, reaching about 6.35 billion by 2000. Most of the births and reduced mortality will occur in the world's poorest countries, but the growth will impact the whole earth. To support all peoples, world agricultural production must increase by some 90 percent. This must come with the addition of only 4 percent more arable land than is presently available to agriculture. Mostly higher food production will need to come from improved practices to increase yields on lands already in use, or by conversion of lands now in forests. Many countries will face great problems in satisfying energy needs, and requirements for fuelwood should exceed accessible supplies throughout the world by 25 percent. Again, the most severe shortages will occur in the poorest countries, which are least able to compete in world markets for expensive oil and coal. Regional water shortages will also intensify as demand by the growing population doubles in nearly one half of the world.

Significant losses of forest area will likely continue, and available volumes of commercial-sized timber will decline. The forest area in many of the less-developed countries could decrease by 40 percent. In addition, because of erosion and other soil deterioration, desertlike conditions could accelerate in many areas. This would trigger important losses of cropland and grasslands. As a result of these and other changes, as many as 20 percent of all plant and animal species on earth could vanish due to loss of habitat.

Expansion of forestry practice will not solve all of these potential problems, but it can help in many areas. Conservation of existing forests and planting of new ones can reduce the worldwide loss of tree cover. This will help in reducing potential soil loss and will slow desertification. It will preserve many vital habitats for plants and animals, and help enhance watershed values and water quality. Greater production and use of wood as a renewable material might slow withdrawals of mineral and petroleum supplies. And wood could even substitute for other materials as a fuel in many countries. Thus, in several important ways, the skills and knowledge of forestry can contribute to the solutions of many serious global problems (Figure 16–1).

This worldwide perspective seems consistent with recent policy changes in the United States and the World Bank. In the past, neither showed much financial support for international forestry programs. But more recently, the World Bank identified about 1 billion hectares it considers critical to the world's watershed, wildlife, recreation, and timber needs. It also increased financial support to developing countries in an effort to strengthen forestry efforts for these critical lands. Likewise, U.S. policy now reflects greater emphasis on worldwide social welfare and the ecologic aspects of forestry, but continues to give attention to the

Figure 16–1.
The skills and knowledge of professional foresters can be an important contribution in addressing many serious global environmental problems and in assuring a steady supply of wood and other forest values into the future. *(U.S. Forest Service.)*

more traditional activities of timber production and wood products trade.[4] Adoption of metric units for measurements, record keeping, industry, and business would represent an important step toward strengthening this international commitment. Decisions by the Congress to promote voluntary use of the International System of Units contribute to the strengthening of these new policies.

These changes in attitude toward worldwide problems must receive support from new national programs. Clearly the emerging global wood shortage seems to lessen opportunities for importation to satisfy anticipated national shortages. Thus the United States may need to step up its timber production and become more of an exporter of wood, even after satisfying growing internal demand. Rapidly rising needs to produce more wood within the continental United States would seem to make the growth of forestry inevitable.

Such possibilities should create great new demands for forestry professionals. Their skills would help enhance the nation's chances for increasing wood production, thereby improving its wood-supply independence. Implementation of sound forestry practices should also ensure continued production of high-quality water, preserve vital habitats for wild creatures, safeguard soils against erosion and other modes of deterioration, and provide continued opportunities for recreation and realization of the many amenity values of forested lands. But what happens in the future depends upon the sale of commodities, and opportunities for forestry-related careers undoubtedly will rise and wane with the changes in the economic climate. Times of prosperity will offer the best opportunities for persons entering forestry practice, and periods of economic tightness may force temporary slowdowns in hiring. This occurred in the past and has characterized forestry for several decades.

Pieced together, these forecasts of growing worldwide needs, changing pressures upon forested lands, and revitalization of wood as a renewable source of material and energy make the long-term prospects for forestry appear bright. At least the need for foresters seems essential. Growth will take place at a rate tempered by an unpredictable economic climate and the influence of demand upon price and costs. As a result the growth of forestry-related disciplines should follow somewhat unpredictable patterns, showing intermittent spurts of activity, interrupted by no-growth periods when employment opportunities are scarce. However, over the long run, the potential looks good and the challenges great.

GEARING UP FOR THE FUTURE

In the long run, the greatest promise of forestry rests with those who master its technologies and devise creative schemes for applying them in the daily routines of managing forested ecosystems. Sound, progressive education will likely stand at the forefront. It will nurture the skilled minds, and the research and transfer of new technology needed to address the increasingly complex problems of forest resources management. This education will not just take its traditional form of residential instruction. It will also expand to offer technologic updating of the women and men who already work within the forestry sector of the world's economy. And it must translate the messages of wise forest resources management into language comprehensible to the many people who have no direct professional involvement in forestry. These lay people play vital roles in owning forestland, controlling its use, and influencing attitudes toward and policy concerning approaches to forest resources stewardship. This three-pronged approach has already begun to emerge, and should gain strength as time passes.

General Education

The general system of forestry education throughout the nation has three broad responsibilities:

1. to accumulate, preserve, and disseminate knowledge of forests as acquired and tested by years of successful application;

2. to apply current developments in supporting sciences, technology, and art to synthesize new concepts of forest care and use, and test these innovations for soundness and utility; and

3. to share with all interested parties information about the state of the art and science of forestry as can serve to enhance human welfare.[6]

The professional schools that specialize in forestry education participate in serving these needs through their combined programs of education, research, and public service.

In some way, all forestry professionals will take part in all of these aspects of forest resources education at some point in their careers. They start with a strong basic education geared toward intellectual development as well as learning the specific technologies of forestry. The two must go together. The appropriate application of forestry technology and the inventiveness necessary to address complex problems depend upon creativity and imagination nurtured by intellectual growth and broadened understanding. Hence forestry education guides students through a learning experience that expands individual capabilities for serious thought and creative invention. As with all educational ventures, forestry works to encourage the attitude that thought and understanding provide the principal expedients for addressing the major problems people face (Figure 16–2).

The art and science of managing forest resources for human benefit draw upon many fields of science, technology, and art. Studies of biology in the broadest sense form the scientific base, as forestry practice strives to manage forested ecosystems so that they better serve human need. Mathematics, physics, chemistry, and engineering provide critical foundations for the biologic sciences, and for studies of forest use, harvest, transport, and products processing. These sciences also serve as the basis for quantifying the supplies and output of wood, water, and several other commodities derived from forested lands. Economics and business administration hold the tools of decision making about effective ways to maximize returns from money and human effort invested in forest improvement. Behavioral sciences and public administration help in developing firm policies for forest use on the basis of diverse human needs and demands, and for resolving conflicts for these services. All play a critical role in the practice of forestry. Its professionals must school themselves in these many disciplines and develop the art of blending them to achieve satisfactory results from their endeavors.

In recognition of these needs, forestry schools require a general preparation in mathematics, physics, chemistry, biology, and social sciences. They also demand skills in oral and written communication, because reporting and communication play such an important role in a forester's routine duties. In addition to the basics, schools provide instruction in dendrology, forest ecology, silviculture, forest soils, forest management, forest protection, forest measurements, biometrics, forestry economics, forest policy, forest administration, and many other related subjects. The elective fields may include wood technology, watershed manage-

Figure 16–2.
Classroom study plus laboratory work, field exercises, and visits to forestry operations help students to develop capabilities for independent thinking and for applying their newly acquired knowledge to solving real problems.

ment, hydrology, range management, photogrammetry, entomology, pathology, timber harvesting, urban vegetation management, forest tree genetics, recreation, forest engineering, and much more. These courses build upon the core requirements, and provide individuals the opportunity to strengthen particular areas of interest.

Although students may have many options from which to choose, schools frequently offer concentrations of study in several common subdisciplines of forestry. These include:

Silviculture
Forest management
Forestry economics
Range management
Wood utilization and technology
Wildlife management
Landscape architecture
Park management
Lumber merchandising and light construction
Forest engineering
Forest mensuration
Forest biology
Forest entomology
Forest pathology
Logging engineering
Paper science and engineering
Forest chemistry
Packaging technology
Conservation of natural resources
Forest soils
Watershed management and hydrology
Forest recreation
Urban forestry
Forest policy and planning
World forestry
Biometrics

These represent several common fields, but not all the options open to undergraduate students. Still they do indicate specialized areas of study that many potential employers recognize in recruiting their staff professionals. Related areas of growing interest include computer science, business management, public administration, communications, personnel management, environmental law, marketing, and commerce.

Accreditation

To guide schools in developing and maintaining effective academic programs, the Society of American Foresters has established standards for forestry education. Along with other organizations, it offers accreditation to those institutions that satisfy their requirements. These standards call for a curriculum including study of basic forestry principles, with specialization through electives that build upon a core of general academic and professional courses. These must include communication skills as well as courses to provide concepts, principles, and working knowledge in the following:

1. Forest biology
2. Forestry in the social context
3. Forest resources management
4. Forest ecosystems management
5. Forest resources administration

Curricula that may qualify for accreditation include forest resources management, forest science, forest engineering, forest recreation, urban forestry, watershed management, wildlife habitat management, range management, and forest products.

Each school of forestry will have its own blueprint for blending the areas of study common to forestry, and the Society of American Foresters encourages imagination and innovation in designing a curriculum to capitalize upon the institution's facilities and capabilities. Consequently programs of study will differ from place to place, and students must explore the offerings of interest and work with a faculty adviser to build a personalized program based upon the courses available. These will include the basic courses that provide a foundation in forestry and general education, plus electives to reflect personal career interests.

Another unifying element in forestry school curricula involves requirements of the U.S. Office of Personnel Management in setting qualifications for persons to fill federal forestry positions. Among their basic requirements are 30 credit hours in biologic, physical, or mathematical sciences, or engineering. Of these, at least 24 must represent courses in forestry taken at an accredited college. The curriculum must also include at least eight semester hours each in management of renewable resources, forest biology, forest resources management, and inventory. As a result, most forestry school core programs will involve courses that satisfy these minimum requirements, while allowing students to strengthen specific areas of interest through use of electives.

FORESTRY RESEARCH AND GRADUATE PROGRAMS

Most professional foresters work directly in resources management, administration, policy development, planning, business, protection, manufacturing, and related fields. In these capacities, they make use of new ideas, technologic innovations, and scientific developments to bring about more effective management of forests and greater realization of values sought through use. Research provides the avenues for improvement, and demand for resources or other economic factors provide the stimuli to precipitate change and advancement.

Innovation in forestry practice comes about through the combination of creative thought and trial by those directly engaged in management, and through implementation of the results of research. This research operates at two levels. Some focuses upon basic questions in biology, physical sciences, mathematics, and similar fields. These provide a better understanding of how ecosystems and organisms grow and reproduce, what factors stimulate change, and how these interact. In wood products, the basic research probes mysteries of wood as material and how to process and use it better. A second level of research addresses problems in applying available basic scientific knowledge to bring about more effective management and use of resources. Progress comes when scientists using both approaches recognize and challenge questions of importance.

This research takes place within the many agencies, firms, and institutions engaged in forestry. Universities play an active role through involvement by faculty, staff, and graduate students. In states with a forestry school associated with the land grant college, the research often is carried out through the agricultural experiment station. Other colleges work through different types of administrative structure. State forestry or conservation organizations often also maintain research and development groups. In these, staff specialists usually undertake applied research and problem solving directed toward questions of immediate importance in the daily operations of the agency. At the federal level, research takes place in formally organized experiment stations, such as those maintained in each region by the U.S. Forest Service. Their research addresses a full range of problems from basic to applied, and of importance to the region served. In addition, several of the larger corporations maintain research departments. They generally explore questions of importance to operations of the company and mostly involve applied research, problem solving, and feasibility studies. In composite, the different organizations provide an elaborate network of forestry research throughout the United States.

Persons seeking careers in research need to pursue graduate education to develop the necessary specialized skills, and the capability for independent thinking needed to address complex problems. This usually requires at least 30 credits of graduate-level study beyond the bachelor of science degree, including a research project leading to a thesis. Many career opportunities await graduates with the M.S. degree who seek positions in research. Such studies lead to a specialized capability in some subset of forestry. Graduates often work in conjunction with senior scientists who hold the Ph.D. degree, which entails study beyond the M.S. It combines course work in preparation for comprehensive examinations to demonstrate depth of technical understanding in a field of specialization, plus completion of independent research that makes a unique contribution to science or practice. Although most schools of forestry offer advanced degrees at the master's level, some do not award the Ph.D., or many have programs supporting only a select few of the specializations of forestry.

Admission to graduate schools will require better-than-average performance at the undergraduate level. A minimum grade point equivalent to B-level work will qualify a candidate for consideration by most schools, though competition among applicants may limit admission to persons with higher grades. Beyond academic performance, graduate schools also look for such personal characteristics as capability for independent thought, persistence and diligence in addressing questions, communication and analytic abilities, and similar traits necessary for successful performance in graduate studies, and afterward.

Some forestry programs also offer advanced degrees, especially at the master's level, geared for persons who do not intend to work in research. These people seek employment in other positions requiring skills in problem solving, advance planning, technical specialization, and administration. These degrees may consist of greater amounts of course work beyond the bachelor of science degree, but without a thesis. Instead students may participate in internships, projects geared toward sharpening problem-solving skills, or cooperative studies to test new ideas or upgrade some aspect of practice. Such programs may require as many as 40–45 hours of advanced-level study beyond the bachelor of science degree and grant to successful participants a professional master's degree. Afterward students can often gain admission to Ph.D. programs or move directly to positions of responsibility with many agency or corporation staffs. These master's degrees also prepare people to work in various technology transfer programs, such as those of cooperative extension information services and conservation education and interpretation.

CONTINUING EDUCATION

Mushrooming developments in the sciences and forestry practice, increased awareness of the interactions of people through the structures of society, innovations in communications and management methods, advent of new technologies, and many other modes of progress make continuing education a necessity in a modern world. New ideas grow from research and development groups at a prodigious rate. This makes professionals in many fields realize that the notions and concepts developed during their early years of academic preparation quickly lose relevance. To compete successfully for advancement, and to capitalize upon the potential for improvements offered by the new ideas, they must actively seek ways to continually update and expand their knowledge. It helps them to sharpen many professional skills and find better ways for putting their understanding to work (Figure 16–3).

Continuing education takes many forms. Usually professionals rely upon a variety of opportunities to help keep current. For example, they have available a

Figure 16–3.
Throughout their years of professional practice, foresters must continue their education to keep abreast of technological innovations and to sharpen their skills.

wide range of professional journals that report on newest developments in forestry science and practice. The *Journal of Forestry, Forest Science, Forest Ecology and Management, Forest Products Journal, Wood and Fiber, Journal of Wildlife Management, Journal of Environmental Quality,* and the *Southern Journal of Applied Forestry* are but a few of the numerous publications. A professional will usually subscribe to one or more of these, and become a member of the professional society that publishes the journal of interest. The societies also sponsor technical meetings for the exchange of ideas, and provide specialized training courses in conjunction with universities, experiment stations, and other institutions.

Most schools of forestry organize short-term workshops or specialized study programs that professionals can attend to learn about new developments or even to earn advanced degrees. Often employers will cooperate by offering financial assistance for tuition and expenses, and by encouraging attendance in other ways. Other colleges have special graduate programs where experienced professionals spend part of their time in residence on the campus, and continue their study or undertake guided projects while on the job between periods of formal instruction. In some cases, public agencies or corporations arrange opportunities for in-service training by specialists within the agency or firm, or aided by consultants. All of these provide working professionals with opportunities for continuing study and technical growth following the formative years of undergraduate education.

The Society of American Foresters has recognized the importance of continuing education and has organized a formal program to encourage and recognize accomplishments by its members.[1] It maintains records of postgraduate education and study and provides a system of certification for members who demonstrate regular involvement in some form of educational endeavor. The study may include participation in technical meetings sponsored by the society or other groups, attendance at formal courses and training programs, guided readings linked to personal study, and other methods of learning and professional development. Such an approach supplements the interests of employers in encouraging continued growth and personal development in forestry skills.

EMPLOYERS OF FORESTRY PROFESSIONALS IN THE UNITED STATES

A survey of employment among members of the Society of American Foresters for the period 1974–1978[3] revealed that private industry and federal agencies employed nearly comparable numbers of foresters, collectively representing 52 percent of those sampled in 1978. State governments employed 13 percent, colleges and universities employed 10 percent, 6 percent worked at consulting or in some type of self-employment, and others served in a variety of other employment categories. Such statistics identify clearly the importance of governmental agencies as employers of professional foresters, especially considering that most schools of forestry receive major support via governmental funding. About 42 percent of foresters work in some form or level of governmental service outside the universities.

The activities of the several agency groups involved in forestry and related resource management together with the types of professionals and technicians employed include:

Federal Agencies

Department of Agriculture: Forest Service, and Other Bureaus

Activities: National Forest administration, forest and forest products research, world forestry. Cooperates with other federal agencies, states, and private forest owners and industries in forestry activities.

Employ: Foresters, wood technologists, pulp and paper technologists, engineers, landscape architects, range managers, wildlife managers, forest pathologists, entomologists, ecologists, and hydrologists, at the professional and technician level.

Department of the Interior: Bureau of Land Management, Bureau of Indian Affairs, National Park Service, Bureau of Outdoor Recreation, Fish and Wildlife Service

Activities: Administer public domain lands, grazing lands, Indian lands, National Parks, wildlife refuges. Cooperate with other federal agencies and states in outdoor recreation planning and coordination, cooperate with states on fish and wildlife administration and research.

Employ: Foresters, wildlife managers, fishery biologists, range managers, landscape architects, and recreation specialists, at the professional and technician level.

Other Federal Agencies

Other federal agencies concerned with managing land, reviewing budgets, and reviewing land policies employ foresters on a full-time or consulting basis. Among them are the Department of Defense, the Agency for International Development, the Office of Management and Budget, the Environmental Protection Agency, the Congress, the General Accounting Office, the Geological Survey, and the Tennessee Valley Authority

Such agencies employ mainly professional foresters, in a variety of capacities.

State Conservation Departments

Activities: Administer state forests and parks, game refuges and management areas. Control fire, insect and disease on state and private lands. Enforce state laws. License motor boats and off-road motor vehicles. Administer in-state waters and their use, plus state forest practice laws affecting private lands. Furnish services to private landowners and industries. Operate state fish hatcheries and game-rearing stations and distribute the fish and reared game. Operate forest nurseries and distribute the stock. Cooperate with sports organizations in fish and wildlife betterment activities. Furnish advice to forest landowners and industries. Conduct research on forestry, fish, and wildlife.

Employ: Foresters, fish and wildlife biologists, law-enforcement officers, forest pest-control officers, landscape architects, and other professionals at professional and technical levels.

Universities and Colleges

Activities: Offer education at the professional and technician levels in forestry, wood technology, pulp and paper technology, wildlife and range management, forest protection, architecture and landscape architecture, hydrology, engineering, biologic science including ecology and environmental science, physical education and outdoor recreation, and many related subject areas. Conduct research and offer postgraduate education in the above fields. Conduct continuing education and sponsor symposiums and conferences in various conservation subjects. Sponsor publications in above fields. Operate college forests and summer camps in conservation.

Employ: Professors, research and extension specialists, and technicians in above-mentioned fields of science and technology.

Consulting Firms

Activities: Perform consulting services for private landowners, industries, investors, associations, and governments.

Employ: Foresters, wood technologists, pulp and paper technologists, forest hydrologists, and experts on environmental sciences, at professional and technician levels.

Forest Products Industries, Primary Manufacturing

Activities: Manage corporate forests, purchase and harvest timber from public and private forest owners, conduct research in forestry, wood technology, and pulp and paper technology. Manufacture lumber, veneer, plywood, particle board, and paper products. Some operate land development subsidiaries.

Employ: Foresters, logging engineers, mechanical engineers, pulp and paper technologists, and landscape architects. Also employ technicans in above specialties.

Forest Products Industries, Secondary Manufacturing

Activities: Manufacture prefabricated houses, mobile homes, millwork, sash and doors, pallets, furniture and fixtures, athletic equipment, turned products, crates and boxes, musical instruments, and numerous other products. Also includes lumber dealers and the light construction industry.

Employ: Wood technologists, business managers, marketing specialists, other professionals, and technicians.

Suppliers to Foresters and Forest Industries

Activities: Supply forestry instruments, logging machinery, sawmill and wood-processing machinery, pulp and paper machinery, chemicals, dry kilns, adhesives, and other equipment and products to the wood processors.

Employ: Foresters and wood technologists at professional and technician levels in product development, and as salespeople and in customer relations.

Associations of Forest Industries and Conservationists

Activities: Provide timely information and promote constructive cooperation among forest industries and users and lay people interested in conservation of nature and environmental improvement.

Employ: Foresters, wood technologists, wildlife biologists, forest ecologists, range experts, landscape architects, and recreation specialists in executive and staff work. Employment limited mainly to professionals and those with writing skills.

Miscellaneous Employers

Various agencies with jurisdiction over land and budgets of forestry agencies, highway departments, power companies, city and county park boards, youth conservation groups, and a few land development agencies employ foresters, landscape architects, hydrologists, and wildlife biologists on a full-time, part-time, or consulting basis. Some may employ forest technicians also.

Self-employment

A few imaginative foresters and wildlife biologists support themselves, at least in part, as writers, lecturers, managers of land, and landowners and operators. Others manage boys' and girls' camps and camps for adults and families. Some are wilderness outfitters and managers of other outdoor activities. Success in such ventures calls for talents and experiences going beyond those included in curricula of forestry colleges.

Other foresters and technicians have become independent loggers, sawmill operators, nursery managers, house builders, and wood fabricators.

THE CHALLENGE AND THE PROMISE

Forestry provides diverse opportunities for rewarding careers to satisfy many interests. Few other professions span such a broad range of activity, including application of biologic sciences, management of resources and people, administration and business management, manufacturing and engineering, sales and service, communication and education, and various other fields. Through wise choices of courses, individuals can prepare for many different career possibilities.

Participation in resources management as an active career brings an adequate salary and good benefits. For individuals who rise through the employment ranks to upper levels of management and administration, or who develop highly specialized skills, the pay levels often prove quite attractive. Most may draw more modest salaries, especially compared with other professionals. For this renumeration, foresters can expect to work hard, assume much responsibility, and work reasonably long hours. During early years, they may live in rural and often remote localities. Many foresters can expect to move frequently, but these moves normally involve promotion to a position of greater responsibility and challenge. Forest resource managers must also demonstrate great self-sufficiency with limited direct supervision in discharging their responsibilities. Such qualities often make them leaders in the communities in which they live and provide many interesting nonprofessional opportunities outside their workplace.

For persons employed in administration and planning, business management, wood products engineering, paper science, technical services and sales, and similar activities the setting will differ. These people work in more urban environments, which house the manufacturing facilities and corporate and agency headquarters. Salaries for specialists in manufacturing, products engineering, and business management may run considerably higher than for field-oriented positions. Qualities of leadership, creativity, communication skills, technical competence, and the like serve individuals well in competing for promotion to upper levels of management and supervisory positions. Opportunities in these fields often provide unique challenges for innovation to keep abreast of rapid developments in processes, methods, and products.

The future offers many exciting challenges for those working in forestry and the many fields related to it. Demand for increased use of wood, needs for product innovation, urgency to make fuller use of renewable resources, and requirements for safeguarding the productive potential and health of our forest ecosystems make forestry-related professions critical to the future. But aside from providing the materials and goods upon which people depend for their daily living and use freely in their recreational activities and leisure, foresters must work hard to extend the creative power of human intellect to safeguard this precious earth for future generations. Such a challenge must draw upon the skills and knowledge of a host of scientific specialties and technologies, and use them in new and creative ways. Foresters must participate in this challenge, and those who choose forestry as a career will make significant contributions to human welfare, business, and prosperity.

REFERENCES

1. R. R. Christiansen, "Continuing Forestry Education—The SAF Role," *Journal of Forestry* **79**(4):200–231 (1981).

2. F. C. Ford-Robertson (ed.), *Terminology of Forest Science, Technology Practice and Products,* Society of American Foresters, Multilingual Forest Terminology Series, no. 1, 1971.

3. H. R. Glascock, "1978 MOS Results: Highlights and Comparisons," *Journal of Forestry* **77**(5):278, 320–321 (1979).

4. J. R. McGuire, "What Role for the United States in World Forestry," *Journal of Forestry* **79**(5):267 (1981).

5. R. B. Phelps, *Timber in the United States Economy 1963, 1967, and 1972,* U.S. Forest Service General Technical Report WO-21, 1980.

6. H. L. Shirley, "Forestry Education in a Changing World," *Proceedings of the Sixth World Forest Congress,* Madrid, 1966, pp. 895–902.

7. Society of American Foresters 1976. *Standards for Accrediting Institutions for the Teaching of Professional Forestry,* Society of American Foresters, Washington, 1976.

8. U.S. Council on Environmental Quality and U.S. Department of State, *The Global 2000 Report to the President. Entering the Twenty-first Century,* vol. 1, U.S. Council on Environmental Quality and U.S. Department of State, Washington, 1980.

INDEX